T0138928

Nanotechnology and
the Resource Fallacy

Nanotechnology and the Resource Fallacy

Stephen L. Gillett

PAN STANFORD PUBLISHING

Published by

Pan Stanford Publishing Pte. Ltd.
Penthouse Level, Suntec Tower 3
8 Temasek Boulevard
Singapore 038988

Email: editorial@panstanford.com
Web: www.panstanford.com

British Library Cataloguing-in-Publication Data
A catalogue record for this book is available from the British Library.

Nanotechnology and the Resource Fallacy

Copyright © 2018 Pan Stanford Publishing Pte. Ltd.

All rights reserved. This book, or parts thereof, may not be reproduced in any form or by any means, electronic or mechanical, including photocopying, recording or any information storage and retrieval system now known or to be invented, without written permission from the publisher.

For photocopying of material in this volume, please pay a copying fee through the Copyright Clearance Center, Inc., 222 Rosewood Drive, Danvers, MA 01923, USA. In this case permission to photocopy is not required from the publisher.

ISBN 978-981-4303-87-3 (Hardcover)
ISBN 978-0-203-73307-3 (eBook)

Contents

Acknowledgments

As anyone who's written one knows, a book is a project that requires the help of many people. Many of the ideas herein appeared first, in a considerably more informal format, as science articles in the magazine *Analog* from the early 1990s through the early 2000s, and I greatly thank Dr. Stanley Schmidt, the then editor, for his enthusiasm about them. I also thank Stan for contributing a cover quote. The immediate predecessor of this book was a white paper I wrote for the Foresight Institute (www.foresight.org), and I thank them for their enthusiasm as well. I thank Eric Drexler in particular for reviewing that paper. I would like to thank my collaborators Prof. Thomas W. Bell, of the University of Nevada, Reno, and his then postdoctoral fellow Masahiro Muraota for their help as I got involved in some actual laboratory research on some of these ideas.

I must especially thank Stanford Chong at Pan Stanford Publishing for his enthusiasm about turning the white paper into this book, and his patience as what was supposed to be a quick project turned into anything but. It's a measure of how fast this field is moving that so much had to be rewritten because of new developments! And last, I'd also like to thank my long-suffering family, in particular my wife Joyce, for their patience. Finally, although I've tried to avoid errors, any remaining are my fault alone.

Stephen L. Gillett
Washoe Valley, Nevada, USA
8 December 2017

Chapter 1

Introduction: The Global Resource Predicament

Civilization exists by geologic consent.
—Will Durant

The spiking of oil prices in 2008 put resource issues into mainstream global headlines for the first time since the 1970s. Moreover, the global communications explosion is causing a worldwide revolution of rising expectations, which puts further pressure on the resource base—not just on energy supplies, but on other raw materials as well. If nothing else, exporters now have other markets for those resources than the long-industrialized countries. Eventually, domestic political concerns may make exporting resources politically impossible. The United States banned exporting oil for 40 years, for example. Although this ban was lifted in December 2015, the precedent has been set, and there seems little doubt that it could be reinstituted should conditions change again.

Resource issues are also being exacerbated by environmental concerns, as the increasing pollution associated with industrialization exacts ever-greater costs, both economic

Nanotechnology and the Resource Fallacy
Stephen L. Gillett
Copyright © 2018 Pan Stanford Publishing Pte. Ltd.
ISBN 978-981-4303-87-3 (Hardcover), 978-0-203-73307-3 (eBook)
www.panstanford.com

and in terms of human well-being. As is well known, present technology is polluting, and a major part of that pollution stems from the production and consumption of resources. For example, little of the material excavated from a mine ends up in finished products. Most is just waste, which is discarded in vast heaps that (at best) must be expensively reclaimed and revegetated. Reaction of some of these materials with water and air can cause further environmental degradation, as in acid-mine drainage. The extraction, transportation, and consumption of petroleum have familiar environmental hazards ranging from oil slicks to photochemical smog. Coal is abundant and cheap, but by-products of its combustion include acid rain and fly ash, not to mention carbon dioxide, and its mining is both dangerous and environmentally disruptive. Even such prosaic commodities as fresh water are in short supply throughout much of the world.

If this were all it would be bad enough. But in fact the stresses will be exacerbated by the sluggish growth of global oil production, coupled with dwindling supplies of other conventional resources such as metal ores. Furthermore, present oil-based technology is fundamental to *obtaining* those other resources. Oil fuels the excavators and front-end loaders; it fuels the haul trucks bringing ore out of a mine; and it fuels the transportation infrastructure that distributes the raw concentrates and finished metal around the world.

Given this oncoming "perfect storm," it is little wonder that many writers foresee the imminent collapse of industrial civilization, accompanied by untold human suffering. Despite the fantasies of a noisy minority, too, the natural environment would be unlikely to prosper in the event of such a collapse. If history is any guide, the increasing desperation of populations will instead put extreme pressure on their environment. A plague of locusts would be more benign. Nor will an oil-starved future be a paradise of "community," as imagined by some other, and equally naïve, writers. Much more likely, in view of historical experience, is a reversion to what Sterling[1] has called the post-Neolithic norm of human history: "starving peasants ruled by bandits."

Even so, some would have us believe that the most dire predictions of doom are inevitable, that humanity must resign itself to a future of scarcity, that not only can't the developing world aspire to the standard of living now enjoyed by the industrialized nations, but that the already industrialized nations must resign themselves to a seriously declining standard of living.[2]

This bleak assessment, however, is wrong.

The Paleotechnical Era

The habit of neglecting or denying the possibility of technological advance is a common problem. . . . Snug limits would simplify our future, making it easier to understand and more comfortable to think about. A belief in snug limits also relieves a person of certain concerns and responsibilities. After all, if natural forces will halt the technology race in a convenient and automatic fashion, then we needn't try to understand and control it.

Best of all, this escape doesn't feel like escapism. To contemplate visions of global decline must give a feeling of facing harsh facts without flinching.

—K. Eric Drexler[3]

What can be done? Well, yes, new technology is key—but with a twist. It's not a substitute or a stopgap but an improvement. A vast improvement, in fact. Not only is present technology

---

[1]Stirling, SM. Introduction: Fusion energy and civilization. In *Power*, Stirling, SM, ed., Baen Books, 1991.

[2]A cursory search will turn up dozens of books and websites. Some are better written than others, but all contain profound flaws of omission, analysis, or both. Potential technologies are overlooked, the limits of particular technical approaches are repeatedly confounded with laws of nature, and the degree to which present technology squanders energy as heat is never mentioned. This book will address such issues in detail.

[3]Drexler, KE. *Engines of Creation*. Doubleday, 1986, pp. 166–7.

exhausting its resource base while polluting the environment, it is also ludicrously inefficient and almost gratuitously dirty. No law of nature demands that things *must* be done this way.

Both the inefficiencies and the pollution come about largely through an overwhelming reliance on *heat*: on energy in its most disorganized and wasteful form. After all, "fuels" are "burned!" Indeed, at least in Western culture, "technology" has been identified with "fire" since the legend of Prometheus. But heat is a hugely messy and wasteful way to use energy. For example, around two-thirds the energy content of a car's fuel tank just goes out the radiator. It is often heard that the energy density of conventional fuels is "irreplaceable." But not only is that not true, the energy density of conventional fuels is merely a brute-force compensation for the inefficiency with which they are used. Furthermore, present technology relies on sources of non-energy raw materials that are *also* intrinsically limited, consisting of anomalous deposits ("ores") where particular desired substances have been concentrated by geologic happenstance.

Other inefficiencies, and associated pollution, come about from what may be termed the "(dis)economies of scale." Because arranging matter with conventional technology is difficult, matter organizers ("factories") have traditionally been large-scale, capital-intensive affairs to which raw materials are brought, and from which finished products are exported. This obviously requires an enormous transportation infrastructure, one of course that now subsumes the entire world. For example, bauxite, a tropical soil, is mined in Jamaica and sent to Norway for processing into aluminum (the electrolytic process for extracting aluminum requires vast amounts of electricity (Box 3.6), and Norway has abundant cheap hydropower). Then the aluminum metal is sent around the world to be further fabricated into finished products. Yet aluminum is the most common metal in the Earth's crust (Table 1.1)! If the promise of distributed fabrication (Chapters 2 and 5) can be achieved—if "matter becomes software"—this energy-intensive transportation network will become largely obsolete.

Table 1.1 Composition of the Earth's crust

Calculated from data in Mason and Moore.[1] Columns don't add to exactly 100% due to rounding and, in the case of weight percentages, due to neglected elements.

Element	Weight %	Mole percentage
Oxygen (O)	46.6	60.4
Silicon (Si)	27.7	20.5
Aluminum (Al)	6.3	8.1
Hydrogen (H)	0.1	2.9
Sodium (Na)	2.8	2.6
Calcium (Ca)	3.6	1.9
Iron (Fe)	5.0	1.9
Magnesium (Mg)	2.1	1.8
Potassium (K)	2.6	1.4
Titanium (Ti)	0.4	0.02

[1]Mason, B; Moore, CB; *Principles of Geochemistry*, 4th Ed. Wiley, 1982, pp. 46–7.

Another result of the present clumsy approaches to matter arranging is the unintended manufacture of unwanted by-products. With little control over occurrences at the nanoscale, conventional synthetic processes yield a suite of molecular products, of which typically only one or two is desired. The others become waste that must be discarded or treated.

To summarize:
- present technology squanders energy, mostly because it's used as *heat*, energy in its most disorganized and wasteful form. (What do we do with fuels? We burn them!) From the smelting of ore to the newest steel minimill to the ubiquitous internal-combustion engine to the sleekest jet, industrial technology is dominated by vast flows of heat. Much is made, as described above, of the high energy density of conventional fuels such as

coal and oil. Indeed, the crisis of oil depletion is thought to be largely due to the impossibility of finding a similar cheap and energy-dense substitute.

- the high energy density of conventional fuels, however, is merely a brute-force compensation for the gross inefficiency with which they are used. We have the "heat" crisis, not the "energy" crisis.

In fact, current "industrial" technology is so primitive and so cumbersome and so inefficient we might call the present the Paleotechnical Era.

The Biological Inspiration

How, then, do we move beyond the Paleotechnical? Let's look at what might be called "the biological inspiration." Living organisms carry out feats of energy collection, energy use, materials extraction, and matter organization that beggar the capabilities of conventional technology—and they do it all with the diffuse and fitful energy of sunlight.

To begin with, we loosely speak of food as "fuel," but biology doesn't do anything so inefficient as "burning" it. The energy of food is converted into metabolic work by a cascade of chemical reactions that minimize the generation of heat at each step. In the language of thermodynamics, their energy efficiencies "approach the reversible limit." (In warm-blooded creatures such as ourselves, some energy can also be turned into heat deliberately to maintain body temperature. But that conversion is to maintain a constant-temperature (isothermal) environment. It is *not* to set up a flow of heat to a cooler environment from which some work can be extracted, the way the heat from burning fuel is used in current technological applications.)

Biosystems excel in the gathering of energy, too. Consider a green plant, quietly turning the fickle energy of sunlight both into stored chemical energy, and into a degree of molecular organization that puts even the most high-tech fabrication facility to shame. Compared to even the humblest weed, the latest integrated circuit is extraordinarily crude.

The organization of matter is basic to all technology, from the simplest handcrafted tools to the newest jumbo jet or the most cutting-edge computer. But, as already noted above, present technology for fabricating matter is clumsy and wasteful in the extreme. First, many of the raw materials have to be obtained from anomalous natural concentrations ("ores"). They are difficult and energy-intensive to separate, largely because separation is traditionally done with lots of heat (*that* again). Of course, the orebodies are not necessarily conveniently located, either, so that once again large-scale transportation costs ensue.

Traditional fabrication also deals only with *bulk* matter: whether it's a structural material such as metal, wood, stone, wallboard, or glass; or a textile such as cotton or nylon, polyester or wool, or indeed anything at all (Box 1.1). At present, the arrangement of atoms inside *any* material is irrelevant, except in the gross sense that the properties of the bulk material result from the average properties of the interactions of the astronomical number of atoms making it up. Steel has the properties it does because of the way all the iron (and alloy) atoms in it are arranged. It would have very different properties if those atoms were arranged differently, but we don't deal with steel at the atomic level. We are only able to influence atomic organization indirectly, via bulk processes such as tempering or cooling rates.

This example also underscores again that one way we fabricate bulk matter is with heat: other heat-based examples are boiling, casting, melting, firing, welding. We also use the raw shaping of bulk matter by grinding, cutting, buffing, bending, and so on. In fact, most industrial fabrication isn't much different from handicraft; it's just faster and at a larger scale, using machines rather than hands to apply the forces necessary. There's little essential difference between a blacksmith's forge and the most modern steel minimill; there's little essential difference between a coppersmith's cutting and bending metal, and a highly paid

Box 1.1 Bulk Matter

Bulk matter, of course, is what modern and ancient technology relies on, everything from wood to steel to the latest polymer plastics. Such materials have a range of properties, things like strength or hardness or flexibility, and those properties are basic to their applications.

As noted later, macroscopic properties such as pressure and temperature result from the average properties of the huge number of constituent particles in a body of matter. Pressure is the average of trillions of individual atomic collisions per second, whereas temperature is proportional to the average energy of individual atoms (Box 2.3). Similarly, one might expect that other macroscopic properties, such as strength, conductivity, or hardness should also reflect an average over the entire ensemble of particles.

It turns out, though, that in general this is not so. Unlike temperature and pressure, bulk properties are often dominated by *second*-order phenomena, because they are *non*-uniform at the nano- and microscale, and those heterogeneities are responsible for the bulk properties. As described in Chapter 6, for example, it's easy to calculate that macroscopic materials strengths are orders of magnitude lower than what would be inferred from the chemical bonds making them up. The strength of everyday materials—metals, for example—is dominated by such things as the boundaries between the crystals making them up. Those boundaries are zones of weakness that fail much more readily than the individual chemical bonds.

The fact that humans routinely have dealt with matter in bulk is built almost subliminally into current engineering practice and behavior. As we move toward finer organization, however, we move toward the atomic scale, in which the nanoscale organization of the atoms becomes ever more important, and present-day intuitions as to how matter "behaves" become ever less relevant. Parenthetically, this also is why "living" matter seems so different; it already has a seething organization that truly extends down to the molecular level.

technician doing the same to high-performance alloy sheets in an aircraft plant.

The chemical industry *does* fabricate things at the molecular level; it makes molecules, of course. But it does so in a clumsy way: only relatively simple substances can be made, often to the accompaniment of lots of unwanted by-products because of the lack of control of the synthesis conditions at anything like a molecular level. And the syntheses are again often driven by (yes!) the application and extraction of prodigious amounts of heat.

As emphasized below, biological systems *do* carry out fabrication at the molecular level, so in a sense agriculture is a way in which we carry out molecular-level fabrication at present. But not really, of course; all we are doing is exploiting fabrication capabilities that already exist. We have to treat the products of agriculture just like more bulk raw material: wood is sawn and ground and clumsily pegged ("nailed") together; cotton or wool is ginned and spun. We can hardly grow a house or a shirt from scratch! Even food, the most important product of agriculture, often gets processed like just another bulk raw material: consider the grinding of wheat into flour, the leaching of corn by lye to make hominy, the drying of fruit or jerky. We even process *food* with heat; consider cooking!

And, because of the vagaries of the conditions demanded by the plants, they are grown in certain areas but not in others. Wheat is picky about the conditions under which it will grow; so is cotton. So are trees grown for timber or for fruit. The consequence is that agricultural products are much like mineral products: they can only be obtained from certain areas, which may not be anywhere near where they are wanted or needed.

Furthermore, because matter organizers ("factories") are so cumbersome, they are big and thus expensive. They must bring in partly fabricated bulk matter ("raw materials") and export the finished products, often around the world. Finally, because the sources of those raw materials often must be located in specific and inconvenient places (mines on those anomalous deposits, agricultural commodities in particular climates), they usually must be transported to the factory as well.

The result is that huge and energy-intensive transportation system, for both bulk and finished commodities, that now subsumes the entire world, and whose inception goes back well before the 19th century. This whole system is so taken for granted—so internalized—that it is embedded into the very substance of conventional economics. Economists speak of "relative advantages," due to the vagaries of the raw materials sources, or of a location on the transportation infrastructure linking the raw materials sources with the matter organizers that use them as inputs. They also speak of "economies of scale" (once you've built a matter organizer, you want to use it as much as possible to pay for it). The story of industrialization has therefore been a story of centralization: of replacing lots of distributed, inefficient handicraft fabricators with one large and considerably more efficient factory. (Of course, this energy-intensive transportation infrastructure we now rely on is itself a substantial cost, but when energy is cheap we haven't cared.) In the 19th century machine-made textiles from Britain could undersell locally handmade textiles in India! (And the cotton used by the British mills was itself brought from overseas, from the American South or from India itself.)

But all this is merely a consequence of clumsy fabrication techniques, not a law of nature. Again the contrast with biosystems is striking. A tree doesn't send off to the root factory, and the leaf factory, and the stem factory. It fabricates itself from ambient materials, according to a molecularly based instruction code (its DNA). And it does all this with a diffuse and fitful source of energy: sunlight.

Finally, the biosphere is certainly sustainable. It recycles itself and its materials indefinitely, using only the energy of sunlight.

So, how are biosystems capable of such feats? It's not due to some sort of "vital force"—living things are subject to the same laws of physics and chemistry as are anything else. It is because they are organized at molecular scales.

Nanotechnology and Resources

The embryonic field of nanotechnology also aspires to the organization of matter at molecular or near-molecular scales, and has attracted much attention recently for its possible applications in such fields as information handling and medicine. However, as these biological examples indicate, its most important near-term application lies in addressing resource and pollution issues.

To summarize with a set of "pre-conclusions": for resources, the broad categories of nanotechnology applications include

Vastly Improved Efficiency of Energy Usage

- *non-thermal energy usage*, via nanostructured devices such as fuel cells, which do not rely on the vast flows of heat that dominate present technology;
- *molecularly tailored catalysts*, for heightened product selectivity, by-product elimination, practical non-hydrogen fuel cells, electrolysis, and so forth;
- *high-strength materials*, which will decrease transportation costs, including access to space;
- *electricity storage and electrosynthesis*, both for portable power sources, such as ultrahigh performance "superbatteries," and for chemical fuel generation;
- *distributed fabrication*, summed up in the slogan "matter as software," as local fabrication of items gradually renders the large-scale transportation of bulk raw material and finished products obsolete. This is already happening with information products: consider the downloading of MP3s or PDFs, and even with physical artifacts, with the growing interest in "3D printers."

Information-Intensive Energy Extraction

- *extensive real-time sensing*, for better conventional resource extraction such as through oil or geothermal reservoir management;
- *cheap nanofabrication*, which will make practical distributed collection from diffuse sources (e.g., artificial photosynthesis and low-head hydropower), as well as better energy management through such devices as "smart windows" and non-thermal light sources.

Solid-State Energy Generation

- *solar power*, via next-generation photovoltaics such as photovoltaic tiles and coatings, as well as artificial photosynthesis;
- *thermoelectric conversion*, the direct solid-state transformation of thermal gradients into electricity;
- *piezoelectric conversion*, for direct conversion of mechanical energy into electricity.

Molecular Separation of Elements and Molecules

- *non-thermal separation of solutes from solutions*, be they leachates, natural brines, or wastewater, which in blurring the distinction between a "pollutant" and a "resource," will render obsolete the geologically anomalous deposits ("ores") now used.

Change of Materials Mix

- *obsolescence of structural metal*, replaced by high-strength covalently bonded materials such as carbon;
- *"waste as resource,"* such as biowaste replacing petroleum as a source of reduced carbon compounds for synthesis feedstock;

- *carbonate rocks (limestone)*, as the backstop resource of carbon on the planet;
- even *silicates*, the very stuff of a rocky planet, will become more valuable than nickel steel.

The rest of this book will look at these various aspects in more detail. First we will look at energy issues in general, including the issues of global oil production, and with a brief sidetrip into the much misunderstood and much misused laws of thermodynamics. We will then survey other resource issues, including pollution, which proves merely to be another aspect of resource utilization. Following this, we will survey nanotechnology itself, with a brief review of potential approaches to fabricating nanostructured materials. After this, we will then return to energy and materials resources, with some specific suggestions as to how nanotechnological approaches can contribute to them. Finally, we will indulge in a look ahead as humanity escapes the Paleotechnical Era. Throughout, background material, including most mathematics, is isolated in sidebars so as not to interrupt the flow of the narrative. A lot of ground will be covered, with a necessarily sketchy overview of a sprawling set of topics, and the author apologizes in advance to all those investigators whose work has been overlooked,

The sustainability of human populations—indeed, the long-term viability of any human populations—is at present a major source of concern to thoughtful people. Nanotechnology holds out the prospect of sustainable "First World" standards of living for billions of people, and indeed it is the only technology to do so.

Chapter 2

The Heat Crisis

Energy, Free and Otherwise

What *is* energy? A quick answer is that it's the fundamental "stuff" that's required to *do* anything. The formal physics definition isn't too different from this everyday one: energy is the ability to do *work*. In physics "work" has a very specific meaning: it means applying a force over a distance. Again, that's not so very different from the everyday meaning. "Work" in the physics sense, however, is not necessarily "useful" in a biological or economic sense. To a physicist an explosion that destroys a building does "work." But all this means for everyday purposes is that "energy" is value-neutral; it can be applied for either constructive or destructive purposes. What energy is used for is a political and social question, not a physical question. In any case, though, it's necessary to start with.

So, what do we use energy *for*? Almost everything. One basic use is for body maintenance; i.e., as *food*! The "calorie" content of food is nothing more than its energy content (Box 2.1), and that energy content represents solar energy stored into a form that humans (and animals) can use. Ultimately the solar energy was trapped by photosynthesis in crops. This is not trivial; agriculture remains fundamental even now, and indeed,

Nanotechnology and the Resource Fallacy
Stephen L. Gillett
Copyright © 2018 Pan Stanford Publishing Pte. Ltd.
ISBN 978-981-4303-87-3 (Hardcover), 978-0-203-73307-3 (eBook)
www.panstanford.com

Box 2.1 Measuring Energy

One of the causes of confusion about energy is that, for historical and traditional reasons, it's measured in a welter of different units. Some are based on heat, whereas others are based on mechanical motion. One of the triumphs of 19th century physics was to show that heat is a form of energy; but we're still stuck with this legacy of measuring energy in these fundamentally different ways. Even worse, still other energy measurement units are based on the amount of energy involved in particular processes: such things as a charged particle moving in an electric field, or the amount of energy obtained by reacting a standard fuel (such as a barrel of oil or a ton of coal) with oxygen.

In the modern scientific system, the so-called "SI" (Système Internationale), which is a highly standardized version of the metric system, the fundamental energy unit is the *joule* (pronounced "jool", abbreviated J[1]). A mass of two kilograms moving at a speed of one meter per second (about a crawl) has an energy of one joule. Obviously, joules aren't very big, so the standard metric prefixes are added for larger amounts: a "kilojoule" is a thousand joules, a "gigajoule" is a billion joules, and so on. A 2000 kilogram (2 tonne, 2.2 ton) automobile moving at 100 km/hr (about 62.5 miles an hour) has an energy of roughly 772 thousand joules (772 kilojoules or kJ).

When talking about global energy usage, the "exajoule" (EJ) is useful. It's one quintillion joules, written more conveniently as 10^{18}, read 10-to-the-eighteenth. Global energy use in 2012

---

[1] A bit of nomenclatural trivia. When a unit of measurement is named after a person, it's lowercase when spelled out in full, but its abbreviation starts with a capital letter. James Prescott Joule was a 19th century British physicist who pioneered measuring the amount of energy in heat—the "mechanical equivalent of heat" as physicists then called it. Similarly, "watt," named for steam-engine pioneer James Watt, is lowercase when spelled out, but is abbreviated "W."

was about 553 exajoules, of which the US alone accounts for a bit less than a fifth (100 EJ).[2]

An older metric unit is the *erg*, which is based on the gram and centimeter instead of the kilogram and meter. A two-gram mass moving at the speed of 1 centimeter per second has an energy of one erg. It's exceedingly small by everyday standards: there are 10 million (10^7) ergs to a joule.

Another traditional metric unit, which is slowly falling out of use, is the *calorie*. It's the amount of heat needed to raise 1 cubic centimeter of water one degree Celsius (centigrade). It's also pretty small, but not so small as the joule: there are 4.18 joules in a calorie. Chemists have traditionally used the *kilocalorie* (1000 calories) in measuring the energy of chemical reactions, though it's gradually being replaced with the kilojoule. The food "calorie" is really a kilocalorie—and so "counting calories" really just means measuring one's energy intake.

A "sort-of" metric energy unit is the kilowatt-hour (kWh), widely used in measuring electrical energy. A watt is a joule per second (Box 2.2), so a kilowatt is a thousand joules per second. Since there are 3600 seconds in an hour, a kilowatt hour is thus $3600 \times 1000 = 3,600,000$ joules $= 3.6$ megajoules (MJ)—somewhat less than five times the energy of the 2-tonne car going 100 kilometers an hour.

Physicists often use yet another unit, the "electron volt." It's the energy gained by a single electron in falling through a potential of one volt. It's tiny indeed: there are 6,242,000,000,000,000,000 (6.242×10^{18}) electron volts in a joule. Electron-volts are traditionally used in talking about the properties of semiconductors, such as for solar-cell applications, because then how much energy a single electron receives is of interest. A single photon—a "light particle"—of yellow light has an energy of about 2.25 eV.

---

[2]These values, and others unless a different source is specifically cited, come from the US Department of Energy's Energy Information Administration, www.eia.gov

There is a set of energy units based on traditional English units, too. The "foot-pound" is the amount of work done in lifting one pound up one foot in Earth's gravity. It's equal to 1.356 joules or 0.324 calories.

The heat-based unit in the English system is the British thermal unit, or BTU. It's defined as the amount of heat needed to raise the temperature of one pound of water one degree Fahrenheit. It's 1054 joules or 252 calories or 777.6 foot-pounds. It probably would sound funny to rate a room air conditioner in foot-pounds, or talk about the BTUs in one's diet, but one could do so perfectly correctly. This is an example of how the welter of energy units just causes confusion!

A quadrillion BTUs (1,000,000,000,000,000 BTUs—i.e., 10^{15}) is often called a Quad, and is also commonly used when talking about global energy usage. A Quad is nearly the same as an exajoule—it's 1.055 EJ, to be exact. So, to rephrase, world energy consumption in 2012 was about 524 Quads, of which the United States' share was about 95 Quads.

We can also measure energy in terms of complete reaction of a certain quantity of a standard fuel. For example, a barrel (42 US gallons) of oil, fully reacted with oxygen, yields about 5.8 million BTU (MBTU) or 6.1 gigajoules (GJ). A short ton (2000 lbs) of coal is equivalent to about 20.7 MBTU or 21.8 GJ. These are approximate values, because the compositions of coal and oil are not all exactly the same. But they're certainly accurate enough for "ballpark" calculations.

In this book, SI units will usually be used for the purpose at hand, but often traditional units will be given to make comparisons easier.

solar energy trapped by photosynthesis was the sole basis for traditional cultures.

Usually, though, when talking about "energy" we are referring to what might be termed "external" energy, energy not in the form of food and usually obtained from sources other than photosynthesis. Its dominant purposes include

---

Box 2.2 Energy vs. Power

"Power" and "energy" are used similarly in everyday language, but they are precisely distinguished in physics. "Power" measures a *flow* of energy: how much energy is supplied in a given time. The basic SI unit for power is the *watt*, which is a joule per second. A "kilowatt" is thus a thousand watts, a "megawatt" is a million watts, and so on. It's often thought that a "watt" is specifically for measuring electricity, but it's not. It just measures power.

The traditional unit of power in English-speaking countries is the "horsepower," which is 550 foot-pounds per second or 746 watts. It was originally defined in the late 1700s by James Watt to give potential customers an idea of how many horses his steam engine could replace.

One could rate automobile engines in watts, but perhaps saying a 400 hp Corvette produces "298 kilowatts" just doesn't have the same cachet.

This also means that one can't sensibly speak of storing "power." The quantity *stored* is always energy. The *power* comes from how fast that energy can be delivered. As we'll see, certain storage systems are much better than others in delivering high energy flows, and that proves critical in applications such as automotive power plants.

---

(1) *moving things around (transportation)*
(2) *arranging matter (fabrication/processing/resource extraction/agriculture)*

This includes such things as cooking and refrigeration as well. Obviously there is also some tradeoff with transportation: if something can be made locally, it need not be brought in. (Why is agriculture in this list? Because modern agriculture uses energy inputs besides raw sunlight.)

(3) *information   processing   (communication/support/recreation/research)*

(4) *environmental maintenance (space heating/cooling)*

Energy occurs in a great variety of forms. It occurs as *kinetic energy*, the energy of motion: a moving car, the spin of the Earth on its axis. It occurs as *potential energy*, the energy of position: a rock poised to fall, the water impounded in a reservoir behind a dam. In particular, *chemical energy* is a form of potential energy: a fuel that can react with oxygen, a battery that can be discharged. Energy occurs as *electricity*, the coherent flow of electrically charged particles. It occurs as *electromagnetic radiation*, that sweep of phenomena subsuming things as seemingly different as visible light, x-rays, microwaves, and radio waves.

And last *and* least, energy can occur in the form of *heat*, its most disorganized and wasteful form (Box 2.3).

Energy cannot be created or destroyed, a principle dignified as the "conservation of energy" or as "the First Law of Thermodynamics."[1] So if energy cannot be destroyed, how can we have an "energy crisis?" Because all forms of energy are not created equal. "High-quality" forms of energy can in theory be completely transformed into work—and into each other, for that matter. Energy that can do work is called "free energy" or "available energy,"[2] and when we use the term "energy" loosely, as we do everyday life, what we are really talking about is *free* energy.

What, then, is low-quality energy? It's heat. Heat is the lowest common denominator of energy, the dregs, the least useful form. Once energy is in the form of heat, it's been partially lost. It can't be converted completely back into work, or into other forms of energy.[3] The only way that some of the heat can be transformed

---

[1] Albert Einstein showed, with his famous equation $E = mc^2$, that matter is also a form of energy. This becomes important only in the nuclear realm.

[2] The ugly and confusing neologism "exergy" is sometimes seen as well.

[3] There seems to be a contradiction here. If energy is the "ability to do work" and heat is a form of energy, why *can't* it be all converted into work? The problem is that complete conversion of heat into work would require it to

Box 2.3 Heat

What *is* heat? (Besides being the most disorganized and wasteful form of energy!) It's the *random* energy of motion of individual molecules. Consider a gas such as air, for example. In a gas, the gas molecules are moving freely and bouncing off each other like billiard balls. The heat of a gas measures the average energy of those moving molecules. What, then, is temperature? It's a measure of the average speed with which the molecules are moving. The higher the temperature, the faster on average the molecules are moving. In fact, the original definition of "absolute zero" is that temperature at which molecular motion ceases. (It turns out that due to the laws of quantum mechanics, molecular motion never can *completely* cease, but that's a detail.)

The molecules making up a solid substance are not moving around freely, but they're still moving. They're jiggling around their average positions, and the higher the temperature, the more violent the jiggling. If the temperature gets too high, the jiggling gets too violent and the molecules separate. The solid melts.

In a liquid the molecules are no longer fixed in place. Even though they're sliding around, though, there's still enough attraction between the molecules to keep them from separating completely. When even more energy is added, however, there comes a point at which the molecules *will* separate from each other. The liquid boils, to yield a gas.

Actually, even below the boiling point the fastest-moving molecules are escaping from a liquid at the surface. That's what evaporation is. That's why, of course, something can be dried without heating it to the boiling point. Or at least it can be dried if the air's not already holding all the water vapor it can. If the relative humidity's 100%, water never evaporates completely, because on average as many molecules of water go back into the liquid as evaporate from it.

Radiant heat

A glowing stove gives off heat, as is certainly familiar. In fact, anything that's at any temperature above absolute zero gives off heat radiation. For reasons that actually are sensible when one looks at the details of the physics,[1] a perfect heat radiator is called a "blackbody." Of course, blackbody radiation can be described mathematically, and is historically important because it provided one pathway to the theory of quantum mechanics. For our purposes, though, the details don't matter. What's important is that like any form of heat, radiant heat, or blackbody radiation, is subject to the Carnot limit (Box 2.5). Since to a good approximation the Sun is a blackbody at a temperature of about 5800 K, solar energy is fundamentally Carnot limited as well. However, because of the high temperature of the Sun, very high theoretical efficiencies, roughly (5800–300)/5800 ~ 95% are possible, assuming an Earth surface ambient temperature of 300 kelvins, or 26.85°C. It's probably superfluous to note that no current application of solar energy remotely approaches such an efficiency!

Parenthetically, sunlight is hardly a "low entropy" source, as is sometimes naïvely claimed in the popular literature. The Sun is generating entropy (Box 2.4) at a prodigious rate, transferring some 4×10^{26} joules to a reservoir of roughly 4 kelvins (the "background temperature" of outer space) every second. This yields an entropy generation rate of some 10^{26} joules per kelvin per second. The Sun is a source of free energy, not "low entropy."

---

[1]Something that's "black" absorbs all radiation that impinges on it. Now imagine a completely enclosed oven with a tiny hole in one wall. Any light—any type of radiation at all—that lands on that pinhole goes right through. It's completely absorbed. But that pinhole also lets out the light from the glowing walls of the oven. That's the textbook example of a perfect "blackbody" radiator.

back into work is to let it flow to a cooler reservoir, and the amount that can be turned into work is proportional only to the *difference* in temperature between the reservoirs.

Furthermore, all *real* processes that transform energy, as opposed to the "theoretical limits" in elementary textbooks, inevitably make some heat as a byproduct, and then that heat is lost, dispersed into a form in which it's unavailable to do work. (On the Earth this waste heat is ultimately radiated to outer space, which acts as an infinite sink.) The upshot is that free energy can't be recycled completely, despite the First Law. Some is always lost on usage.

Hence, ongoing sources of free energy—or just "energy," loosely—are required for all biological and technological activity. Sunlight is the source for the biosphere. In the case of conventional technology a whole suite of sources is used, but the dominant source is fossil fuels, which will be discussed below. These have an energy density considerably higher than that of sunlight. Their high density, however, merely turns out to be a brute-force compensation for the inefficiency with which they are used, as will be seen.

Formally, scientists measure the "unavailability" of energy with a physical quantity called *entropy*. The "Second Law of Thermodynamics" states that *total* entropy can never decrease in spontaneous processes, and in real-world processes—as opposed to the theoretical "reversible limit"—it invariably increases.[4] As befits its fundamental relation to heat, entropy has units of energy divided by temperature: joules per kelvin, in SI units. That said, entropy has been wildly misunderstood and misused,

flow to a "cold" reservoir at absolute zero, and one consequence of the Third Law of Thermodynamics—which otherwise we aren't concerned with—is that absolute zero can't be reached, only approached. From a practical standpoint, of course, even this is irrelevant because achieving a lower temperature for a cold reservoir than for the ambient environment would require refrigeration—and that requires energy!

[4]Many texts on thermodynamics and physical chemistry discuss the formulation of the thermodynamic laws at length. See Moore, WJ. *Physical Chemistry*, 4th edition, Prentice Hall, 1972, pp. 77–115, for example.

and not just in the popular literature (Box 2.4). In particular, the qualifier "total" is key. *Local*, spontaneous decreases of entropy are not just possible, but ubiquitous, and the box contains a worked example.

Box 2.4 Entropy and its Misuse

A little learning is a dangerous thing.
—Alexander Pope

Some scientific concepts have more than their share of confusion surrounding them: relativity, quantum mechanics, evolution … and entropy. Entropy is a precise thermodynamic concept, but it has been wildly misunderstood and misapplied. Many qualitative statements of the form "the laws of thermodynamics say that x must happen," particularly in the popular literature, are simply wrong. At the very least such pronouncements should be buttressed with numerical values, and it furthermore should be a red flag when such values are not presented. Indeed, although it's certainly true that the laws of thermodynamics, like any physical laws, constrain human affairs, when evaluated quantitively the constraints prove to be considerably more permissive than often thought.

So, in an attempt to clear the air, here are some facts about entropy.

First, entropy is not a mystical notion, and it's not a vague epithet for "things I don't like about industrial civilization" or some such. It is a real physical quantity. It is deeply related to heat, and as noted in the main text it has units of heat divided by temperature; joules per kelvin, in the SI system. Values of entropy can be, and routinely *are*, measured on real substances, and those values can be looked up in standard references such as the *Handbook of Chemistry and Physics*. (Cf. the values quoted below.) The amount of entropy generated in transferring a given amount of heat Q to a (cooler) reservoir at absolute temperature R is simply:

$$S = Q/R$$

where S is the entropy. And in fact, determining the entropy of real substances involves measuring their heat capacities, such measurements having been carried out since the latter 19th century.

Second, the second law of thermodynamics states that the *total* entropy of a *isolated system* (i.e., a system closed with respect to matter and energy) must not decrease in a spontaneous process. Note the emphasis. The second law does *not* state that *local* spontaneous decreases of entropy can't occur. They just must be compensated for by an increase of entropy elsewhere in the system.

And in fact, such local, spontaneous decreases of entropy are ubiquitous in real processes. An everyday example is ice (frost) forming on a cold winter's night by condensation from the atmosphere. The entropy of water ice, per mole (Box 2.7), at 0°C (273.15 K) is 38.1 joules per kelvin (J/mol K), which is considerably less than that of the vapor (224.9 J/mol K) from which it condensed. However, when it condensed the water vapor released its heat of vaporization, some 51000 J/mol, into the environment, and that leads to an entropy increase of some 51000 J/273 K = 186.8 J/mol K. As it must, this balances the entropy lost on the formation of the ice.[1] In fact, even more entropy than this is generated because heat is in turn radiated to the even colder reservoir of outer space.

---

[1]Sources of data for calculation of the heat of sublimation from the Clausius-Clapeyron equation: Weast, RC; Ed. *CRC Handbook of Chemistry and Physics*, 50th ed., p. D-136-7, 1970; For the absolute entropy of water at the triple point: Meyer, CA; McClintock, RB; Sylvestri, GJ; Spencer, RC, Jr. *Thermodynamic and* Transport Properties of Steam, Am. Soc. Mechanical Engineers, New York, 1979; Giauque, WF; Stout, JW. The entropy of water and the third law of thermodynamics. The heat capacity of ice from 15 to 273°K. *J. Am. Chem. Soc.*, *58*, 1144–50, 1936. Note that the Steam Tables define the entropy of water at the triple point to be 0.0, so this offset must be accounted for.

This leads to the third point: the Earth is not an isolated system! Neither is Earth's surface. Nor is the biosphere. Naïve authors often give the impression that entropy "accumulates" on the Earth in some fashion. This is simply not so. The Earth is warmed by energy received from the Sun, that heat generating entropy in the process (irrespective of whether life's present or not), and then the heat is later re-radiated to space, as in the example above. Thus the "thermodynamic system" is the Sun, Earth, *and* the Universe. In fact, the Earth lies between two effectively infinite thermal reservoirs at constant temperature: the Sun and outer space. (They're "effectively" infinite because we can neither cool down the Sun nor warm up space.) The flow of energy between these reservoirs powers life, as well as weather, ocean currents, and so on. Note especially that it's as critical that heat be re-radiated to space as that new energy arrive from the Sun. In the language of *non*-equilibrium thermodynamics, the biosphere is an example of an "energy dissipative system." Life maintains its highly ordered systems with this energy through-flow. Thus, it is just not true that "one blade of grass now means one fewer in the future" as one naïve "ecological economist" put it.[2]

Fourth, entropy is fundamentally defined in terms of *heat*; that is, in terms of the unavailability of energy to do work. It has no direct relation to any other form of energy. As already implied, though, entropy has a deep connection to disorder, because heat is *random* molecular motion (Box 2.3).

Or, more correctly, entropy has a deep connection with *probability*. The increase of total entropy in closed systems

---

[2]Georgescu-Roegen, N. *The Entropy Law and the Economic Process*, Harvard University Press, 1971, p. 296. Although Georgescu-Roegen was certainly correct that physical laws fundamentally constrain human activity, this book garbled the thermodynamics beyond recognition. (An amusing exercise would be to have a bright undergrad or first-year physics graduate student highlight the howlers in the narrative.) Unfortunately, he is largely responsible for the misuse of entropy in the "ecological economics" literature, from which it has seeped into popular accounts.

does not reflect a mysterious, perverse drive to disorder. It merely measures a drive to the most *probable* state. *Other things being equal*, disorder is simply a more likely way for things to be. A deck of cards has millions of poker hands, but only 4 royal flushes; and even small everyday systems contain (typically) a trillion trillion atoms. Thus the possible combinations beggar the term "astronomical." A uniformly warm body does not spontaneously become ice cold at one end and boiling hot at the other simply because it's a grotesquely improbable thing to happen.

A system at "thermodynamic equilibrium" is incapable of further spontaneous change. But as in the frost example above, a system at equilibrium may include highly ordered phases (i.e., the ice crystals); it is *not* necessarily completely disordered, and in fact usually it is not. Some authors have thought that matter, over time, becomes uniformly dispersed, and this also is simply wrong.[3]

The reason is that other things are *not*, in general, equal. Enormous forces act between atoms (Box 4.5), and this leads to large energy differences among the possible configurations of the atoms. Thus, in the case of the ice crystals, there is a tradeoff between the increase in disorder resulting from scattering those atoms, versus the energy needed to tear the ice crystal apart. In water ice, for example, the hydrogen atoms in one water molecule are strongly attracted to the oxygen of another (a so-called "hydrogen bond," see Box 4.5), so that the lowest energy arrangement is an ordered, rigid array: i.e., a crystal. Indeed, as discussed in Chapter 4, such "self-organization" is both ubiquitous and important.

Here's an everyday analogy. Consider dropping a mixture of steel and cork balls into a bucket of water, and shake the bucket up. Now what's the probability of finding all the steel balls on the bottom and all the cork balls on the top? It's obviously 100%. The balls' positions aren't going to

---

[3]E.g., the bogus "fourth law of thermodynamics" proposed by Georgescu-Roegen.

be randomized, no matter how much the bucket is sloshed around. Strong forces in the system push the steel to the bottom and the cork to the top.

Only in the case where the forces between atoms are negligible—that is, in a gas—do the atoms tend to a maximally disordered configuration. However, even such a random arrangement is not necessarily at equilibrium. A 2:1 mixture of hydrogen and oxygen gases obviously contains a great deal of free energy!

It's worth repeating: the laws of thermodynamics, just like other fundamental physical laws such as gravitation and electromagnetism, place basic constraints on human activity. However, thermodynamics is both more subtle and more quantitative than many popular expositions recognize, and in any case the thermodynamic limits must be evaluated quantitatively.

The Promethean Paradigm

Prometheus, they say, brought God's fire down to man
And we've caught it, tamed it, trained it, since our history
began.

—Jordin Kare, "Fire in the Sky"

Since heat is the lowest form of energy, it follows that the generation of heat should be avoided at all costs. Yet how is energy used in conventional technology? "Fuels" are "burned"; that's what fuels are *for*. Indeed, "energy" often means just heat—a generally unexamined assumption that might be termed the "Promethean paradigm." After all, in Western culture "fire" has been a metaphor for technology at least since the legend of Prometheus.

That's why current technology is so vastly inefficient. In fact, present-day energy efficiencies lie anywhere from factors of several to orders of magnitude below the thermodynamic limits.

And *this* is why the high energy densities of conventional fuels are merely a brute-force compensation for the inefficiencies with which they are used.

Let's look at heat in more detail.

Heat Engines

When fuel is burned to run an engine, the chemical energy of the fuel has been converted into heat. Now only *part* of that heat can perform useful work, and it must do so by flowing to a cooler environment. Furthermore, the ultimate efficiency with which heat can be converted into work is set by the *difference* in temperature between where the heat was created and that cooler environment. That's worth emphasizing: the amount of work possible depends, not on the temperature itself, but on the *difference* in temperature. The Sun, with a surface temperature of some 5800 K, obviously contains considerably more energy per unit area than the Earth. Nonetheless, that high energy content could not be used to run a heat engine unless there was a cooler reservoir to which the heat could flow (Box 2.5).

On the Earth, the "cooler environment" to which the heat flows is that at the Earth's surface. In turn, heat in the environment, whether the result of technological or biological activity, or merely from heating by the Sun, is eventually radiated to space.

In practice, the heat created by burning a fuel drives the expansion of gases: gaseous combustion products, in the case of an internal combustion engine; steam, in the case of a conventional boiler; or another so-called working fluid. Mechanisms—pistons, turbines, or whatever—then convert that expansion into mechanical motion. Nearly all current transportation technology uses this paradigm, from the quaintest steam locomotive (Fig. 2.1) to the latest jet turbine. Of course, any other application using a fuel-burning engine also shares this inefficiency, such as mechanical processors (grinding, crushing,

Box 2.5 Sadi Carnot and His Limit

This maximum possible efficiency of a heat engine is called the Carnot efficiency. It's named after Sadi Carnot, a French engineer who worked it out in 1824 (!—that's not a misprint), and it's a direct consequence of the second law of thermodynamics.[1] Let the Carnot efficiency be e_c, T_h the temperature of the "hot" reservoir and T_c the temperature of the "cold" reservoir to which the heat flows. (Again, a heat engine always involves a flow of heat between reservoirs at different temperatures.) Then the efficiency is simply:

$$e_c = (T_h - T_c)/T_h.$$

T_h and T_c are so-called "absolute" temperatures, i.e., temperatures measured from absolute zero, $-273.15°C$. The Celsius (centigrade) scale in which absolute zero is taken as zero is called the Kelvin scale, and its units are "kelvins" (formerly called "degrees kelvin"), abbreviated K. So, for example, $1000°C = 1273.15$ K, $100°C = 373.15$ K, and $0°C = 273.15$ K.

In principle, then, if heat is flowing from a reservoir at $1000°C$ to one at $0°C$, the maximum possible efficiency is

$$e_c = (1273 - 273)/1273 \approx 79\%$$

whereas the Carnot efficiency for heat flowing from a reservoir at $100°C$ to $0°C$ is only

$$e_c = (373 - 273)/373 \approx 27\%,$$

where "$\approx$" means "approximately equal."

Obviously, other things being equal, the hotter the "hot" reservoir, the more efficient the conversion.

By the way, there is also a "Fahrenheit" temperature scale in which absolute zero is taken as zero. It's called the "Rankine"

---

[1]E.g., Moore, Ref. 4 (in main text), pp. 78–85.

scale and its units are degrees Rankine (°R). So, for example, −459.59°F = 0°R; 32°F = 491.59°R, 70°F = 561.59°R, and so forth. The Rankine scale is still used by American engineers for thermodynamic calculations, but is hardly used otherwise. Obviously, the same efficiencies are calculated whichever scale is used.

Figure 2.1 Steam locomotive.

A 19th-century apotheosis of the thermal paradigm. Not only is a steam engine Carnot-limited, it is exceptionally inefficient even so, typically less than 10%. This observation suggests a generalization: "The more spectacular a technology, the cruder it probably is. A truly mature technology will be as exciting as watching the grass grow."

shredding, threshing, etc.) or, most especially, the steam turbines driving electrical generators. It's also immaterial whether the heat is made by chemical reactions or by nuclear reactions or by whatever means.

In fact, even the Carnot limit is seldom approached in real applications (Box 2.6). Moreover, heat has other practical disadvantages. Fuel must be dry—impurities such as water limit the temperature that can be attained by absorbing some of the released energy. And that's, of course, if the fuel can be ignited at all. Burning is also intrinsically polluting, as combustion products are automatically generated. The very waste heat can itself become a pollutant, as when heat discarded into a river

Box 2.6 Entropy Costs: a Worked Example

Let us look at the free energy that could be extracted from the reaction of hydrogen (H_2) and oxygen (O_2) gases in stoichiometric proportion (i.e., in the exact ratio necessary so that no H_2 or O_2 is left over) to form water vapor at standard temperature and pressure (STP, 1 atmosphere and 25°C):

$$H_2 + \tfrac{1}{2} O_2 \rightarrow H_2O \text{ (g)}. \qquad (2.6\text{-}1)$$

where the "(g)" means the resulting water is in the gaseous (i.e., vapor) phase.

For processes at constant pressure and temperature, typical of those at the surface of the Earth, the maximum useful work that can be obtained (or alternatively, the minimum energy cost of a thermodynamic transformation, the so-called "reversible limit") is given by the change in the so-called "Gibbs[1] free energy:"

$$\Delta G = \Delta H - T \Delta S \qquad (2.6\text{-}2)$$

Here ΔH (read "delta H") is the change in *enthapy*, the so-called "heat content" of a substance, whereas T is the absolute temperature and ΔS is the change in entropy. The $T\Delta S$ term, which has dimensions of energy, represents the "tithe" paid to the second law. It is a measure of the energy that is lost. Hence the maximum thermodynamic efficiency is given by $\Delta G/\Delta H$.

Obviously, ΔG depends on temperature. Since ΔH and ΔS are functions of temperature and pressure, ΔG is a function of pressure as well. Furthermore, because the $T\Delta S$ term increases with temperature, the irreversible losses also are larger the higher the temperature at which reaction is carried out.

For reaction (2.6-1), the Gibbs free energy at STP is −228.6 kilojoules per mole (kJ/mol; Box 2.7), whereas the enthalpy is −241.8 kJ/mol. (The minus sign just means that the reaction *releases* energy. By convention, positive values are used when

---

[1]Named after Josiah Willard Gibbs (1839–1903), who carried out a thorough mathematical formalization of the laws of thermodynamics toward the end of the 19th century.

a reaction absorbs energy.) In the thermodynamic limit, therefore, some $-228.6/-241.8 \approx 95\%$ of the total reaction energy is available to do useful work.

This is far more efficient than conventional processes. For example, the heat from the stoichiometric combustion could be used to form 3.8 moles (68.4 grams) of steam at 550°C (823 K) at a pressure of roughly 2.1×10^7 pascals (3000 psi), which are typical operating conditions for a modern steam-turbine electrical generating plant.[2] The limiting Carnot efficiency (Box 2.5), assuming discharge to a reservoir at 25°C (298 K), is then only:

$$e_c = (823-298)/298 = 63.8\%.$$

In practice, however, the Carnot limit doesn't apply even theoretically. For engineering convenience the steam will be allowed to expand with as little increase in entropy as possible, as part of a so-called Rankine cycle, and this expansion will do work by spinning the turbine blades. The entropy of the steam at high pressure is 177.0 joules per kelvin per mole (J/mol K). Its enthalpy (which is just its heat content, as no chemical reactions are occurring) is 63.8 kJ/mol.[3] If this steam is expanded to a pressure of 3386 pascals (1 inch of mercury, a typical value) the enthalpy at the same entropy is 38.0 kJ/mol. The limiting efficiency is then only $(63.8-38.0)/63.8 \sim 40\%$. As can be seen, the Rankine cycle is considerably less efficient than even the Carnot cycle.

This underscores that a great deal of the free energy potentially available has been sacrificed by converting the chemical energy of the fuel into heat, and this conversion is *not* required by the laws of thermodynamics. These are limits set by particular technological approaches.

It's worth repeating: we have had the "heat" crisis, not the "energy" crisis.

---

[2] E.g., Peltier, RV. Steam turbines. In *Handbook of Turbomachinery*, Logan, E, Jr. ed. Dekker, 1995.
[3] E.g., Meyer et al., Box 2.4, Ref. 1.

Box 2.7 Of Moles and Mass

Physical scientists and engineers routinely express properties of substances, such as their energy content, in terms of the number of molecules or atoms involved. This usage is so convenient we'll also follow it, albeit sparingly. The phraseology is "such-and-such a quantity per *mole*," where a "mole" consists of Avogadro's number of atoms, or molecules, or whatever. For example, the heat content, or *enthalpy*, of a mole of methane is −74.8 kJ at STP (standard temperature and pressure), the minus sign here meaning that it *takes* this much energy to break a mole of methane molecules apart (or alternatively, that this much energy is released on forming methane from carbon and hydrogen). Methane, CH_4, is the simplest hydrocarbon (Box 2.8) and the main constinuent of natural gas.

The *gram-molecular-weight* (or gram-formula-weight, or just molecular weight) is the weight, in grams, of one mole of a substance. Since the GFW of methane is 16 grams, that's how much a mole of methane is. Gram-formula weights are calculated from the so-called atomic weights of the atoms making up the substances. They're tabulated in standard references.

By everyday standards Avogadro's number is huge, roughly 6.02×10^{23}, or almost a trillion trillon. Its size reflects the fact that everyday quantities of substances contain an absolutely enormous number of atoms. For example, the gram-molecular weight of water (H_2O) is 18 grams. That means there are 6.02×10^{23} water molecules in 18 grams of water.

The sheer size of Avogadro's number proves to be a critical consideration when we turn to the issues involved in nanotechnology, because nanotechnological fabrication, where we aspire to "a place for every atom, and every atom in its place" involves somehow positioning numbers of atoms that beggar the term "astronomical." There are a number of approaches to this problem (and indeed, biology illustrates a set of solutions), and they will be considered in more detail in Chapter 4.

makes it too warm for the indigenous fish. The inefficiency of burning of "fuels" to release heat also forces technology to use concentrated energy sources, and this is one reason diffuse sources such as solar energy have been largely impractical.

Finally, it is worth re-emphasizing that when a fuel is burned, a great deal of the free energy potentially available is thrown away by converting the chemical energy of the fuel into heat. In the case of an automobile, for example, some two-thirds of the gas tank goes right out the radiator! This wastage, furthermore, is *not* required by the laws of thermodynamics. It is a consequence of particular technological approaches. We have had the "heat" crisis, not the "energy" crisis.

Work Smarter, Not Hotter

The trick in efficient utilization of energy is to avoid turning it into heat at any step in the process. This is the secret of the efficiency of metabolic processes.
—Dickerson, Gray, and Haight[5]

Well, obviously the opposite of thermal technology is non-thermal technology. Simple mechanical motion is non-thermal; the efficiency by which a turbine can be spun by wind or flowing water, for example, is not subject to the Carnot limit. Electricity is also highly organized energy. Such devices as electric motors, for example, are not heat engines, and so their efficiencies are not limited by Carnot's cycle. Fuel cells, which are finally receiving serious attention as mobile power sources, convert fuel directly to electricity without generating heat first. In theory, they also could run on any fuel, not merely hydrogen.

Let's first turn, though, to the biological inspiration again. As already mentioned, although we loosely speak of animals (and humans) "burning" food, in fact biology does not do anything nearly so inefficient. Organisms are not heat engines

---

[5]Dickerson, RE; Gray, HB; Haight, GP, Jr. *Chemical Principles*, W.A. Benjamin, 1970, pp. 601–2.

and do not rely on flows of heat. All energy conversion takes place isothermally; that is, at constant temperature, in stark contrast to most technological systems. Of course, organisms do generate heat, but it is a by-product, or else generated to maintain a particular isothermal environment, as in warm-blooded animals. Biosystems most emphatically do *not* set up a heat gradient and then convert some of that heat to work as it flows toward a cooler environment. Furthermore, aside from trifling exceptions such as the microbial communities at subsea spreading centers, the biosphere is powered completely on solar energy. Global "primary productivity," the solar energy trapped into biomass, has been estimated as about 3450 EJ,[6] about 7 times the global consumption of non-biological energy.

This shows that the common belief that "sunlight is simply too diffuse" to run a technical civilization is simply wrong. Sunlight is only too "diffuse" to run a paleotechnical civilization, one that relies on "burning" fuels to generate vast flows of heat.

But we will defer further discussion to Chapter 5. Let's look now at conventional energy sources.

Conventional Energy Sources: The Numbers and the Problem

> *Analysis without numbers is just opinion.*
> —Robert A. Heinlein

> *Figures don't lie.*
> —Anonymous

Primary Sources of Energy

These are those in which the energy already exists in the environment, and is merely collected for human use. In

---

[6]Calculated from values in Field, CB, Behrenfeld, MJ, Randerson, JT, Falkowski, P. Primary production of the biosphere: integrating terrestrial and oceanic components. *Science 281*, 237–40, 1998.

other words, a primary source does not merely represent the transformation of some other kind of energy into a more convenient form.

Renewable Energy Sources

Renewable sources of energy, as is familiar, are not exhausted when gathered and used. They instead persist as ongoing flows over geologic intervals of time, and are thus effectively "eternal" on human timescales. Nearly all rely on energy input from the Sun.

Biomass

This is all once-living matter, and it ultimately reflects sunlight stored by plant photosynthesis. It obviously includes foodstuffs, plus such other agricultural commodities as wood and fiber. A great deal of this material, however, is just waste, including substances as diverse as waste paper, agricultural debris, timber slash, sewage sludge, feedlot debris, and so on. Although wood and dung have been used as fuels since time immemorial, their contribution to the energy mix in the modern world is small. Indeed, at present waste disposal is an energy *cost*. Nonetheless, waste biomass represents a largely untapped resource of both energy and materials, as is discussed further in both Chapters 5 (p. 242) and 6 (p. 331).

Hydropower[7] and wind

Heating by the Sun drives the global systems of weather and ocean circulation. Hence, both wind and flowing water represent conversion of a small percentage of the Earth's solar energy inputs into mechanical motion. In effect, the Sun is driving a

---

[7]Obviously, as per Box 2.2, the term should be "hydroenergy," but we'll stick with the traditional phrase.

huge, inefficient heat engine, and the fluid flows generated by this heat engine represent a source of energy that can be tapped. To a much greater degree than other natural sources, however, they suffer from what might be termed "storm risk." The vagaries of weather demand that power-extracting infrastructure must be highly overdesigned to withstand the rare freak storms or floods. This obviously greatly increases the capital expenditure.

The fall of water originally raised by solar evaporation has been used as an energy source for centuries. Nearly all is now used for electricity generation, accounting for something like 17% of electrical capacity worldwide, although there are great variations among countries. It has disproportionate importance despite this relatively small percentage, because the generating capacity is easy to turn on and off. It is much easier to open a valve leading to a turbine than to fire up a thermal generating station. Hence, hydropower is now nearly all "peaking power," used at the time of highest demand. Indeed, hydraulic head in a reservoir is one of the few ways to "store" electricity at present.

Of course, hydroelectric dams have serious environmental consequences. They have profound and generally deleterious effects on riparian ecosystems, including destruction of fisheries, and through interruption of sediment transport. They also permanently flood farmland, and in any case reservoir lifetimes are finite, on the order of a few centuries at most, due to ongoing sedimentation, so that silting in of the reservoirs will be a growing problem in the coming years. Finally, few sites remain for hydropower dams even if building them were not politically difficult. It thus is somewhat ironic that hydroelectric energy is the single largest renewable energy source at present.

Wind, of course, also results from the stirring up of the atmosphere by solar heating. Its use is also ancient, but in recent years it has become a competitive source of electricity, at least in certain cases.

Direct solar energy

Of course, direct use of solar energy exists, and some such applications (e.g., drying food, evaporating seawater for salt) go back to antiquity.

Solar thermal *electric* energy, ironically, has recently become economic locally.[8] Such installations use sunlight to generate steam, which then powers a conventional turbine. Thus, the energy of the sunlight is turned into heat, which then powers a Carnot engine. It's an odd combination of 19th- and 21st-century technology, but at least for now such plants have proved to be more economic than photovoltaic installations.

Further discussion of the large-scale *direct* use of solar energy is deferred to the "Alternatives" section below.

Tidal energy

The gravity of the Moon raises a bulge in the sea surface that is directed both toward and away from the Moon, and that follows the Moon on its orbit around the Earth. As the Earth rotates through this bulge, sea level rises and falls as seen at a given place on the Earth, at roughly 12.5 hour intervals. The change of sea level as the tide flows in and out can be used to drive a turbine or analogous device. Although tidal energy is of course not derived from the Sun—the energy comes from the rotation of the Earth—the same sorts of issues arise as for surf power and low-head hydropower (p. 74). Although some such installations exist, their contribution to the current energy mix is negligible.

Geothermal energy

The interior of the Earth is very hot, and so the thermal difference between the interior and the surface is potentially a vast energy resource. It is also an energy source that does not rely ultimately

---

[8]E.g., Nevada Solar One, outside Boulder City, Nevada, USA: www.acciona.us/projects/concentrating-solar-power/nevada-solar-one

on the Sun. Unfortunately, it has been impractical in all but unusual situations because the thermal gradient is generally so low. Prohibitive depths would need to be reached to obtain a thermal difference that could be exploited with conventional technology. The typical geothermal gradient in the crust is only ~25°C/km, so that even reaching the boiling point of water (at standard temperature and pressure) requires impractical depths. Moreover, conventional geothermal power generation relies on steam as a working fluid, and to get the volumes required the steam must have access to a large volume of rock. Thus the rock must be reasonably porous, with cracks if nothing else, but the deeper the level, the greater the degree to which any open space has been closed up by the confining pressure.

Therefore, for conventional geothermal power the geothermal gradient must be considerably elevated, so that temperatures of 100–200°C can be attained within a kilometer or so of the surface. In addition, the rock must both be permeable and be thoroughly saturated either with steam or with water sufficiently hot that a significant fraction will vaporize on exiting to atmospheric pressure. A further engineering difficulty ensues because hot water in contact with rock tends to be highly mineralized, so that scaling from the precipitation of solid compounds is a serious problem. The hot water is typically groundwater of ultimate meteoric origin, so the degree to which it is mineralized tends to reflect the average residence time of the water. The longer recharge takes, the more mineralized the water.

The requirement of a working fluid leads to further difficulties. First, water loss by extraction is often a serious issue. Hence many geothermal plants replenish the water by reinjecting the spent, cooled water back into the field. This has the added advantage of disposing of the water, which is commonly too mineralized or saline to release to conventional wastewater treatment without special pretreatment. Second, many areas of high geothermal gradient contain little groundwater: so-called hot dry rocks. There have been ongoing experiments on drilling and fracturing such rocks, followed by injection of water to make steam. The US Department of Energy in particular has been conducting such studies since the late 1970s. However, the process is not yet practical.

Finally, many geothermal fields are too cool to produce much steam. If conveniently located, these can be used as a source of low-grade heat for space heating or for such applications as drying food.[9] Alternatively, in so-called two-phase generation systems, the hot water is used to boil a working fluid with a lower boiling point, which is used to drive a turbine. Obviously, however, this both increases complexity and decreases net power output.

The upshot is that at present the geothermal contribution to electrical generation is extremely small, being less than 0.5% worldwide (as of 2012). Geothermal production is also not completely "renewable," even on human timescales, because the rocks cool down, and the thermal conductivity of rock is so low they take decades to heat up again. This furthermore means that it's easy to overproduce the reservoirs.

Finally, conventional geothermal energy, even more than hydropower dams, often involves a tradeoff with aesthetic values, because many of those areas with anomalously high heat flow are also places of great scenic beauty. Yellowstone National Park in the United States is an outstanding but hardly contrived example.

Nonrenewable Sources: Fossil Fuels

Conventional fossil fuels, which include coal, oil, and natural gas, constitute the bulk of energy usage today worldwide, around 86% of that 553 EJ used in 2012. Furthermore, the percentage has only marginally declined in recent decades: in 1970 fossil fuels amounted to some 88% of global use. And, of course, the *total* amount of fossil fuel energy being used today is much greater, about twice as much as in 1970, even though it makes up

---

[9]E.g., at Brady's Hot Springs, about 80 km east of Reno, Nevada, USA.

a slightly smaller fraction of the total. Moreover, fossil fuels are being depleted even while demand for energy grows worldwide, as people in developing nations aspire to a lifestyle like that in the wealthy, industrialized countries.

This, of course, is the problem. One way or another, this trend will change sharply in the near future, probably within the couple of decades or so.

Fossil fuels are *fossil* biomass. They represent ancient solar energy trapped by ancient photosynthesis, and buried, processed, and preserved by grotesquely inefficient geologic happenstance. Hydrocarbons and carbon, because of their stability, are the ultimate product of processing organic material with heat and pressure under anoxic conditions. They are clearly limited on a human timescale. Because of their tremendous importance, fossil fuels also have an enormous literature on their formation, their characterization, the geologic settings in which they occur, and exploration for them.

Also, because of the localized nature of fossil fuel deposits, a great deal of production occurs in political jurisdictions different than where it is consumed. Only a handful of industrialized nations (e.g., Canada, Norway) are self-sufficient in oil. This can lead to political conflicts and even outright military clashes. It is widely believed, for example, that the deep US involvement in the Middle East is the result of the oil resources there.[10] Moreover, as global oil consumption continues to increase, such conflicts are unlikely to abate.[11]

Oil: the global thirst

What *is* it about oil? It's both the most convenient and most limited fossil fuel—and, of course, that's not a coincidence. It has high energy density, necessary for fueling Carnot-limited

---

[10]E.g., Klare, MT. *Blood and Oil*. Metropolitan - Henry Holt & Co. 2004; McQuaig, L. *It's the Crude, Dude: War, Big Oil, and the Fight for the Planet*. Doubleday Canada, 2004.

[11]E.g., Klare, MT. *Resource Wars*. Metropolitan/Owl, 2001.

devices. Being liquid, oil is also easy and convenient to store and transport, and furthermore makes fueling an engine easy. The upshot is that nearly all transportation, worldwide, is fueled with oil.

Oil consists mostly of a mixture of hydrocarbons (HCs, Box 2.8). The most important HCs, because they are liquids at room temperature, are those with chains of roughly 6 to 30 or so carbons. When a HC reacts completely with oxygen (i.e., burns completely) it yields carbon dioxide (CO_2) and water (H_2O), and

---

Box 2.8 Hydrocarbons

Hydrocarbons (HCs), as their name implies, are chemical compounds of the elements carbon (C) and hydrogen (H). The simplest is methane (CH_4), the main component of natural gas, consisting of a carbon atom surrounded by hydrogen atoms in a perfect tetrahedron. The next simplest is ethane (C_2H_6), which contains two carbon atoms bonded to each other. Since carbon atoms can link to form chains and branched structures of almost indefinite length, hydrocarbons can become very complicated indeed as the number of atoms increases. Gasoline, for example, consists mostly of molecules with 4–12 carbons in the chain. (A HC with 8 carbons is "octane", so it's obvious where the "octane" rating of gasoline came from historically. Nowadays, though, the nominal rating has little to do with the actual content of octane.) Kerosene and jet fuel contain 6–16 carbons, whereas diesel and fuel oil contain from 8–21 carbons. Obviously, there's a certain amount of overlap among these fuels, but the short chain HCs make up a much smaller proportion of the heavier fuels than they do in gasoline.

Even longer HC chains are found in asphalt and tars, which are nearly solid. The longer the carbon chain, the better the molecules stick to each other through van der Waals forces (Box 4.5). Hence they're much more difficult to deal with, since they neither flow nor react easily. Processing such long chains into useful forms requires breaking them up ("cracking"), and that traditionally has required heat(!) as well as catalysts.

Very long hydrocarbon chains don't occur naturally, but make up the plastic polyethylene.

Because of their importance, hydrocarbons are encrusted with more than their share of jargon. Hydrocarbons containing only single carbon-carbon links are called *alkanes*, of which ethane is the simplest. They make good fuels but are relatively unreactive otherwise. For reasons that will become obvious, these compounds are termed *saturated*.

Carbon atoms can also link with a *double* bond, the simplest such compound being ethylene, C_2H_4. It has the structure:

$$H_2C = CH_2.$$

Such hydrocarbons are called *alkenes* or *olefins*. They're also termed *unsaturated* HCs, because "opening up" the double bond—i.e., making it a single bond—provides bonding capability for additional atoms. In other words, their bonding capacity is not "saturated."

Alkanes and alkenes, collectively, are *aliphatic* hydrocarbons. (Carbon is capable of making triple bonds, too, the simplest such compound being acetylene, C_2H_2. These "alkynes" are also considered unsaturated aliphatics. However, alkynes don't occur in petroleum as they're much too reactive.)

Finally, *aromatic* HCs, or *arenes*, contain rings of carbon atoms in which electrons are "smeared" (delocalized, in the language of quantum mechanics) around the ring, somewhat in the matter of a metallic bond (Box 4.5). The most familiar example, if not quite the simplest, is benzene, C_6H_6, which contains a 6-carbon ring.

Table 2.1 Energy contents

Approximate energy contents of some common or illustrative sources. These are *total* energy contents and make no allowance for efficiency losses (e.g., the lost energy in raw fission products).

Energy source	Energy content (traditional units)	Energy content (SI)
Oil	6.1 GJ/bbl	38.36 GJ/m^3
Natural gas	1.055 GJ/1000 cubic feet	37.26 MJ/m^3
Coal	26.57 GJ/short ton	29.23 GJ/tonne[1]
Hydropower		9800 J/m^3 per meter drop
Wood (dry)	18.2 GJ/short ton	20 GJ/tonne
Nuclear energy (fission)		
^{235}U fission		81.5 GJ/gram
(assuming fission products in rough 4/3 mass ratio)		
Fusion energy (theoretical)		
proton-proton (Sun)		644.2 GJ/gram
deuterium-deuterium (assuming ^{3}He as product)		97.6 GJ/gram
Solar constant -		1 kW/m^2
(full direct sunlight at Earth's surface)		

[1]Bodansky, D. Appendix: Energy Units. In *The Energy Sourcebook*, Howes, R, Fainberg, A, eds. American Institute of Physics, 1991.

lots of energy. The energy content (i.e., enthalpy, Table 2.1) of a barrel[12] of oil is roughly 6.1 gigajoules.

Oil has been cheap to produce, because locally it has been formed and collected by geologic happenstance (Box 2.9). All that's been required is to find it and gather it up—although, to

[12]For historical reasons a "barrel" of oil, abbreviated "bbl," is 42 US gallons, or 159 liters. Because traditionally oil usage is measured in barrels, that convention will be followed in this book.

Box 2.9 Conventional Oil Formation

Oil is a fossil fuel, and thus is a product of the processing of once-living material over geologic time. Not only that: it's a product of processing once-living material under a stringent set of geologic conditions. Most sedimentary rocks contain a trace of oil, but exploitable oil deposits require an unusual set of circumstances. They require a *source* of the organic matter. In fact, in many cases large oil deposits can be traced to particular organic-rich horizons through "fingerprint" compounds in the oil itself.[1]

Formation of oil occurs by thermal processing of buried organic matter in the so-called "oil window" of temperature and pressure. Too cool, and the organic matter remains embedded in the rock as a tangle of high molecular-weight, largely aliphatic polymers termed *kerogen*; too hot, and all the long-chain hydrocarbons break down into graphite and natural gas (mostly CH_4). Of even more importance, however, the rock "plumbing" must be just right. The oil, once formed, requires porous rock in which it can accumulate, a "reservoir"; and it requires a "trap", a geologic structure or structures that keep the oil from flowing away and being lost. Source, reservoir, and trap—that's the traditional trinity of the oil geologist. It's a stringent set of conditions, and many times all the factors don't come together just right, with the result that there's no oil.[2] (In some cases, production by hydraulic fracturing—"fracking"—may occur directly from a source rock where the hydrocarbons have never flowed out.)

There has been much discussion in the popular press and on the Internet about the notion, popularized by the late astronomer Thomas Gold, that oil is abiogenic—that is, not formed by the processing of once-living matter. For a host of reasons this idea is not taken seriously by petroleum geologists

---

[1] See Ref. 15 (in main text), pp. 14–39, for some nice examples.
[2] Ref. 15 (in main text), pp. 40–69, gives an excellent review accessible to the non-specialist.

and explorationists. All of the purported "unexplained" occurrences of oil have much more prosaic explanations.[3] In some cases degraded drilling lubricants were taken as evidence of natural hydrocarbon occurrences!

---

[3]Apps, JA; van de Kamp, PC. Energy gases of abiogenic origin in the Earth's crust. In Ref. 24 (in main text), pp. 81–132; Deffeyes, KS. *Beyond Oil*, Hill & Wang, 2005, pp. 62–3; Glasby, GP. Abiogenic origin of hydrocarbons: an historical overview. *Resource Geology 56*, 85–98, 2006.

be sure, that's become *much* more difficult over the years, as ever-more challenging areas (the polar regions, the deep continental shelf) are explored.

Oil consumption

As of 2014, global usage stood at about 93 Mbbl/day (including NGL), of which the United States alone consumes about 20 million barrels. Global production, after increasing by about 1.5% a year from the mid-1990s through 2005, remained roughly flat from then till 2009. Since then, however, global production has again been increasing by about 1.5% per year. This is a direct result of new technology: hydraulic fracturing, or "fracking," in which hydrocarbon-bearing rocks too impermeable for HCs to flow out are fractured in place. In the jargon, such rocks have "porosity"—voids filled with oil or gas—but no "permeability"—that is, the voids are not connected. The cracks then provide pathways by which the HCs can escape and be collected. Indeed, because of fracking, since 2012 the United States has broken annual oil production records that had stood since the early 1970s. Of course, the technology is controversial and has been blamed for groundwater contamination and even minor earthquakes.[13]

---

[13]Drogos, Donna L., ed. *Hydraulic Fracturing: Environmental Issues*. ACSSS, 1216, 2015.

Natural gas

Natural gas is akin to oil in its chemistry, formation, and occurrence; indeed, one common mode of formation is from oil that was "overcooked" geologically. It is mostly composed of the simplest hydrocarbon, methane (formula CH_4), with an admixture of some heavier HCs such as ethane (C_2H_6), propane (C_3H_8), and butane (C_4H_{10}). The heavier HCs are high-value products and are typically separated from the raw gas. These so-called natural gas liquids (NGLs) are typically considered part of oil production. The remaining "dry" natural gas, which also has had impurities such as carbon dioxide (CO_2) removed, is what is usually meant by otherwise unqualified "natural gas."

The good news is that there is still lots of gas. Even in the developed part of North America, which had had declining production up into the early 2000s, production has increased sharply in recent years. As with oil, this is due to fracking of nearly impermeable unconventional source rocks, as well as to other unconventional sources such as coalbed methane. Unlike oil, too, there is still *lots* outside North America. Even Europe, which in recent years has become ever-more reliant on imports from Russia—which in turn has led to political conflicts—may have similar large unconventional resources, although their possible development remains controversial.

Gas has many desirable attributes as an energy source. Not only is it clean-burning, it produces the least amount of carbon dioxide per unit of energy output of any natural fuel, because it has the highest ratio of hydrogen to carbon. Indeed, some approaches toward hydrogen-powered vehicles use natural gas as the hydrogen carrier, stripping hydrogen out of the methane, in so-called "reforming," to leave the carbon behind. Although this eliminates CO_2 emissions, it obviously adds complexity, not to mention the loss of energy from oxidizing the carbon.

Gas does have some problems, though. One is a relatively low energy content per volume, due to its low density: 1000 cubic feet of gas—a cube 10 feet (3 meters) on a side—has an energy content of only 1.055 GJ, less than 1/6 that of a barrel of oil.

The single biggest problem, though, is that most of the world's gas is "stranded"; that is, remote from its potential markets. Indeed, even now some gas, produced as a byproduct during oil extraction, is wastefully "flared" (i.e., burned off) because of the absence of a local market. The problem is that natural gas is hard to transport. It is transported most conveniently by pipeline, but even so it's not moved as efficiently as oil. Natural gas is a *gas* after all; it's a lot less dense than oil. So the same diameter of pipe can transport much less natural gas, proportionately, than it can oil, and thus the pipeline is proportionately more expensive. Of course, in any case it takes time and money to build pipelines, and obviously, the farther from markets, the more expensive the pipelines are.

It's especially hard to ship gas overseas, again because it's so much less dense than oil. It *can* be done; it's compressed and cooled to temperatures like cryogenic rocket fuels. At ordinary pressures, the boiling point of methane, the main constituent of natural gas, is −164°C (−263.2°F). For comparison, the liquid oxygen in the Space Shuttle external tank boils at −183°C (−297.4°F). This "liquified natural gas" (LNG) can be transported in special tankers, a remarkable technological feat in itself, but in 2015 only ~3% of US gas came from overseas.

A tanker full of liquified natural gas makes a dandy terrorist target, too.

An alternative to cryogenic cooling and shipping as LNG is to convert the gas into a more transportable form. Such "gas to liquids" technology is the focus of much current research, and is discussed in more depth in Chapter 5. This will be another application where greatly improved catalysts (discussed below) are critical.

Coal

First of all the good news: There is *lots* of coal. As of 2011, global conventional reserves amount to roughly 891 billion metric tons. These include all grades of coal recoverable with current technology. Of these, over 26% is in the United States alone, with nearly 18% in Russia and 13% in China. Even though recently the world has been going through coal at the rate of about 8.1 billion tonnes a year, that's still enough for a while yet.

For "Old King Coal" is not just a relic of the 19th century. In 2012 it still accounted for over 25% of the global energy budget. In fact, it's the largest single contributor to electricity production, in recent years accounting for nearly 40% of the electricity generated in the United States, making up about the same percentage in the energy mix as it does worldwide.

The bad news? Coal is a very dirty fuel—as is hardly news. Not only does coal smoke contain a lot of noxious compounds, as any resident of a 19th-century industrial city could attest, but it releases the most carbon dioxide per unit of energy produced of any fuel at all. Coal is mostly just carbon, and when carbon is burned completely it turns into carbon dioxide, of concern for its large role in the greenhouse effect (Box 2.10). (If carbon *doesn't* burn completely, it yields highly toxic carbon *mon*oxide—which is not only inefficient but leads to even more serious pollution issues.)

In addition, coal contains impurities that are also caught up in the combustion, and some of those combustion products are also serious pollution sources. The combustion of impurity sulfides is a major source of acid rain, for example (Box 2.11). Other impurities in the coal are also distributed far and wide. Coal burning is now the main source of atmospheric mercury, for example.

Coal mining is also extremely dangerous for the miners.

For many purposes coal is also exceedingly inconvenient. Oil and gas, of course, flow. Coal does not. It cannot be pumped

Box 2.10 The Greenhouse Effect

Carbon dioxide (CO_2) is a "greenhouse gas"—it absorbs heat radiated from the Earth, inhibiting its escape into space. The molecule has a couple of broad absorption bands in the infrared, one with wavelength near 4.2 micrometers[1] and another centered around 15 micrometers. These absorb infrared radiation radiated by the surface, after it's warmed by the sun, so that the atmosphere near the surface of the Earth in turn warms up.

A certain amount of greenhouse effect is a good and healthy thing—Earth would be frozen over completely if it had no greenhouse gases in its atmosphere—but one can always get too much of a good thing. It's known that burning fossil fuels has increased the CO_2 content of the atmosphere by about 40% since the mid-19th century, and there's now little doubt that this increase has led to an overall increase in global temperatures. *How* much warming will ultimately result, and what all the effects will be, are still points of dispute, but the warming is not. On the one hand, making large, inadvertent changes to the global environment is probably not a good idea; on the other hand, there will be winners, as well as losers, in all reasonable global warming scenarios. Beachfront property in Florida is at risk of flooding by rising sea levels, for example, due both to melting of glacial ice and to simple thermal expansion of seawater. Alternatively, however, the wheat-growing season in (say) northern Saskatchewan will be getting longer.

Moreover, the Earth has also gone through some extraordinary climatic excursions in the geologic past. Such *natural* climate stresses also indicate that the more lurid predictions of the outcome of global warming are grossly exaggerated.

---

[1]A micrometer is a millionth of a meter, and was formerly (and still commonly) called a "micron."

Although CO_2 is the most notorious greenhouse gas, it's not the only one. Trace atmospheric gases of both natural and artificial origin, such as methane (CH_4), and chlorofluorinated hydrocarbons (CFCs) such as the refrigerant Freon, are even more effective greenhouse gases, although present at vastly lower concentrations. But the most important greenhouse gas on Earth is simply water vapor. The overall water vapor content of the atmosphere, however, is set by the evaporation of liquid water, and *that* in turn is largely set by evaporation from the oceans. So regulating the H_2O content of the atmosphere isn't practical.

Ultimately the CO_2 content of the atmosphere is set by its reaction with surface rocks to yield limestone in the so-called "Urey reaction" (Box 6.7). Unfortunately, this regulating reaction occur much too slowly to keep the CO_2 content of the atmosphere from rising over the next century or so.

out of the ground; it must be mined, a cumbersome, energy-intensive, and dangerous procedure.

The coal must then be transported to where it's burned. Even the mere fueling of an engine by solid coal is cumbersome. At the end of the steam-locomotive era, for example, the fireboxes were fed by a large screw drive running the length of the tender that pushed coal chunks into the firebox. The time had long passed when a shoveling fireman could keep up with the locomotive's demands. Many modern coal-fired power plants solve the transport problem by transporting the coal as a slurry—a mixture with water that can be pumped and piped like a liquid. Of course, the required crushing and grinding of the coal, as well as the water to make the slurry, all are additional costs.

Coal *can* be converted into other fuels, including hydrocarbons, and these technologies will be touched on in Chapter 5. Such conversion, however, yields even more carbon

Box 2.11 Acid Rain

Much coal, such as that in the eastern US, contains a little pyrite,[1] an iron disulfide (formula FeS_2). One product when pyrite burns is sulfur dioxide (SO_2), which eventually forms sulfuric acid in the atmosphere via the following reactions with oxygen (O_2) and water vapor (H_2O):

$$SO_2 + \quad\quad \tfrac{1}{2} O_2 \quad\quad \rightarrow \quad SO_3$$
sulfur dioxide   oxygen   sulfur trioxide

$$SO_3 + \quad\quad H_2O \quad\quad \rightarrow \quad H_2SO_4.$$
sulfur trioxide   water vapor   sulfuric acid.

Sulfuric acid is a so-called "strong acid." When dissolved in water, it ionizes essentially completely to hydrogen ions (H^+) and sulfate anions (SO_4^{2-}); indeed, the presence of free hydrogen ions, which are highly reactive, is what makes acids "acidic." This acidity can accelerate weathering of vulnerable materials such as building stone, especially limestone. The hydrogen ions also tend to displace other cations from their compounds, putting them into aqueous solution. Depending on the metal, the consequences can range from unattractive (e.g., corroded statuary) to hazardous, in the case of dissolved toxic metals.

---

[1] Big chunks of pyrite are probably more familiar as "fool's gold," but the mineral occurs in many other circumstances.

dioxide per unit of energy, because of the additional overhead in the conversion.

These engineering inconveniences are reasons *why* there is still lots of coal left. Coal was supplanted in so many uses in the first place *because* oil and gas are so much easier to handle.

It *is* a backstop energy resource, but it's hardly optimal.

Nuclear fission energy

[Due to nuclear power] a few decades hence energy may be free—just like the unmetered air.

—John von Neumann[14]

All commercial power reactors use the *isotope* ^{235}U, which *fissions* (Box 2.12) to yield energy, largely in the form of heat(!), and that heat is used to drive a thermal plant. (Why this particular isotope? It's fissionable by slow neutrons, a few of which are thrown off when a uranium-235 nucleus splits, so that *they* can split further nuclei, and so on. This is the famous "chain reaction.") Since ^{235}U constitutes only about 0.7% of natural uranium, fuel must be enriched in the fissionable isotope, a cumbersome and capital-intensive procedure.

Unfortunately, not all the energy ends up as heat. Some remains in the fission products themselves, which remain radioactive (Box 2.14) on timescales ranging from fractions of a second to millennia. This, of course, is the problem of nuclear waste (Box 2.13).

Nonetheless, despite its dangers, highlighted most recently by the disaster at Fukushima in Japan, and the very real problems of nuclear waste disposal, nuclear fission remains a major contributor to the energy mix worldwide, contributing some 9.8 EJ of electrical energy in 2012. Its contribution, moreover, will almost certainly increase in the future if the alternative is blackouts.

A Note on Electricity

Electricity is *not* a primary energy source. It is an energy *carrier*. For many applications it's one of the most convenient forms of energy, and because it's not heat, it can be used with great

---

[14] von Neumann, J. Can We Survive Technology? In *The Fabulous Future: America in 1980*, E.F. Dutton, 1956, p. 37. (Reprinted from *Fortune*, 106 ff., June 1955.)

Box 2.12 Nuclear Energy

Nuclear energy involves exploiting the forces holding the atomic *nucleus* together. To oversimplify, nuclei consist of protons (positively charged particles) and neutrons (neutral particles). The *nuclear strong force* acts between protons and neutrons both, and holds the nucleus together. In addition, though, the protons repel each other due to their positive charges, and so they're trying to tear the nucleus apart. The balance of these two forces determines the stability of the nucleus. The *atomic number*, usually termed Z, is the number of protons, whereas the *mass number*, termed A, is the number of protons and neutrons ("*nucleons*," collectively) together. The "atomic" number is so called because the number of protons determines the chemical properties of the whole atom; that is, it determines the element (Box 4.3). In modern notation, the mass number is written as a superscript *before* the element symbol; e.g., ^{16}O for oxygen-16.

Because the neutrons have no electric charge, they do not directly affect the *chemical* properties of an atom, because those properties depend on the number of surrounding electrons, and that in turn depends on the number of protons. Atoms having the same atomic number but different numbers of neutrons are called *isotopes*, from the Greek for "same place" because they have nearly the same chemical properties. Most chemical elements less massive than lead ($Z = 82$) have more than one *stable* isotope; i.e., one not subject to radioactive decay (Box 2.14). (It should be emphasized that *nuclear* properties, as opposed to chemical properties, *do* depend on the number of neutrons.) Another term commonly seen is "nuclide," meaning a particular nucleus with a specified number of protons and neutrons. For example, ^{16}O is the nuclide with 8 protons and 8 neutrons, and it is also the isotope of oxygen ($Z = 8$) with 8 neutrons.

For very short distances (~1 femtometer, 10^{-15} m), the strong force between two protons is something like 100 times stronger than their electromagnetic repulsion. However, the strong force drops off much more quickly with distance than does the electromagnetic repulsion. The result is that very large nuclei become unstable because the repulsion among their protons overwhelms the strong force. Hence they will release energy if broken apart.

Conversely, very small nuclei can be fused together to release energy, because then the strong force overwhelms electromagnetism. In between there's a region of "maximum binding energy" where nuclei are as stable as they can be. It takes energy to break them up, but it also requires energy to build them up into anything bigger. This region occurs around iron-56 (^{56}Fe, $A = 56$), which has 26 protons (i.e., $Z = 26$) and 30 neutrons.

Therefore, nuclear reactions can yield energy either by (a) merging very light nuclei into heavier ones, termed *fusion*; or (b) breaking apart very heavy nuclei, termed *fission*.

Conventional nuclear energy uses the second process. Fusion, however, powers the Sun, as well as thermonuclear ("hydrogen") bombs. Controlled fusion has been a technological goal since the 1950s, but success has so far been elusive.

efficiency. But it takes another source of energy to make it. Most electricity is generated in thermal plants, in which heat is converted, with modest efficiency (usually about 60% or so), into electricity. In fossil-fuel plants the chemical energy of the fuel is converted into heat by burning. In nuclear plants, the source of heat is nuclear reactions.

In 2012 the world consumed about 77.5 EJ (roughly 21.5 trillion kilowatt hours) of electricity. Subtracting the roughly 17 EJ that's generated by renewable sources, mostly hydropower, leaves some 60.5 EJ that are thermally generated. Formally, this is equivalent to about 27 million barrels of oil a day, or about 29% of the daily usage of oil. However, this ignores the fact that

Box 2.13 Nuclear Waste drom Fission Reactors

When a uranium or other heavy nucleus fissions, the two fragments are nearly always highly radioactive. Why is this? It's ultimately a result of that proton-proton repulsion in the original nucleus. Large nuclei have an excess of protons over neutrons: the nucleus of ^{235}U, for example, contains 92 protons but 143 neutrons. In effect, the extra neutrons are "glue" to help hold the nucleus together.

Smaller nuclei, however, have more nearly equal numbers of protons and neutrons. Hence, the fission products have too many neutrons for the size they now are, and they must beta-decay (Box 2.14) to reach stable forms. Some fission products, such as iodine-129, can have half-lives of millions of years. However, in general the longer the half-life the less energetic the decay. Hence, the most hazardous nuclei are the ones with half-lives of a few decades, because they're still highly energetic and yet persist for a long time on human timescales.

Not all neutrons go into maintaining the chain reaction, either. Some are absorbed by other nuclei in the environment, with the usual result that those nuclei become radioactive. Additionally, uranium nuclei (usually ^{238}U) that don't fission build up into yet heavier nuclei. Even ^{235}U, for example, every now and then just absorbs a neutron to yield ^{236}U, rather than fission products plus energy. These "transuranic" or "actinide" nuclei are alpha-active (Box 2.14) and typically long-lived on a human timescale.

Fission waste is thus a very complex mixture, and it would be difficult to separate even without its seething radioactivity. Nonetheless, it is a potentially valuable resource, not least for its content of unconsumed nuclear fuel. This point is considered further in Chapter 6.

Box 2.14 Radioactivity

Radioactivity is a phenomenon of the atomic *nucleus*, and reflects that nucleus giving up energy to reach a more stable state. For historical reasons, the three common types of energetic particles emitted in radioactive decay are termed "alpha," "beta," and "gamma." Alpha particles are helium-4 (^4He) nuclei consisting of 2 protons and 2 neutrons; beta particles are highly energetic (i.e., fast-moving) electrons, and gamma rays are high-energy photons. They're electromagnetic radiation even more energetic than x-rays.

Alpha decay

The nucleus consists of protons and neutrons held together by the nuclear strong force (Box 2.12). When a nucleus becomes too big, the mutual electrostatic repulsion of the protons starts to tear it apart. Such a nucleus ejects alpha particles, as well as a gamma ray. Alpha decay, therefore, is typical of very large nuclei.

Beta decay

In addition, for a given *mass number* (Box 2.12), only at most two combinations of protons and neutrons have the lowest energies (or alternatively phrased, the lowest mass). Nuclei not having this most stable combination will change into it via "beta decay." If there are too many neutrons, a neutron will change into a proton, while a highly energetic electron (i.e., a beta particle) is ejected to keep charge balance. Often a gamma ray is emitted as well. For example, ^{90}Sr (strontium-90, with 38 protons and 52 neutrons), decays to ^{90}Y (yttrium-90, 39 protons and 51 neutrons) with a half-life (see below) of some 29.1 years. In turn, ^{90}Y decays to ^{90}Zr (zirconium-90, 40 protons and 50 neutrons) with a half-life of only 2.67 days. ^{90}Zr is stable—it contains the lowest-energy combination of protons and neutrons.

All radioactive decays can be described by a characteristic *half-life*, the time taken for one-half the original quantity to decay. This is characteristic of exponential decay:

$$N = N_0 \exp(-\lambda t)$$

where N_0 is the initial number of atoms, N the number remaining after time t, λ (Greek letter lambda) is the so-called decay constant, and *exp* the exponential function, i.e., the power to the base e. It's easy to show that the half-life is related to the decay constant by:

$$t_{1/2} = (\ln 2)/\lambda$$

where *ln* is the natural logarithm. In general, the longer the half-life the less energetic the radioactive decay.

Radioactivity is hazardous to biological systems because of the energy of these emitted particles. Nuclear energies are typically so much larger than the energies of chemical bonds that the emitted particles act much like a bowling ball rolling along a shelf full of bric-a-brac. Disruption of DNA molecules, in particular, can lead to irreparable genetic damage because the information carried by the molecular arrangement is destroyed.

other forms of energy cannot be converted into electricity with 100% efficiency. Assuming conservatively a conversion efficiency of 50% yields a minimum equivalence of 54 million barrels. On the other hand, the average efficiency with which petroleum is used in transportation is only on the order of 30%. Even a primitive electric car will be a lot more efficient than an internal-combustion engine. Nonetheless, the comparison illustrates the sheer magnitude of the energy represented by petroleum, and the scale of the challenge involved in replacing it for transportation applications.

The Alternatives: A Brief Survey

There are plenty of energy sources other than fossil fuels.
Running out of energy in the long run is not the problem.
—Kenneth Deffeyes[15]

What can replace oil and other conventional fossil fuels? Lots of things, and we will briefly survey them here. Many are obvious candidates for nanotechnological development, and *that* will be discussed in considerably more depth in Chapter 5. Furthermore, non-thermal approaches to energy use usually require nanostructuring, as is already suggested by the examples of biosystems. For example, fuel cells require highly structured control of the reaction of fuel and air to avoid thermalizing the chemical energy as it's released. Unfortunately, though, many alternatives are not nearly so well developed as they could be— one consequence of low oil prices.

Efficiency

This is least glamorous but most important, and one, of course, that's already been foreshadowed by the emphasis on how present clumsy, thermal-based technology squanders energy. If we have had a "heat" crisis, not an "energy" crisis, then obviously energy should not be used as heat! But the use of heat goes well beyond Carnot-limited engines. Process heating, the widespread application of heat to cause desired chemical and physical changes, pervades current engineering, ranging from the simple drying of foodstuffs, to the smelting of steel and other metals, to the ferocious heat of an industrial kiln baking limestone into lime. It is also the basis of conventional extractive metallurgy, as discussed further in Chapter 3.

---

[15]Deffeyes, KS. *Hubbert's Peak*, Princeton University Press, 2001, p. 176.

Better catalysts

A "catalyst" makes a chemical reaction proceed under milder conditions than it otherwise would, without itself being consumed in the reaction. "Milder conditions" usually means at lower temperatures, but it can also mean under less extreme chemical conditions, such as lowered acidity. The catalyzed reaction must be thermodynamically favorable, but it does not occur because of a so-called activation energy barrier (Box 4.6). In effect, a catalyst "builds a tunnel" through the barrier so that much less energy is required to make a reaction happen.

For example, a mixture of hydrogen and oxygen at room temperature does not react. It, of course, explodes if heated, since the heating furnishes the needed activation energy, but it also explodes if simply contacted with a platinum wire. Here the platinum is a catalyst, because it causes the reaction to proceed without itself participating, and without any change in the ambient conditions.

In many applications, another basic function of catalysts is to increase the yield of the desired product. This not only is more efficient, but it minimizes the synthesis of by-products that are often simply waste, and that in any case must be separated from the primary product. Indeed, it's no exaggeration to say that catalysis pervades potential nanotechnological approaches to energy usage, and we will see many examples throughout this book. Detailed discussion is deferred to Chapter 5, but it's worth noting that the "biological inspiration" is especially cogent in the case of catalysts. Biological catalysts, termed "enzymes," have a selectivity and activity that puts most industrial catalysts to shame. Enzymes are structured proteins, sometimes including additional molecules, that work by having a "lock and key fit" of reactants and enzyme, something like a molecular-scale clamp and template. They provide an extreme example of *molecular recognition* (Box 4.7), another phenomenon that will be encountered repeatedly throughout this book, and are how living things can carry out their seething multitude of chemical reactions at body temperatures.

Molecular separation technologies

This set of applications, which includes things as seemingly diverse as pollution control, environmental remediation, purification, and resource extraction, is so important it receives its own chapter (Chapter 6). The fundamental point is that separation is not an intrinsically energy-expensive problem, despite the widespread impression conveyed by conventional pyrometallurgy and resource extraction. In fact, that impression comes about because the use of vast flows of heat dominates resource extraction to an even greater degree than it does transportation and motive power.

Minimization of waste heat

Waste heat doesn't just come from heat engines, though. Of course, any real process produces by-product heat, but in some cases the technology is so clumsy that the by-product heat produced is utterly disproportionate to the service rendered.

Perhaps the most egregious example is traditional incandescent lighting. Less than 10% of the electrical energy consumed by an ordinary light bulb shows up as visible light; most of the radiance is invisible infrared, which simply shows up as environmental heating. In many climates, this dumped heat has the further consequence of increasing air-conditioning costs, so that there is a "cascading inefficiency." Such considerations, of course, have motivated the push to ban incandescent lamps in the last few years. It is noteworthy, moreover, that their replacements, such as white light-emitting diodes (LEDs) and compact fluorescents, involve a significantly greater degree of low-level organization.

Electricity transmission is another area where efficiency losses are significant, averaging around 6% in the US—about as much as is consumed in air conditioning.[16] Local generation, as by widespread use of solar arrays, would help minimize transmission losses because transmission distances would be minimized. Alternatively, high-temperature (HT) superconductors could be used for transmission. However, not only is it now unknown how to build such superconductors, particularly at the scale required, they will undoubtedly require molecular-scale nanostructuring.

Superstrong materials

The stronger a material, the less of it needed for a given structure, and the lighter that structure can be. And, in the case of a moving part, the less energy needed to move it. Hence, stronger materials lead to energy savings, vehicle fuel economy being an obvious example. One promise of nanotechnological fabrication is materials that approach the limiting strengths set by the chemical bonds in them. This is treated at length in Chapter 6.

Smart materials

"Smart" materials react to their environment, to such factors as temperature, lighting, and so on. An energy-related example would be self-darkening windows for passive solar-energy management. Of course, such approaches grade into the "information intensive" efficiencies noted below.

Distributed fabrication

As noted in Chapter 1, biology already carries out what might be termed "distributed fabrication." Plants assemble themselves

---

[16] www.eia.gov/tools/faqs/faq.cfm?id=105&t=3.

out of ambient materials according to a molecular "instruction set" (their DNA), using the diffuse energy of sunlight. Similar technological capabilities would render that enormous globe-spanning transportation infrastructure, both for raw materials and finished products, largely obsolete. Of course, this is unlikely to happen completely in the near term, but it is already happening with purely information products. Music sold as downloads, rather than as embodied in a physically shipped form such as a CD, is already a substantial part of the market, and in the last few years e-books have made huge inroads into the traditional book market. Most of the technical and trade journals have now migrated online, with articles available as downloads. In the US, many government agencies (e.g., NASA, USGS) are also making their publications available online. Not only does this save on physical paper, publication, and storage costs, but it makes it easy to "virtually" republish out-of-print publications. Indeed, at present the legal challenges are a bigger inhibition to broader online distribution than the technical issues, due to the ongoing concerns about copyright. Another contemporary example of distributed "information" fabrication is in the vanishing of specialized letterhead paper. With current printers, it's now easy to include the letterhead as part of the text—and unlike pre-printed paper, it's easy to update with new telephone numbers or other information.

Indeed, the distributed fabrication of physical objects is now beginning to occur, via the spread of so-called 3D printers,[17] which have garnered a great deal of attention in the last few years.

In the somewhat longer term, distributed energy generation will become important. Solar energy collection, for example, will be intrinsically distributed because of the diffuse nature of sunlight. Another example might be generation via many

---

[17] E.g., www.3dprinter.net/reference/what-is-3d-printing; Ambrosi, A; Martin Pumera, M. 3D-printing technologies for electrochemical applications. *Chem. Soc. Rev. 45*, 2740–55, 2016; Barnatt, C. *3D Printing*, 3rd ed., self-published.

local fuel cells, perhaps those powering vehicles, instead of by centralized thermal generating plants.

Electricity storage

Electricity, as has already been noted, is an extremely convenient energy source for many devices. What is *not* convenient about electricity is that conventional storage, in rechargeable batteries, is both bulky and extremely expensive. Next-generation batteries, as well as new technological approaches such as "supercapacitors," however, will require nanostructuring. The push toward electric vehicles is also providing a major incentive for new electric storage technologies.

It should be noted that since a battery/electric motor combination is not Carnot limited, a battery need not hold the entire energy content of a tank of gasoline to yield comparable performance to a conventional IC engine. Moreover, with an electrically powered vehicle, the kinetic energy of the moving car can be recovered during braking, as the motor acts as a generator when spun backwards. This is so easy to do that even present, "first-generation" hybrid automobiles do so.

An alternative means of storage would be the electrosynthesis of chemical fuels. A simple example is the electrolysis of water to hydrogen. Efficient synthesis and electrolysis, however, once again involves new catalysts. There are also synergies with fuel cell development, as noted in Chapter 5.

Information technologies: getting more out of matter

The explosive growth in computing power over the last few decades is surely familiar. That box on your desktop is vastly more powerful than the Pentagon's biggest in the 1960s. Among the host of applications of cheap and widespread computing are many directly related to energy usage and management.

One less obvious example is modeling petroleum or geothermal reservoirs. Such techniques as three-dimensional seismic modeling are extremely computation-intensive, but have led to a considerably better success rate in locating producing wells. And, of course, computation is much cheaper than drill-rig time.

New Energy Sources

Let's now move on to a quick survey of new energy sources. After all, ultimately *some* energy source is required, no matter how efficiently it's used. Many of these will be further discussed in Chapter 5 as we look at nanotechnological applications.

Unconventional fossil fuels

As might be expected, conventional fossil fuels represent a tiny fraction of "reduced" (Box 2.15) organic matter, ultimately of biological origin, that is held in sedimentary rocks. By far the bulk of this material is much too dispersed to be realistically considered a fuel prospect.[18] Several types of relatively concentrated deposits, however, have excited interest as potential energy sources over the years. Moreover, the development of fracking certainly underscores that new extraction technology can have a profound effect on hydrocarbon production. Of course, like all fossil fuels they have the serious disadvantage of generating carbon dioxide when fully oxidized, unless they are "reformed" to oxidize only their hydrogen content, as mentioned above and as briefly discussed in Chapter 5 (p. 196). Of course,

---

[18] The non-specialist tends not to realize that even so-called "dry holes" typically encounter hydrocarbons, just not in sufficient quantity to be economic. Indeed, hydrocarbons are ubiquitous in most sedimentary rocks. For example, for his dissertation the author had occasion to dissolve limestone samples, from rocks very similar to those in Fig. 6.8, in formic acid to extract the other minerals present. Invariably there was a sheen of oil floating on the solution when dissolution was complete. What is unusual about oil deposits is not the mere formation of petroleum but its collection into exploitable deposits.

> **Box 2.15** Redox Reactions
>
> When electrons are transferred between atoms in a chemical reaction, the atom that loses electrons is "oxidized", and the atom that gains electrons is "reduced." Overall, chemists speak of "redox" reactions when such electron transfer takes place. (As usual, the world is not so stark: few reactions involve the *complete* transfer of an electron from one atom to another. But the formalism is useful nonetheless.) In the example of the perfect ionic crystal of sodium chloride in the next chapter (Box 3.3), for example, sodium has been oxidized by chlorine.
>
> "Oxidation" originally referred to just oxygen, because oxygen is very good at taking electrons from other atoms. But as the example shows, other oxidizers exist. Indeed, there are better oxidizers than oxygen(!), elemental fluorine being an example. Fluorine will displace oxygen from oxides. Another example is the "holes" in a semiconductor (Box 2.17), which will be discussed in more detail in Chapter 5.
>
> "Reduced" compounds, as might be surmised, are those capable of being oxidized. Hence it common to speak of unoxided carbon compounds, including fuels and biomass, as "reduced." Indeed, under ordinary conditions carbon dioxide (CO_2) and *carbonates* (cf. Box 6.7) are the only carbon compounds that can't be further oxidized.

such sources are ultimately limited as well, but they may help in transitioning to long-term technologies.

Better oil recovery

Conventional oil recovery techniques still leave some two-thirds of the oil in the ground.[19] Hence, "oil-in-place" does not equal

---

[19] E.g., Montgomery, CW. *Environmental Geology*, 5th ed. WCB McGraw Hill, 1997, p. 305.

"ultimately recoverable oil" unless some major new technology is developed. "Enhanced recovery" techniques aim to mobilize oil that would otherwise remain in the ground. For example, one form of "tertiary recovery" involves injecting steam into the field to mobilize the oil. In one case,[20] monitoring reservoir temperatures with downhole fiber-optic thermometers while tracking the flow of oil and steam with ultraminiaturized flow sensors inserted into the stream led to substantial (~20%) savings. Less injected steam was required, and it could be targeted more effectively. This is an example of how nanotechnology will provide incremental improvements of techniques already carried out in embryonic form with microtechnology, and also is an example of using information technologies to optimize energy management.

Biotechnological approaches have also been under investigation for tertiary oil recovery[21] and are likely to be both much cheaper and far less disruptive environmentally. Obviously, injecting hot fluids itself requires energy, and, depending on the source of the heat, may require more energy than the oil is worth. This is another example of the "net energy" (Box 2.16) problem.

Oil "shale"

This is a double misnomer. "Kerogen marl" would be a better name, but try selling *that* to investors! This is a sedimentary rock that contains ~15% kerogen.[22] As noted (Box 2.9), kerogen is "proto-petroleum"; heating kerogen causes it to break up into shorter-length hydrocarbon chains that can then flow more freely. Natural oil formation is thought to involve the thermal

---

[20] Paul, D. Chevron Oil Co. Presentation at Mackay School of Mines, University of Nevada, Reno, August 2001.

[21] E.g., Yen, TF, ed. *Microbial Enhanced Oil Recovery: Principle and Practice*, Boca Raton, Fla: CRC Press, 1990.

[22] Hinman, GW. Unconventional petroleum resources. In *Energy Sourcebook*, eds. Howes, R, Fainberg, A, Am. Inst. Phys., pp. 99–126, 1991.

Box 2.16 Net Energy

If one spends more energy developing an energy resource than one gets out of it, it's not a resource. This seems like common sense, right? And such considerations have been dignified as "net energy" analysis, or more grandiosely "energy returned on energy invested" (EROEI).

However, such analysis is not so straightforward as it seems. First, the *form* of the energy is critical. Obviously spending 10 barrels of oil to recover 5 barrels makes no sense. However, it may well make sense to use 10 barrel-equivalents *in another form of energy* to extract 5 barrels of oil. For example, using 10 barrel-equivalents of oil *shale* to yield 5 barrels of *oil* may well be a sensible thing to do. Similarly, gathering 10 joules' worth of sunlight to yield 1 joule of electricity also makes sense.

Secondly, "net energy" analyses assume, usually implicitly, particular technical approaches to extracting the energy. Such approaches are not laws of nature. Hence at best net energy analyses can finger unpromising approaches to energy development, but at worst they degenerate into assaults on straw men. Alas, many published analyses seem to be of the latter sort—a preformed conclusion dressed up as the consequence of sober evaluation.

Here are a few examples. From a net-energy standpoint, conventional photovoltaics—i.e., solar cells—have been claimed to be marginal at best, because their fabrication is so energy-intensive. But that energy cost is hardly fundamental. It merely reflects the cumbersome and heat-intensive way semiconductor-grade silicon is currently made. Silicon dioxide is "cooked" with coke—essentially pure carbon—at 1200°C to make raw, impure silicon. This silicon is then purified by repeated cycles of melting and recrystallization, called "zone refining." No law of nature says that silicon *must* be extracted and purified in this way. The fundamental energy cost is set by the free-energy difference between silicon and silicon dioxide, some 856.3 kilojoules per mole, and *that's* a tiny fraction of the present cost.

Furthermore, even with conventional fabrication, there's an amortization issue: how long do the solar cells last? If a cell takes two years to collect the energy required to make it, but lasts for 25 years, it's still a net energy source. So net-energy analyses also depend critically on the amortization and maintenance costs assumed.

Another example of a "straw man assault" is ethanol purification by distillation. Ethanol is a potentially attractive alternative fuel that even now is used a gasoline additive (Box 5.8). It is conventionally obtained by fermentation of starch, such as from corn or wheat. Thus, not only does grain raised for ethanol compete with grain raised for food, but it's been claimed[1] that the heat required to purify the ethanol, by distillation, makes fuel ethanol a net energy loser.

However, ethanol doesn't have to be made from starch. It can be made from cellulose—wood scraps, vegetation, and so on. Cellulose is even a major component of sewage sludge. So material that otherwise requires expensive disposal can become fuel.

A more fundamental assumption is that *distillation* is needed to purify the ethanol in the first place. That's just another clumsy thermal process. Semipermeable membranes are one obvious alternative; effectively the ethanol is "strained" out of the water, as noted in Chapter 3 and further discussed in Chapter 6.

Indeed, even more fundamentally: why purify the ethanol in the first place? It needs to be water-free if it's to be *burned*, for fuel; but as emphasized repeatedly throughout this book, "burning" fuels throws much of their energy away off the top. An ethanol-water solution would work fine in a next-generation fuel cell. After all, biosystems metabolize ethanol that's mixed with water. Ethanol is the "alcohol" of alcoholic beverages.

For a last example, it's known that uranium (among many other things) is present in seawater at low levels, and groups

---

[1] Ref. 26 (Chapter 5) in main text.

(mainly in Japan) have carried out research on extracting that uranium. These efforts have been derided on the basis that pumping seawater through an extraction plant would cost more energy than could be gotten from the uranium.

But that's a very naïve approach to extracting *anything* from seawater! Instead, carry out extraction the way corals and other such sessile sea life do. Don't spend the effort to pump; use the currents and the surf. One possibility would be racks of highly adsorbent material in the surf zone, passively straining the water. Every few weeks the accumulated material is removed and the racks replaced. (Of course, with different adsorbing materials other materials could be extracted similarly.)

All these examples, of course, are amenable to obvious nanotechnological approaches, as will be discussed elsewhere in this book.

---

degradation of kerogen in the "oil window" of temperature and pressure.

Oil shale has been proposed as a "backstop" oil resource for decades and remains a subject of interest[23] to this day. The burgeoning of fracking has also led to some terminological confusion, as the rocks subject to fracking are commonly shales, and so the hydrocarbons released are often called "shale oil." However, these are *not* the same as traditional "oil shale," because the hydrocarbons in the fracked units are already present as oil or gas and so can flow without further processing.

Traditional attempts to process oil shale in essence carry out "artificial maturation": the rock is heated to "crack" (i.e., cleave) the kerogen into shorter hydrocarbon chains, which can then flow out. Hence the rock must be fractured extensively and "cooked" in bulk. Approaches that have involved mining the oil shale have been particularly dirty and clumsy as might

---

[23] Ogunsola, OI, Hartstein, AM, Ogunsola, O, eds. *Oil Shale: A Solution to the Liquid Fuel Dilemma*. ACSSS, 1032, 2010.

be expected, generating a great deal of waste. Additionally, of course, the concentration of kerogen must be great enough to make up for the energy expended in mining and heating.

An alternative approach is to break up the rock in place, via fracking, and then heat it in place. This is potentially both more efficient and far less disruptive.

Tar ("oil") sands

These consist of a gummy mess of very high-molecular-weight hydrocarbons dispersed in unconsolidated sand or in sandstone. They represent the "dregs" left behind when a conventional oil reservoir was breached by erosion so that all the low-molecular weight compounds escape. Thus they are "fossil" oil fields. (This, by the way, is the ultimate geologic fate of all oil deposits. Most oil is less than 100 million years old, which is quite young in a geologic sense. Oil deposits are ephemeral, geologically speaking.)

Tar sands in northern Alberta are currently being exploited. They are mined, heated to cause the thick tar to flow out, and then these long-chain aliphatics are cracked catalytically to yield lighter hydrocarbons typical of conventional oil. The technology, although still crude, works well enough that in 2003 Canada's proved reserves of oil were boosted from 4.9 billion barrels to 180 billion barrels by addition of the tar sands. Although the degree of cracking is not so drastic as would be required for oil shale, the exploitation of tar sands involves similar environmental issues.

Both oil shale and tar sands seem to be candidates for microbial processing, especially since bacterial strains capable of metabolizing kerogen have been reported.[24] Such organisms, or the enzyme systems derived from them, may be able to carry

---

[24] Petsch, ST, Eglington, TI, Edwards, KJ. ^{14}C-dead living biomass; evidence for microbial assimilation of ancient organic carbon during shale weathering. *Science* 292, 1127–31, 2001.

out low-temperature cracking of kerogen or of the long-chain aliphatics in tar and asphalt, which could provide considerably cleaner and less energy-intensive approaches to processing. As further discussed in Chapter 5 (p. 196), catalysts for breaking up aliphatic hydrocarbons at low temperatures will be necessary for hydrocarbon-fueled fuel cells, and they may have application in processing the products of oil shale and tar sands as well. For this to be practical, however, cheap, high-volume synthesis of such catalysts will be required, which is probably not a near-term prospect.

Molecular separation technologies, which are discussed in Chapter 6, also may provide cheaper amelioration of the environmental impacts from exploiting these sources, both of which generate a great deal of contaminated wastewater.

Clathrates ("gas hydrates")

These consist of a form of water ice having a very open crystal structure that can accommodate small guest molecules such as methane.[25] Clathrate deposits are locally abundant beneath the sea floor in high latitudes, but they present formidable problems for exploitation. Deep water drilling is difficult and expensive, and the degree of drilling needed to tap a deposit is uncertain. Conversely, on depressurization some such deposits may undergo catastrophic release, with potentially serious consequences. Naturally triggered catastrophic release has been speculated to have caused sudden warming events in the geologic past,[26] as methane is an excellent greenhouse gas.

---

[25] Kvenvolden, KA. A primer on gas hydrates. In *The future of energy gases*. U. S. *Geological Survey Professional Paper* Howell, DG, Wiese, K, Fanelli, M, Zink, LL, Cole, F, eds. 1570, pp. 279–91, 1993.

[26] E.g., Nisbet, E. Climate change and methane. *Nature 347*, 23, 1990; Hesselbo, SP, Gröcke, DR, Jenkyns, HC, Bjerrum, CJ, Farrimond, P, Morgans Bell, HS; Green, OR. Massive dissociation of gas hydrate during a Jurassic oceanic anoxic event. *Nature 406*, 392–5, 2000.

Solar energy

Solar energy has obvious advantages that have been recognized for over a century.[27] It is free, ubiquitous, amounts to an enormous energy input, and is virtually eternal (several billion years) on a human timescale. The Sun is powered by the fusion of protons, the nuclei of ordinary hydrogen, into nuclei of ^{4}He, as is described below. At the surface of the Earth, full direct sunlight delivers about a kilowatt per square meter (Table 2.1), so a patch of desert 15 meters on a side receives the equivalent of a barrel of oil in about 7.5 hours, and the 93 MMbbl of oil the globe uses in a day is equivalent to a piece of desert about 140 km on a side. Now, obviously there will be significant efficiency losses in any real system for trapping that energy (Box 5.14 has a more realistic calculation). Still, it's clear that there is no long-term "energy crisis!" It's just a matter of gathering and distributing the energy that's arriving already.

Of course, sunlight has obvious disadvantages: it is intermittent, unreliable, and diffuse, and even if clouds aren't present, the Sun does not shine continuously in the sky. Because of its diffuse nature, moreover, the area that must be devoted to energy collection also becomes an issue. Land is a high-value commodity in most of the places where energy is in high demand. Nonetheless, the Sun powers the biosphere and has done so for billions of years—and we've already repeatedly marveled at the capabilities of biosystems.

Solar energy, indirect use (the solar heat engine)

Hydropower

As mentioned already, electricity generation from large hydropower dams is certainly a mature technology, but it is unlikely to be expanded significantly due to the social and environmental

---

[27] Ciamician, G. The photochemistry of the future. *Science* 36, 385–94, 1912.

issues surrounding dam construction. Running water nonetheless still represents a large potential energy resource. Even in the US, where essentially all conventional hydroelectric sites have been developed, conventional hydropower accounts for only ~20% of the total potentially available.[28] Of course, the remaining untapped hydropower is so dispersed that little could be tapped by conventional damming in any case. Over the last few decades an alternative approach to hydropower generation has attracted attention: "low-head hydropower" or "microhydro." This ranges from turbines driven by low dams ("weirs") to "run of river" systems. Like wind power, the latter use the natural currents to generate power, typically with small turbines or water wheels.[29] It also has the advantages of not interfering with fish passage, nor creating permanent flooding.

Wind

Direct electrical generation from wind is a maturing technology, with a growing number of commercial installations. The issues are much as with distributed hydropower: the energy is diffuse, with large capital investments required, and with the added risk that not only the speed but the very direction of the current to be tapped is variable.

Wave energy

The surf raised by the wind contains a great deal of energy, and tapping it has intermittently excited interest over the years. It is another diffuse resource, however, and the "storm risk" factor is particularly severe.

---

[28] Dowling, J. Hydroelectricity. In *The Energy Sourcebook*, Howes, R, Fainberg, A, eds. Am. Inst. Phys., pp. 225–38, 1991.

[29] Fraenkel, P. Flowing too slowly - Performance and potential of small hydropower. *Renewable Energy World*, March 1999.

Ocean thermal energy conversion (OTEC)

In tropical areas there is a significant temperature difference between deep seawater (~4°C) and the surface (~25°C). This difference, of course, is maintained by solar energy. Deep ocean water is cold because it sinks at the poles and flows along the bottom toward lower latitudes, while the sun directly warms up the surface water. Although the temperature difference is small, it has attracted much interest because the reservoirs are both large and fluid. Thus the issue of dealing with cubic kilometers of unyielding rock, as in geothermal applications, does not arise. Sites for OTEC, however, need access to deep water, and so must be either situated on a steep shoreline or in the open sea.

Waste biomass

This reflects energy stored by natural photosynthesis, and although it's "waste," it still contains free energy. This, of course, is why bacteria can live on it, metabolizing it to carbon dioxide and water. Indeed, its disposal is now an expensive problem. Aside from its infection hazard, the main deleterious effect of sewage dumped back into streams is its "biological oxygen demand." Because it *is* chemically reduced, micro-organisms can oxidize sewage to derive energy, and that consumes the oxygen in surrounding water. This process, termed "eutrophication," can destroy an aquatic environment for other oxygen-using aquatic life such as fish. Hence sewage treatment focuses on oxidizing such material before its discharge into the environment.

Waste biomass, however, is a readily available feedstock. Even if not used as a basis for fuel (p. 242), it will eventually replace oil as a chemical feedstock, probably within a few decades, as discussed futher in Chapter 6 (p. 331). But for now, at least in the industrialized world, there exists elaborate urban infrastructure to gather together reduced-carbon waste and destroy it as fast as possible—even as there's a *global* infrastructure to expensively gather *other* reduced carbon compounds (i.e., petroleum and coal) from deep in the Earth. The irony often goes unremarked.

Solar energy, direct use

Photovoltaics for electrical generation

This, of course, is commonly thought of as *the* "solar energy." Conventional photovoltaics devices, "PVs" or "solar cells," consist of an illuminated semiconductor interface that generates an electrical current. Restricted to niche markets for decades, they now seem to be crossing the threshold into large-scale commercial applications as costs have dropped.

Photovoltaic devices are based on semiconductors (Box 2.17 and Fig. 2.2), usually elemental silicon. Absorption of a photon with energy greater than the semiconductor bandgap promotes an electron across the bandgap into the conduction band, leaving behind a vacancy, or "hole" in the valence band. If this electron–hole pair is formed in the vicinity of a space charge, such as at a p–n junction, charge separation takes place due to the electric field. With the proper provisions for collecting the opposite charges, this generates a potential that can drive an electric circuit.

Present-day PVs are useful, without question, but the prospects for "next-generation" photovoltaic devices, discussed at length in Chapter 5 (p. 244), seem especially bright. PVs that are unrolled like carpeting, or applied as paint, would utterly revolutionize the economics of energy generation. However, even with such devices the storage of electricity remains an issue (cf. note above). Even in the sunniest areas the sun doesn't shine at night.

Storage as chemical energy

It's remarkable that solar energy is so often thought of in terms of turning sunlight into electricity. That's not how biology does

Box 2.17 Semiconductors

As discussed in the sidebar on chemical bonding (Box 4.5), because of the laws of quantum mechanics the electrons in atoms have very specific energy levels, or "orbitals," from the old idea that that's where the electrons "orbit." When a substance is built up from individual atoms the separate orbitals blend into what are called *bands*, which actually are composed of a vast number of individual levels with minutely different energies. The electrons then occupy these individual levels, starting at the lowest energy. The energy level where the electrons run out is called the "Fermi level."[1] In most substances, too, there are several distinct bands of different energies, which results from the different sets of blended orbitals in the original atoms. A zone of "forbidden" energies between bands is, sensibly enough, termed a "bandgap."

If the Fermi level ends up in the middle of a band, a metal results. Metals are electrical conductors because there are still unfilled energy levels in the band into which electrons can move; indeed, this is also the basis of metallic bonding (Box 4.5). On the other hand, a "semiconductor" results when the Fermi level falls within a bandgap, so that the lower band (the "valence band" or "V.B." is filled) and the upper one ("conduction band" or "C.B.") is empty.

Strictly speaking, though, they're "filled" and "empty" only at absolute zero. At ordinary temperatures, some electrons are in the C.B. due to thermal excitation, with corresponding vacancies ("holes") left behind in the V.B. These account for the (small) "intrinsic conduction" of a semiconductor; indeed, it's why they were called "semi" conductors in the first place. The electrons in the C.B. can conduct, because there are lots of unfilled levels available. The "holes" in the V.B. can *also* conduct. In fact they act just like single positive charges. Of course, it's still electrons that are moving, but when electrons shift into a vacancy one after another, it looks just as though

---

[1]Named after Enrico Fermi (1901–1954), one of the leading physicists of the first half of the 20th century.

it's the hole that's moving. It's so convenient to treat the holes as single positive charges that it's always done.

Of course, the amount of charge carriers that are present depends both on the energy of the bandgap and the ambient temperature. If there's a big bandgap, there's no significant excitation and so no conduction. In fact, an insulator is just a semiconductor with a *really* wide bandgap. For example, the bandgap of silica (silicon dioxide, SiO_2), an excellent insulator, is about 1.76 attojoules (aJ; 10^{-18} J or 11 eV[2]), while the average energy of molecules at room temperature is only about 0.004 aJ or 0.026 eV. By comparison the bandgap of silicon is only 1.12 eV.

Conversely, as the bandgap gets smaller and smaller (or as temperatures get higher and higher), a semiconductor becomes more and more like an ordinary conductor, because the electrons can slosh more and more freely across the bandgap.

Adding impurity atoms—"doping"—can change the properties of a semiconductor drastically and usefully. Impurity atoms with extra valence electrons donate those electrons to the conduction band, to make "n" (negative)-type semiconductors. Alternatively, "p" (positive)-type semiconductors have doping atoms with too few electrons, so that holes are present in the V.B. In either case, we can look at these impurities as shifting the Fermi level up or down.

What happens at a boundary between two substances with different Fermi levels? Electrons tend to spill from the substance with the higher Fermi level to that with the lower. This continues until the overcoming the electrostatic attraction to remove another electron requires more energy than is gained by having the electron drop to the lower Fermi level. The result is a region around that boundary—a "Schottky barrier"—that has permanent electric charge due to the shifted electrons, the "space charge region."

---

[2]Solid-state physicists still traditionally measure bandgaps in electron volts.

Schottky barriers are *extremely* important technologically, but those details are well beyond the scope of this book.[3] Our interest here is that a Schottky barrier is also critical to the *photophysics* of semiconductors. When a semiconductor absorbs a photon with energy greater than the bandgap, the photon generates a "hole-electron pair." An electron is promoted into the C.B. while a hole is left behind in the V.B. Normally the hole and electron then just recombine, and the energy is dispersed as heat.

If the electron-hole pair is formed in the space charge region, however, charge separation takes place. The hole and electron go in opposite directions due to the opposite electrostatic attractions they experience. This separation is just what has to happen to convert the energy of that photon into useful forms.

---

[3]Any number of texts on solid-state physics, such as C. Kittel's classic (*Introduction to Solid State Physics*, 6th ed., Wiley, 1986), have a detailed treatment.

it. Photosynthesis stores the energy of sunlight as chemical bonds. That not only makes a lot more sense biologically, it may make a lot more sense technologically as well. After all, conventional fossil fuels are products of photosynthesis, just with a little additional geologic processing. So-called *artificial photosynthesis* is another embryonic technology, with serious research now stretching back decades, that seems immediately amenable to nanotechnological approaches, as is discussed in Chapter 5 (p. 254).

Sunlight as a distributed power source

Another unexamined assumption is perhaps the reflexive way in which we immediately consider converting sunlight into another form of energy. Of course, it's been used directly as a source of heating since ancient times, but it's not usually considered

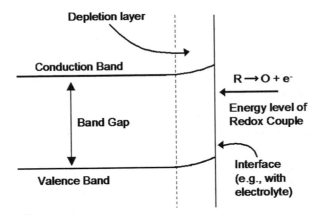

Figure 2.2 Semiconductor bandgap.

Any semiconductor contains a "valence band" (V.B.) separated by an energy gap (the "bandgap") from a higher-energy "conduction band." (C.B.) At absolute zero the valence band is completely filled with electrons, while the conduction band is completely empty. In the immediate vicinity of an interface with another substance, electrons will flow toward the substance with the lower Fermi level (highest filled energy level) until the increasing electrostatic charge makes such flow energetically unfavorable. This so-called depletion layer (space-charge region, Schottky barrier) then causes the energy levels of the bands to change, resulting in "band bending" near the interface.

Absorption of a photon with energy greater than the bandgap promotes an electron to the C.B. while leaving a positive "hole" in the V.B. In the space-charge region, the electric field there causes charge separation to occur. This is the basis both of photovoltaic devices and of semiconductor mediated photochemical reactions, as the hole is a powerful oxidizing agent while the promoted electron is a powerful reducing agent. A redox couple whose energy falls in the bandgap can be oxidized by the hole ($R \rightarrow O + e^-$), or reduced by the electron ($O + e^- \rightarrow R$).

for other purposes. However, light itself can be a "reagent," as suggested in the "photoswitched adsorbers" considered in Chapter 6 (p. 305). Some further possible applications will be considered in Chapter 5 (p. 265).

Nuclear fusion

Fusion is unquestionably of vital importance: it's what powers the Sun! In the Sun, 4 nuclei of ordinary hydrogen (which is just an isolated proton) are turned into 1 nucleus of helium-4—with the release of a great deal of energy. One gram of protons, when fused, yields 0.9999713 grams of helium-4; and that miniscule 0.0000287 grams lost corresponds to the energy of over 100 barrels of oil, according to Einstein's relation $E = mc^2$.

Fusing ordinary hydrogen nuclei, however, is *exceedingly* difficult, and occurs only in the extreme environment of stellar interiors. Certain other nuclei can be fused more easily (Box 5.16), and this has been exploited in thermonuclear bombs ("hydrogen bombs"). "Easier" to fuse does not mean "easy," however. For two nuclei to fuse, they must get close enough that the nuclear strong force overwhelms their mutual electrical repulsion due to their positive charges. A simple but crude approach for doing this is extreme heat: if things are hot enough, where "hot enough" means millions of degrees, occasionally nuclei will strike hard enough to fuse just by sheer happenstance. This the basis of fusion both in the Sun and in hydrogen bombs, which use a fission bomb(!) to achieve the extraordinary temperatures required.

Unfortunately, however, *controlled* nuclear fusion still seems remote, although it's been a "holy grail" of energy research for over 50 years. Conventionally, controlled fusion has been attempted by the "brute force" technique of magnetically squeezing a hot thermal plasma. An alternative using high-powered lasers to compress pellets has also been the subject of much interest. In all cases the result of fusion will be heat, which would power a heat engine. But, to return to a thrust of this book, why use heat to initiate fusion? That's just the thermal paradigm rearing its head again. There are potential alternative approaches, to be touched on in Chapter 5.

In the next chapter, we turn to a consideration of conventional material resources.

Chapter 3

Matter Matters

If it can't be grown, it has to be mined.
        —A common bumper sticker in northern Nevada

All human societies require matter as well as energy. Or rather, they need particular *kinds* of matter. Metals, gems, commodities such as salt, agricultural products such as food and flax—such things have been staples of human economies since there have been humans. Moreover, the fact that even now such materials are obtained from localized sources underpins that enormous global transportation infrastructure—again, an infrastructure that in embryonic form goes back to the beginnings of trade.

There are, at present, two fundamental ways of obtaining such primary materials. The first is *agriculture*: the use of plants to arrange raw matter into food, fiber, and structural materials such as wood. The second traditional way of obtaining primary materials is *mining*, which may be defined as the extraction and processing of particular useful substances from natural enhancements in the Earth. These will be discussed briefly in turn.

Nanotechnology and the Resource Fallacy
Stephen L. Gillett
Copyright © 2018 Pan Stanford Publishing Pte. Ltd.
ISBN 978-981-4303-87-3 (Hardcover), 978-0-203-73307-3 (eBook)
www.panstanford.com

Agriculture

Agriculture is the mother of all arts.

—Xenophon

Agriculture, of course, was the basis for traditional societies, and it remains fundamental even in the high-tech present. All food is still ultimately derived from green plants, and nearly all of those plants are deliberately cultivated by human beings. Agriculture is also a sterling example of the biological inspiration: highly organized matter is assembled out of simple ambient materials using only the energy of sunlight, ultimately via the well-known process of photosynthesis. Moreover, sunlight also becomes stored as chemical energy as a result of this process.

Of course, agriculture's limitations are well known. Crop-growing is highly localized and area-intensive due to the vagaries (water usage, climate, etc.) of the crops' requirements. In turn, this is why the shipment of foodstuffs accounts for a major part of that global transportation infrastructure.

Moreover, although raw agricultural products are molecularly organized, in any further processing they are treated like bulk matter (Box 1.1). Wheat is ground; wood is sawn; and, of course, foods are routinely subjected to low levels of thermal processing, i.e., cooking. In addition, modern industrial agriculture achieves its high yields by large additional inputs of energy (for tilling, transportation, and so forth) and materials (e.g., fertilizers, pesticides), so that it does not rely *only* on sunlight and ambient materials. Finally, agricultural waste products are molecularly organized, to be sure, but at present they are low-value items nonetheless. Although some have value as fuel (e.g., wood by-products), others are noxious waste indeed. Hence, molecular organization per se does not guarantee value.

Nonetheless, the capabilities of agriculture, and of plants in general, provide both an example of what's possible and an inspiration to nanotechnologists.

Minerals and Ores

> *Rich mineral deposits are a nation's most valuable but ephemeral material possession.*
>
> —T. S. Lovering[1]

Historically, all non-agricultural materials (save water and air) that humanity uses are obtained from geologically anomalous enhancements—"ores"—that occur locally on Earth. Ore is an economic term: it is a natural deposit "from which a desired material can be won *at a profit.*" Thus what constitutes an ore is a function of demand and technology as well as of geology. Ores are commonly worked for particular elements, as with most metals (iron, silver, gold, copper, zinc, etc.), but they include nonmetallic commodities as well, such as phosphate rock for fertilizer. They need not even be elements, but can be particular "pre-organized" materials, such as diamonds or other gems, mineral fibers such as asbestos (a valued "strategic material" until quite recently), or chemical compounds such as salts or limestone.

Ore deposits typically represent occurrences in which by happenstance some geologic process, or set of processes, has run to an extreme (Box 3.1). There are three dominant considerations in whether a deposit is "ore":

- the grade, or the concentration of the desired substance;
- the quantity ("tonnage") of the deposit;
- the access, which typically includes legal as well as physical constraints.

Of these, the grade is nearly always the most important. Aside from legal proscriptions, the only exceptions are high-volume, low-value commodities such as sand and aggregate, for which

---

[1]Lovering, TS. Mineral resources from the land. In *Resources and Man*, National Academy of Sciences—National Research Council, W.H. Freeman, 1969, p. 110.

Box 3.1 Ore Formation

Ore deposits are products of a geologic process, or set of processes, running to an extreme. They have a host of different origins, ranging from density separation by flowing water to crystallization of salts from an evaporating brine. Most, however, are products of natural thermal processing on a huge scale and over a very long time. For example, a "porphyry copper" deposit, the source of most of the world's copper, is a large granitic body in which copper-bearing sulfides are dispersed. Because sulfides have quite different physical properties from the common rock-forming silicates that make up the bulk of the granite, it's practical to use physical separation techniques to separate the (useful) sulfides from the (useless) silicates (the "gangue") (Box 3.2).

How did these porphyries form? Granite is an igneous rock; that is, it crystallizes from molten rock, a so-called magma. As the granite crystallized, a sulfur-rich phase fluid phase "exsolved," much as carbon dioxide comes out of a freezing can of soda pop. Because of its chemical properties, copper, as well as certain other metals, were extracted preferentially into that sulfur-rich fluid, and so ended up in sulfides dispersed throughout the larger mass of rock.[1] However, through the phenomenon of solid solution (Box 3.3), there is still a great deal of copper in the rest of the granite, mostly substituting for iron in the iron-bearing silicates. It is just not practical to extract with current technology.

As emphasized in the main text, traditional extractive metallurgy largely relies on the application and extraction of enormous amounts of heat. It is ironic, then, that the natural deposits we seek out usually represent the application of huge amounts of heat to huge amounts of material. (A large granite body can take several million years to cool.) Natural processes themselves thus underscore the limitations of thermal-based extraction, and why resource extraction must move to non-thermal technologies.

---

[1]The details are actually fairly well understood; see Burnham, CW. Magmas and hydrothermal fluids. In *Geochemistry of Hydrothermal Ore Deposits*, 2nd ed., Barnes, HL; ed. pp. 71–136, Holt, 1979), for example.

Box 3.2 Beneficiation

As mentioned in the text and Box 3.1, orebodies typically consist of ore minerals dispersed in a much larger volume of gangue minerals. Thus, another major part of extraction costs is separating out the ore minerals from the rest of the rock. This first involves crushing the ore to detach the minerals from each other, and then mechanical separation of the crushed mixture. "Beneficiation" is simply the catch-all term for the set of processes by which this is carried out. Although there is a fundamental energy cost in crushing, that of creating new surface energy, most (>90%) of the applied mechanical energy is simply thermalized.[1]

The ore minerals are separated out by exploiting differences in their physical properties vs. the gangue. A simple one is density, by which the (typically) denser ore minerals tend to sink to the bottom of a suspension, usually in water. "Panning" gold, to separate the metallic gold particles from the lighter gravel, is a traditional example, but density-based separation is now both much more sophisticated and much more widely applied.

More commonly, differences in surface properties are exploited, usually by *froth flotation*. Added reagents (*surfactants*, Box 4.8) preferentially bind to the ore mineral surfaces, so the ore particles stick to the surfactant bubbles and float up with the froth, which is then skimmed off. (A charming legend has it that flotation was discovered accidentally by a woman washing miners' dirty clothes, who noticed that flecks of ore minerals stuck to the soap bubbles and were carried off in the suds. Alas, this tale isn't true.)

Since its introduction more than a century ago, flotation has been extended to an ever wider range of ore minerals by the design and synthesis of new, "tailored" surfactant molecules. Although this has obviously greatly increased the separations that are now practical, such reagents are also a source of additional costs, particularly when they require elaborate syntheses.

---

[1]Ref. 7 in main text.

Box 3.3 Ionic Crystals and Solid Solution

Elementary chemistry classes usually teach that atoms of individual chemical elements combine to form "molecules," which contain those atoms in a precise ratio and hooked together in precise ways. In turn, the linkage between the atoms—the "bonding"—results from the "sharing" of electrons between the atoms.

However, this is an (over)simplified explanation of only one sort of chemical bonding, so-called "covalent" bonding (Box 4.5). Many important substances are *not* put together in that way. For example, sodium chloride, ordinary table salt, has the chemical formula "NaCl"—one atom of chlorine ("Cl") to one atom of sodium ("Na"[1]). There is no such thing, however, as an NaCl molecule. In salt, the chlorine atom has gained an electron to form the negative chloride ion, written Cl^-. Similarly, the sodium atom has lost an electron to form a positive sodium ion, Na^+. Solid salt, then, consists of indefinitely repeating arrays of alternating Na^+ and Cl^- ions, the whole thing sticking together by electrostatic attraction. In such an "ionic crystal," instead of the atoms' "sharing" electrons, one atom has effectively "stolen" the electron from another, to leave both with a net electric charge. In the terminology of Box 2.15, the chlorine has been reduced, while the sodium is oxidized. For historical reasons, positive ions are termed "cations," whereas negative ions are "anions." ("Cations" were attracted to a negatively charged electrode, or "cathode." Conversely, "anions" were attracted to a positively charged electrode, or "anode.")

There are also ionic crystals in which one or more of the ions is itself a covalently bonded group. In sodium sulfate, Na_2SO_4, for example, Na^+ ions are packed with SO_4^{2-} ions. The sulfate group, SO_4^{2-}, is a covalently bound anion. It consists of a tetrahedron of oxygens with the sulfur atom in the middle, the whole possessing a net negative-two charge. Silicates,

---

[1]From the German "Natrium," a pseudo-Latin coinage by the pioneering chemist Ludwig Wilhelm Gilbert in the early 1800s.

the most common compounds in the Earth's crust (Box 3.5), consist of a huge variety of silicon-oxygen anions that have partial covalent character and that act as structural units in silicate crystals.

In an ionic crystal, the identity of the ions doesn't matter too much. All that matters is their size and electric charge. Therefore it's possible for other ions of similar size and charge to substitute into the crystal. Obviously, the possible degree of substitution is greater the more similar the ions are. Potassium (K^+) substitution into NaCl is limited because the K^+ ion is significantly bigger than the Na^+ ion. On the other hand, magnesium ion (Mg^{++}) and ferrous ion (Fe^{++}, an iron atom with 2 positive charges) are similar in size (0.72 vs. 0.77 picometers[2]), and substitute relatively freely for each other in natural systems. The common mineral olivine, for example, with chemical formula $(Mg,Fe)_2SiO_4$, can range in composition all the way from pure magnesium olivine (Mg_2SiO_4, *forsterite*) to pure iron olivine (Fe_2SiO_4, *fayalite*). (Fayalite is common in iron-bearing slags, such as those from copper smelting (Box 3.8).) Few examples of such "solid solution" are this ideal. However, a great many elements that can exist in a +2 ionic form, including such important metals as manganese and copper, can substitute for iron and magnesium to some degree in such minerals. Such limited solid solution has great practical significance for the occurrence of conventional ore deposits (Box 3.1), and also places limits on separations based on phase changes (p. 108).

[2]Shannon, RD; Prewitt, C. Effective ionic radii in oxides and fluorides. *Acta Cryst. B 25*, 925–45, 1969.

local sources are required because transportation costs become dominant. With present separation techniques, letting nature do as much of the separation as possible is so cost-effective that it is worth spending the effort—worldwide!—to seek out highly unusual deposits in which the concentration of a desired element

Box 3.4 The Composition of the Earth's Crust

The "Top 10" most abundant elements in the Earth's crust were shown (Table 1.1), both as weight percentages and as atomic percentages. The lists overall are very similar. The biggest discrepancy is for the light hydrogen atom, which makes up nearly 3% of the crust on an atom basis but less than 0.2% on a weight basis. Most of this hydrogen is present in water.

Oxygen is far and away the most common crustal element, followed by silicon and then aluminum. This abundance accounts for the overwhelming dominance of *silicates* (Box 3.5) in common rocks. Indeed, certain aluminosilicates, the feldspars, are the most common chemical compounds in the Earth's crust. The overwhelming abundance of silicates also may have profound significance for future resources, as discussed in Chapter 6 (Box 6.6).

Furthermore, although the names of the elements in the Top 10 are all familiar, it's striking how many well-known elements are *not* present: sulfur, carbon, chlorine, and a great many important metals such as copper, silver, zinc, nickel, lead, and so on. Indeed, a number of these latter metals have been known since antiquity. Although they are rare in the crust, their chemistry is sufficiently different from those of the common silicates that they have locally made separate compounds. This phenomenon underpins conventional approaches to ore exploration and extraction, because those compounds can be separated relatively easily from the silicate background by simple physical means (cf. Box 3.1).

is as enhanced as possible. Such orebodies are also difficult to find because they are highly localized, so that "exploration" for mineral deposits remains an important endeavor. "Prospecting" for ore is still carried out, and even though the techniques are substantially advanced beyond a burro, pick, and gold pan, the degree to which the principle remains unchanged is surprising. Sheer luck still plays a larger role than might be imagined.[2]

Even crustally common elements such as aluminum or iron are obtained from their own ores, not from common rocks. Aluminum is the most common metal, and the third-most common element, in the Earth's crust (Box 3.4; Table 1.1). Common rocks typically contain a few percent to a few tens of percent aluminum, but they are not commercial aluminum sources. (Indeed, the aluminum is in *aluminosilicates* (Box 3.5).) As mentioned in the Introduction (p. 4), aluminum is currently extracted from a tropical soil (*bauxite*), from which it is refined into metal in areas having cheap electricity, which commonly lie halfway around the world from where the mining took place (Box 3.6).

Ore deposits also are obviously limited, and the trend has been to exploiting lower and lower grade sources over time. For example, since the 1880s the average grade of US copper ores has dropped from ~3% to ~0.5%.[3]

A further issue is looming here: even though the desired metal (copper, say) may make up only 0.5% of the rock, in conventional ores it is nonetheless segregated into specific phases—so-called "ore minerals—that can be separated from the valueless host rock—the "gangue" (pronounced "gang")—by physical means, through so-called beneficiation (Box 3.2). The gangue consists nearly all of *silicates* (Box 3.5), the most common compounds in the crust, making up by far the bulk of common rocks.

In the case of copper, for example, the typical ore minerals are sulfides, usually containing iron as well as copper, as for example chalcopyrite ($CuFeS_2$). There is a fairly well-defined threshold concentration at which these separate phases do not form.[4] In

---

[2]Anyone who has worked around the minerals industry for any length of time has a collection of anecdotes about major deposits that were discovered accidentally by misinterpretation or misreading of sophisticated data sets.

[3]Gordon, RB; Koopmans, TC; Nordhaus, WD; Skinner, BJ. *Toward a New Iron Age?* Harvard University Press, 1987, p. 42.

[4]E.g., Ref. 3, pp. 26–7.

Box 3.5 Silicates

As silicon and oxygen are the most common elements in the Earth's crust (Table 1.1), it's hardly surprising that compounds of silicon and oxygen, so-called *silicates*, make up most of crustal rocks. Silicon can form four bonds with surrounding oxygens to form a tetrahedron, the "silicate tetrahedron," a strong, compact unit that is the fundamental building block of silicate chemistry (Fig. 3.1).[1]

Some silicates, such as the mineral olivine, $(Mg, Fe)_2SiO_4$, contain the SiO_4 tetrahedron as an isolated anion. The metal cations and silicate anion then stack alternately to make up an ionic crystal (Box 3.3) analogous to other oxyanion salts like sodium sulfate (Na_2SO_4), where the sulfate anion also acts like a single unit.

If this were all, silicate chemistry would be both simple and boring. The enormous variety of silicate chemistry comes from the fact that these tetrahedra can share corner oxygens to make a disiloxy bond (Si-O-Si). Such sharing, termed "polymerization," results in a huge number of silicate anions. In such structures cations still fit between the silicate anions to form an ionic crystal. However, the silicate anions range in form from clusters of a few atoms to chains, sheets, and even entire frameworks.

If each tetrahedron shares two corners, a chain structure results, the *pyroxene*[2] minerals being one example (Fig. 3.2a). Two common pyroxenes are enstatite, $MgSiO_3$, and diopside, $CaMgSi_2O_6$. Double-chain structures, consisting of two parallel single chains in which every other tetrahedron shares a third oxygen with the adjacent chain, are basic to the *amphiboles*, another widespread group of minerals (Fig. 3.2b). Hornblende is a common example.

---

[1] At the (geologically) low pressures typical of the Earth's crust. At very high pressures silicon coordinates with 6 oxygen atoms in a so-called octahedral configuration. At ordinary pressures certain organic reagents, such as catechol (Fig. 6.9), will also form octahedral complexes with silicon, with three catechol groups to each silicon.

[2] Pronounced "PEER-ux-een" by geologists and mineralogists.

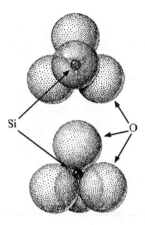

Si   O

Figure 3.1 The silicate tetrahedron.

The silicate tetrahedron, SiO_4. In polymeric structures, adjacent tetrahedra share an oxygen at a vertex to form a disiloxy bond.

The *silicones* (more properly, "siloxanes") are a technological version of a silicon-oxygen chain structure. In this case, instead of negatively charged chains with intercalated metal ions, organic sidegroups such as methyl (CH_3) groups are attached to the chains in place of the unshared oxygens. A simple silicone is polydimethylsiloxane, $(CH_3)_2SiO$.

If three oxygens on *every* tetrahedron are shared, an infinite sheet structure results. These are also ubiquitous in the natural world, making up the micas and clay minerals. The platy structures of these minerals directly reflect their molecular structure. They split ("cleave") easily along the plane of the sheets. The micas, for example, can be visualized as having a layer of tetrahedra all sitting on a face and sharing all three corners on that face, while the unshared corner sticks up, to form a hexagonal pattern that looks a bit like chicken wire (Fig. 3.2c). A corresponding tetrahedral sheet above has all the unshared tetrahedral corners pointing down, while metal cations between the unshared corners cancel most of the negative charge to yield a kind of "sandwich" structure. The residual charge is canceled by large cations between the "sandwiches." They're usually potassium cations in natural minerals, and they tuck into those hexagonal holes.

This plane of weakness, where the potassium cations hold the sheets together, is where the cleavage occurs. In fact, carefully cleaved mica sheets can be atomically flat over distances of centimeters. That is, the surface of the very same(!) silicate sheet is exposed over macroscopic distances. This extraordinary phenomenon has been exploited in surface science studies for years, and such atomically perfect surfaces are an obvious basis for studies in proto-nanotechnology (Chapter 4).

If all four corners of each silicate tetrahedron are shared, an infinite three-dimensional (3D) framework results whose composition is simply that of silica, silicon dioxide (SiO_2). Silica occurs in a number of different structures depending on how the tetrahedra are linked. The most common form is quartz, but there are others (cristobalite, coesite, etc.) One of the most interesting is an artificial form called *silicalite*, in which the structure is so open that it contains channels and tunnels into which other molecules can fit—that is, it is a pure silica molecular sieve (Box 6.3).

If this were all to 3D silicate frameworks, there would be little compositional variety. However, it turns out that certain other cations can substitute for silicon in the framework, aluminum being an important example. Since aluminum has one fewer protons than silicon, the framework ends up with an overall negative charge, one negative for every aluminum atom present. To cancel this charge, other cations need to be incorporated into holes in the framework. The result is 3D frameworks ("tectosilicates") that aren't *just* silica, and that furthermore can accommodate other cations as well. (Some substitution for silicon can also occur in 2D structures, in the micas for example.)

Aluminum substitution is ubiquitous in natural silicates, because aluminum is the third most abundant element in the crust after oxygen and silicon themselves (Table 1.1). Geochemists thus refer to "aluminosilicates," combined silicon-aluminum framework minerals. One group of these minerals, the feldspars, are the most abundant chemical compounds in the crust. Potassium feldspar, for example, has the formula $KAlSi_3O_8$. Since the potassium cation has a single charge while aluminum has three, they exactly compensate for the one replaced framework silicon.

In current technology, besides the obvious uses as bulk fill, rip-rap, and so on, silicates have long played an important role in glass and ceramic manufacture, a role that indeed goes back to antiquity. As described in Chapter 6, too, the role of silicates as structural materials is likely to be far broader in a mature nanotechnology.

Silicates already have important roles in proto-nanotechnology, however. The importance of cleaved mica sheets was mentioned above. More importantly, however, tectosilicates with open frameworks, the so-called *zeolites*, are the prototypes of the enormous variety of *molecular sieves*, which already have abundant applications in catalysis and potentially in molecular separation and nanostructuring.

---

that case the copper remains dispersed in the gangue minerals, into which it substitutes for more common elements such as iron and magnesium, a phenomenon termed "solid solution" (Box 3.3). This "background" concentration is the form in which most elements are present in the crust of the Earth, and it is a *considerably* more difficult form to deal with, especially because silicates are difficult to process. No current technologies separate elements from their background values.

Conventional exploitation of an orebody also leads to serious environmental issues. First, as is obvious from the example of copper, most of an ore does not consist of the desired element, and the crushed piles of discarded gangue minerals are at best an eyesore that must be expensively reclaimed. At worst they become a significant source of pollution in their own right, through such phenomena as dust hazards or acid-mine drainage

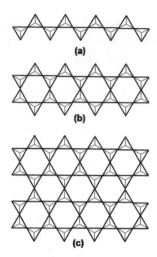

Figure 3.2 Silicate structures.

(a) Chain structure in pyroxenes; on replacing the oxygen at the unshared corner of each tetrahedron with a methyl (CH_3) group, it's also the structure of polydimethylsiloxane ("methyl silicone"). (b) Double chain in amphiboles. The structures in (a) and (b) extend indefinitely both right and left. (c) Sheet structure in mica and clay minerals. This structure extends indefinitely in two dimensions. Individual silicate (SiO_4) tetrahedra are shown diagrammatically; where tetrahedra are joined at a vertex, the oxygen is shared between two tetrahedra.

(AMD; Fig. 3.3). The latter occurs when sulfide minerals newly exposed to air and water begin to oxidize (Box 3.7).

Even the processing of the extracted ore minerals is wasteful. Copper ores, for example, nearly always contain more iron than copper (Box 3.8), but the iron is not recovered. Iron is extracted from its *own* ores, in which it is even more strongly concentrated.

Furthermore, it bears repeating that most of the industrialized nations depend on foreign sources for the bulk of their mineral commodities. In part this is because they have exhausted their high-grade deposits. In part, too, it reflects higher environmental standards, as mining is traditionally a messy business. In any case, the products of mining underpin a large part of that global transportation network.

Figure 3.3 Acid-mine drainage.

Rusty stain along Red Mountain Creek, Colorado, USA. Iron mobilized by acidity due to sulfide oxidation has oxidized.

Separation: A Fundamental Technological Problem

Resources are but one aspect of a single, fundamental technological problem, that of separating out one type of atom or molecule from a huge (and arbitrary) background of other types. If we *want* what is extracted, then it is a resource. But viewed in another way, this is also the fundamental problem of pollution control, as well as of purification processes such as desalination. The difference is a matter of context: if we *don't* want what is extracted, it's not resource extraction.

Usually, however, separation has been regarded as a host of different problems. Resource extraction especially has generally been seen as distinct from (say) pollution control or purification, because it's traditionally been carried out with a spectacular suite of processes. It's also thought of as intrinsically energy-intensive, from those inchoate impressions of gigantic open pits,

Box 3.6 Aluminum Extraction: Bauxite and the Hall–Héroult Process

As already noted, aluminum is the most abundant metal in the Earth's crust, and the third most abundant element. Most common rock-forming minerals contain aluminum as a major constituent. Yet, as emphasized, common rocks are *not* sources of aluminum. Aluminum ore is *bauxite*, a tropical soil. It is a relatively low-iron *laterite*, the typically brick-red soils found in tropical areas that receive a lot of rain. Over time all components are leached out by rainwater, leaving only the least soluble oxides, those of aluminum, iron, and titanium. Although "pre-purified" by nature, the bauxite still must be further purified before processing. It is "digested" in alkali to remove the residual iron, silica, and titanium oxides, yielding so-called "red mud" as a waste product. Unsurprisingly, at present red mud is not a source of iron or titanium, although experimental efforts have been made to determine its suitability as a ceramic feedstock.[1]

The result of the processing is alumina (Al_2O_3, aluminum oxide). Extracting aluminum metal from this oxide requires a large energy expenditure:

$$2\,Al_2O_3 \rightarrow 4\,Al + 3O_2,\ \Delta G = +3165\ \text{kJ/mol at STP},$$

where the plus sign means that an energy input is required for the reaction to proceed as written, and STP is standard temperature and pressure. On a per-atom of metal basis, this is over 10 times the energy input required to separate iron metal from magnetite (Box 3.10). Moreover, it is not surprising that inefficiencies in the actual extraction process yield values significantly higher than this limit.

---

[1]Sglavo, VM; Campostrini, R; Maurina, S; Carturan, G; Monagheddu, M; Budroni, G; Cocco, G. Bauxite "red mud" in the ceramic industry. Part 1: Thermal behavior. *J. Eur. Ceram. Soc. 20*, 235–44, 2000; Sglavo, VM; Maurina, S; Conci, A; Salviati, A; Carturan, G; Cocco, G. Bauxite "red mud" in the ceramic industry. Part 2: Production of clay-based ceramics. *J. Eur. Ceram. Soc. 20*, 245–52, 2000.

The alumina is reduced to metal by the Hall–Héroult process, named after the inventors who independently discovered it in the latter 19th century. Alumina is dissolved in a mixture of molten cryolite (trisodium aluminum fluoride, Na_3AlF_6) and aluminum fluoride at temperatures in excess of 1000°C. Although exceedingly hot by everyday standards, this is far less than the melting temperature of pure alumina, which is over 2000°C. (The purpose of the cryolite is to lower the melting point. At the molecular level, the fluoride ion breaks up oxygen linkages between aluminum cations. That is, it acts as a flux.) The alumina-cryolite melt is then electrolyzed to yield molten aluminum at the cathode, where it pools at the bottom of the reaction vat. Oxygen is not released at the anode, however; the anodes are made of carbon, which oxidizes to carbon dioxide. Obviously this thermodynamically wasteful, but is done for safety and engineering convenience.

The Hall–Héroult process is why aluminum extraction is so electrically intensive. It's why at present it's economic to ship bauxite ore halfway around the world to sources of cheap electricity.

fiery smelters, and armies of heavy equipment hauling tons of dirt (Fig. 3.4). Even in the US, for example, primary metals and non-metals production still accounted for some 2450 EJ of energy consumption in 2010, about 2.5% of total US energy usage[5]—despite the fact that the US, like the other industrialized countries, imports a great deal of its primary materials.

In turn, this profligate energy usage, nearly all in the form of heat, is typically rationalized by vague appeals to the second law of thermodynamics, especially since those processes are applied to deposits that are *already* anomalously enriched. This misconception, moreover, is not confined merely to naïve popularizers, but is occasionally echoed by scientists, such as

---

[5]www.eia.gov/consumption/manufacturing/data/2010/pdf/Table 1_1.pdf.

Box 3.7 Acid-Mine Drainage: The Details

Acid-mine drainage (AMD) results from the oxidation of sulfide minerals that are now exposed to water (H_2O) and atmospheric oxygen (O_2) as a result of mining operations. The most common such mineral is pyrite, iron disulfide (FeS_2), although other sulfides react very similarly. As mentioned (Box 2.11) pyrite is much more widely distributed mineral than its occurrences as "fool's gold" might suggest. Pyrite also is normally discarded as part of the gangue. The overall reaction is:

$$4 \, FeS_2 \; + \;\; 10 \, H_2O + 15 \, O_2 \;\; \rightarrow \;\; 4 \, FeOOH \; + \; 8 \, H_2SO_4$$
$$\text{(pyrite)} \quad \text{(water)} \qquad\qquad \text{(goethite)} \quad \text{(sulfuric acid)}$$

(Note the similarity of this reaction to that forming acid rain from sulfide impurities in coal combustion (Box 2.11).) The mineral goethite is the main component of common rust. It is bright orange and accounts for the stains commonly seen along streams in mining areas (Fig. 3.3). The sulfuric acid formed decreases the pH (i.e., increases the acidity) of the water. Not only is this directly harmful to aquatic life, but the increased acidity also leaches metals into solution where they become a further source of pollution.

In many cases, these reactions are carried out by so-called chemolithotrophic bacteria, which derive energy from oxidizing the sulfides. This underscores the irony that many *waste* products from current technological processes are still energy-rich, and indeed that often accounts for much of their hazard.

with Preston Cloud's assertion[6] that the increase in extraction energy with lower ore grade "follows inexorably from the entropy law." (Box 3.9 analyzes the true entropy costs of mixtures because of this pervasive misconception.) These rather nebulous

---

[6]Cloud, P. *Cosmos, Earth, and Man.* Yale University Press, 1978, p. 321.

Figure 3.4 Open pit mine.

Active gold mine at the site of Rawhide, Nevada, USA. The scale of the operation may be gleaned from the haul truck visible at the left. It is much larger than a standard, "street-legal" dump truck. The tires on the vehicle are roughly 10 ft (3 m) in diameter. This is an example of a mine recovering gold from rock at, in some cases, parts-per-million levels. Photo by the author.

beliefs are further buttressed by the practical difficulties with pollution control and remediation, which involve extracting low-concentration components from a background of other substances.

The Biological Example

As will be demonstrated below (Box 3.8), it's easy to show these perceptions are mistaken by actually working out the limiting thermodynamic costs quantitatively. This conclusion, however, could have been anticipated by again considering the biological inspiration. Biosystems carry out considerably more spectacular feats of element extraction with considerably less sound and fury, all the while using considerably less energy. The kidneys of higher animals extract certain solutes from the blood at high efficiency out of a background of many other solutes while working isothermally at near-ambient temperatures. "Shell builders" such as mollusks extract calcium carbonate from

Box 3.8 Roasting of Sulfide Ores

The ores of many important metals, such as copper and zinc, consist largely of sulfides, which conventionally are first "roasted" to convert them to the oxides.[1] This is a simple but wasteful step, as it both discards the free energy released by the oxidation of the sulfide, and puts the metal into a form (the oxide) in which an additional energy input will be required to reduce it to metal.

In the case of zinc, for example,

$$ZnS + 1.5\ O_2 \quad \rightarrow \quad ZnO \quad + \quad SO_2.$$
zinc sulfide　　　　　　zinc oxide　　　sulfur dioxide

Thermodynamically it is possible to extract zinc metal at a net energy *gain*:

$$ZnS + O_2 \rightarrow Zn + SO_2,\ \Delta G = -100.5\ kJ/mol\ at\ STP.$$

Ironically, chemolithotrophic bacteria nearly manage this feat, although they do not extract the free metal but oxidize it as well. For example,

$$ZnS + H_2O + 2\ O_2 \rightarrow ZnO + H_2SO_4,\ \Delta G = -573.7\ kJ/mol\ at\ STP.$$

Note the similarity of this reaction to that causing acid-mine drainage from pyrite oxidation (Box 3.7). Formerly, the sulfur dioxide released by roasting was itself a major pollutant, but it is now recovered as sulfuric acid (H_2SO_4), an important industrial chemical.

The zinc oxide is then converted to metal vapor (!) and carbon dioxide by carbon monoxide at high temperature, the carbon monoxide being made by partial combustion of carbon:

---

[1]Ref. 8 (in main text) pp. 277–86.

$$2 C + O_2 \rightarrow 2 CO \text{ (gas)}$$
$$ZnO + CO \rightarrow Zn \text{ (gas)} + CO_2 \text{ } (\sim 1000°C).$$

The zinc vapor is then condensed to separate it from the carbon dioxide.

In contrast, copper ores, consisting mainly of iron-copper sulfides such as *chalcopyrite,* $CuFeS_2$, are melted to form a *matte.* Iron is then separated by oxidizing the matte with atmospheric oxygen,[2] the process being carried out at about 1200°C:

$$2 CuFeS_2 + 4 O_2 + SiO_2 \text{ (silica)} \rightarrow 2 Cu \text{ (liquid)} + Fe_2SiO_4 \text{ (slag)}$$
$$+ 4 SO_2 \text{ (gas)}.$$

As with zinc, the byproduct sulfur dioxide (SO_2) is no longer discarded but is now recovered for sulfuric acid (H_2SO_4) production. In contrast to zinc, however, the copper metal is recovered directly from this melt as so-called "blister copper," instead of requiring a further reduction step with carbon monoxide. However, further purification steps are required for most applications, such as electrical wire.

Note that an additional reagent, silica (SiO_2), must be added to cause the iron to separate into the slag. Other impurities are also extracted into the slag at this point. Furthermore, although nearly all copper ores contain more iron than copper, the iron is not recovered but merely discarded with the slag.

---

[2]Ref. 8 (in main text) pp. 324–41.

the surrounding water to build their shells. Diatoms, a type of single-celled alga, are particularly impressive: their shells are a low-temperature silica glass, made from silica (silicon dioxide, SiO_2) extracted from the ambient water (Fig. 3.5). Silica is barely soluble in water, with a maximum solubility of only a few parts per million (ppm). Nonetheless, the diatoms manage to extract it, molecule by molecule, from the water around them. Plants

Figure 3.5 Diatoms.

Rock ("diatomite") composed almost entirely of shells ("tests") of single-celled alga (*Melosira granulata*) from the Pliocene of western Nevada. This organism constructs its test out of a hydrous silica glass by extracting dissolved silica (as H_4SiO_4), which is present at levels of a few ppm in its aqueous environment. (Scanning electron microscope photo by the author with assistance of L. Grasseschi, University of Nevada, Reno.)

extract atmospheric CO_2 (concentration ~350 ppm) to build their intricate organic structure, and their roots extract nutrient elements at low concentrations, some (e.g., molybdenum, a critical component of the photosynthetic mechanisms) also at parts-per-million levels. All this, moreover, is carried out using a diffuse and intermittent energy source—sunlight. And, of course, the example of agricultural products has been mentioned.

Box 3.9 Mixing Entropy

As noted (Box 2.4), there is a deep, albeit often misunderstood, connection between disorder and entropy; or more correctly, between *probability* and entropy. Since entropy is a real physical quantity, it follows that we should be able to calculate the entropy of a disordered mixture. Mixing entropy is given by:

$$S_m = - R \sum_{i}^{N} x_i \ln x_i, \tag{3.9-1}$$

where R is the gas constant (in SI units, 8.314 J/K-mol), x_i the mole fraction of component i, ln is the natural logarithm, and there are N components,[1] with the summation over all components. Note that mixing entropy is independent of temperature.

It is illuminating to use this relation to put numbers on "entropy costs." For extraction of one component from a background of other components, we may treat the mixture as made of two components, with $x_1 \equiv x =$ the mole fraction of the desired component and $x_2 = 1 - x_1 \equiv 1 - x =$ "everything else." Equation 3.9-1 then becomes:

$$S_m = - R [x \ln x + (1 - x) \ln (1 - x)]. \tag{3.9-2}$$

Eq. 3.9-2 gives entropy per mole of *mixture*. In the case of the desired component, we would like to determine its mixing entropy per mole of *component*, S_m'. To get this divide 3.9-2 by its mole fraction x:

$$S_m' = - R [\ln x + [(1 - x)/x] \ln (1 - x)]. \tag{3.9-3}$$

As would be expected, this *normalized* entropy goes to infinity as x goes to zero. As the solute concentration goes to zero, more and more solution must be processed to extract a mole. However, we're not interested in the cost to extract a *mole* of

[1] E.g., Moore, Ref. 4 (Chapter 2) (in main text), p. 239.

impurity; we're interested in the costs to purify a mole of the desired component. This is illustrated in the example treated below.

In the limit of non-interacting particles, the free energy of the solution reduces to the mixing entropy alone, multiplied by the absolute temperature:

$$\Delta G = -T\Delta S.$$

Let us naïvely examine, for example, a 1 part-per-million (ppm) aqueous solution of copper sulfate ($CuSO_4$), assuming it is a molecular mixture of non-interacting particles. Such a solution might arise from a sulfuric acid leach of ore rock or from acid-mine drainage, and so is typical of "real world" dilute solutions that arise repeatedly in both resource extraction and pollution control. A concentration of 1 ppm corresponds to a 6.3 micromolar (μM) solution, and this in turn corresponds to a mole fraction of $\sim 1.13 \times 10^{-7}$. Evaluating equation 3.9-3 then yields $S_m' = 141.3$ J/K mol, so that $-T\Delta S$ at 25°C is -42.1 kJ/mol.

This is a trivial amount. For comparison, it is about the energy needed to evaporate 5 moles (~ 90 g) of water at STP; and along with 1 mole of pure $CuSO_4$ one has also extracted 999,999 moles of pure water.

Of course, the assumption of non-interacting particles is incorrect: the ions in the solution in fact strongly interact with the water solvent, and hence the enthalpy is non-zero. The correct expression for the free energy that takes this into account is:

$$\Delta G = \Delta G_0 + RT \ln \{Cu^{++}\} \{SO_4^{2-}\}$$

where ΔG_0 ($= -16.6$ kJ/mol[2]) is the free energy of formation of the aqueous ions in their standard states from solid $CuSO_4$, R is the gas constant, T is absolute temperature, and the braces indicate the activities of the aqueous ions. Using the

[2] Thermochemical data from: Robie, RA; Hemingway, BS. Thermodynamic properties of minerals and related substances at 298.15 K and 1 bar (10^5 Pascals) pressure and at higher temperatures. *USGS Bull. 2131*, 1995.

Debye–Hückel formula[3] for aqueous activities, we find $\{Cu^{++}\}$ = $\{SO_4^{2-}\}$ = 6.16×10^{-6}, and hence $\Delta G \sim -76.1$ kJ/mol $CuSO_4$. Therefore, even on taking the interactions into account, the limiting thermodynamic cost is still less than double the value that assumes the mixing of non-interacting particles only.

Macroscopic entropy

Thus mixing entropy is small, even for truly molecular mixtures. However, most things are *not* molecularly mixed, and in such cases the true thermodynamic entropy becomes utterly negligible, as will now be demonstrated.

Elementary textbooks[4] assert that the entropy of a macroscopic mixture is "zero." The reason is simply combinatorial arithmetic: molecules and atoms are so much smaller than even "small" particles that the number of their possible arrangements swamps the number possible with the particles.

This can be illustrated with some crude examples. Consider a 1:1 mixture of 1 micrometer (μm) grains of salt (sodium chloride, NaCl) and quartz (silicon dioxide, SiO_2). We can use equation 3.9-2 to calculate the mixing entropy:

$$S_m = -2 R (0.5 \ln (0.5)) = 5.76 \text{ J/mol K,}$$

but this is per mole of *grains*. We now need to find the entropy per mole of compound. Each grain contains $(1 \mu m)^3 d_i/g_i$ moles of compound i, where d_i is the density and g_i the molecular weight. For salt $d_i = 2.165$ g/cm^3 and $g_i = 58.44$ g/mol, and so there are 3.70×10^{-14} moles/grain; for quartz ($\rho_i = 2.660$ g/cm^3 and $g_i = 60.08$ g/mol) there are 4.43×10^{-14} moles/ grain. Hence there are $\sim 4 \times 10^{-14}$ moles/grain on average, so a mole of grains contains $\sim 6.02 \times 10^{23} \times 4 \times 10^{-14} \sim 2.4 \times 10^{10}$ moles of compound. The correct mixing entropy per mole of *compound* is then $\sim 5.76 / 2.4 \times 10^{10} \sim 2.4 \times 10^{-10}$ J/mol K, which is utterly negligible for all practical purposes.

---

[3] Ref. 2 in Box 3.12.
[4] E.g., Broecker, WS.; Oversby, VM. *Chemical Equilibria in the Earth.* McGraw-Hill, 1971, p. 240.

For another illustrative example, consider an orebody containing 0.1% of an ore mineral that consists of randomly distributed 1 μm grains. These could be copper-bearing sulfides in a copper orebody. As described elsewhere, such grains need to be extracted for processing into copper metal, so let us estimate the true entropy cost of doing so.

Using (3.9-2) with $x = 0.001$ yields $S_m' = 65.74$ J/mol K, and again this is per mole of grains. If the grains are chalcopyrite ($CuFeS_2$, a common ore mineral for copper) there is ~2.3×10^{-14} mole/grain, and a mole of grains contains $6.02 \times 10^{23} \times 2.3 \times 10^{-14} = 1.4 \times 10^{10}$ moles. The resulting entropy is then ~$65.74 / 1.4 \times 10^{-10} \sim 4.8 \times 10^{-9}$ J/mol K, which is again utterly negligible.

Extracting ore minerals from such orebodies is energetically intensive, but that energy cost results from the technologies used, not from the laws of thermodynamics.

How do biosystems manage these feats? Not by extravagant applications of heat. Instead, they literally move individual atoms or molecules using specialized molecular mechanisms; e.g., the binding of nutrient elements by specialized proteins. This is not only vastly less costly energetically but allows separation from considerably lower concentrations. (Organisms also are commonly effective in concentrating pollutants out of the environment, which often leads to a new suite of problems.)

To anticipate the conclusions of this book, biosystems are capable of such feats because they work at molecular scales. Such biomimetic—"life-mimicking"—approaches are exactly what nanotechnology makes accessible. They will be examined in much more detail in Chapter 6.

Fundamental Costs of Element Extraction

Let us first, however, examine quantitatively the fundamental thermodynamic cost involved in any separation. Analogously to the maximum amount of work that could be extracted from

---

Box 3.10 Iron Smelting

Iron smelting has not changed in essence since antiquity. An iron oxide is reacted at high temperature with carbon (C) to yield carbon dioxide (CO_2), which wafts off to leave molten iron behind. In essence the carbon displaces the iron from the oxide. Here's the reaction for the common oxide magnetite (Fe_3O_4), for example:

$$Fe_3O_4 + 2\,C \rightarrow 3\,Fe + 2\,CO_2, \Delta G = 223.9 \text{ kJ/mole,}$$
$$\text{or } \sim 1300 \text{ kJ/kg Fe.}$$

Here ΔG is again the Gibbs free energy of the reaction, the plus sign indicating that an energy input is required to make the reaction go in the direction written. The reaction takes place at temperatures around 700–1200°C. Furthermore, any silicate impurities form a molten glass (a *slag*) which can be poured off separately and discarded. Molten iron and molten slag do not mix, just like oil and water. Usually, however, a so-called *flux* is also added to lower the melting point of the slag. Examples include limestone (calcium carbonate, $CaCO_3$), fluorspar (fluorite, calcium fluoride, CaF_2), borax (a sodium borate), and so on. Fluxes not only save on heating costs by lowering the melting point, but also lower the melts' viscosity so they flow more easily. They also have a role in removing impurities, which tend to partition into the melt. Of course, there is a tradeoff, as these additional reagents increase both costs and complexity.

---

a chemical reaction (Box 2.6), this cost is set by the Gibbs free energy difference between the initial and final product. An example is the extraction of iron (Fe) from magnetite (Fe_3O_4), a magnetic oxide that is a common ore mineral (Box 3.10). In this case, where the initial phase (i.e., the ore mineral) is a metal oxide, this true thermodynamic cost is only a few percent of the actual energy expended.[7] Moreover, that energy is *only* the process cost for the chemical transformation. The energy expended for crushing and beneficiation (Box 3.2) is not included.

Sulfide ores

But in fact, though, such analyses[7] drastically understate the inefficiencies in many cases, because many ore minerals are *not* oxides but sulfides. Metal sulfides *release* energy when they react with oxygen. Thus, according to the laws of thermodynamics, we could use sulfide ores as fuels and get metal as a by-product! Here again, biosystems are far ahead of current technology. So-called chemolithotrophic bacteria, which derive energy by oxidizing inorganic materials, nearly manage this, although they do not usually extract the free metal but oxidize it as well to yield additional energy. Indeed, such bacteria are important in the generation of acid-mine drainage (Box 3.7).

Instead, current technology begins by "roasting" the sulfide to convert it to the oxide(!), thereby not only throwing its free energy away but generating noxious sulfur dioxide (SO_2) as a by-product. The metal oxide is now converted to the free metal by adding new reagents—usually also with the application of additional heat. The detailed reactions for two important metals, copper and zinc, are shown in Box 3.8.

The processes are grossly energy intensive, but not because of the laws of thermodynamics.

The Promethean Paradigm, Revisited: Pyrometallurgy and Phase Separation

Conventional resource extraction is so energy-expensive because it relies on using *phase changes*—melting, crystallization, vaporization, and so on—and those phase changes are usually driven by the application and extraction of vast quantities of

---

[7]Kellogg, HH. Sizing up the energy requirements for producing primary materials. *Eng. Mining. J.*, 61–5, April 1977.

heat. The self-organization (Box 3.11) of huge numbers of atoms is exploited as the desired element separates ("partitions") into one of the phases. In other words, the atoms of a particular element will distribute themselves among coexisting phases depending on which environment the atom "prefers." Such "preferences" can be expressed by "partition coefficients:" the ratio of the element concentration in one phase (one crystal, say) to another (say, the melt). For practical separation, this ratio obviously must be very different from 1; i.e., the element must *really* prefers one phase to another. This is the basis for phase-based separation: arrange the system so that the desired element is extracted into one particular phase; then use physical means to separate off that phase while discarding the rest. And, in most cases, those phase changes are thermally driven. For this reason conventional extractive metallurgy is termed "pyrometallurgy," the heat treatment of ore concentrates being termed "smelting."

The examples in Boxes 3.8 and 3.10 have illustrated these points. Loss of sulfur or carbon compounds as a gas is common. So is the formation of an immiscible silicate melt—a "slag"—into which many impurities separate, and which can be decanted and discarded separately. Moreover, by products containing geochemically abundant elements are usually uneconomic and discarded as waste, as noted in the case of the iron content of copper ores (Box 3.8). In fact, a beneficiated copper sulfide feedstock typically contains more iron than copper.

As was also seen in the examples, reagents are commonly added. These can cause new phases to become stable (e.g., silica in copper processing), or simply change the physical properties in useful ways, such as to make the slag fluid at lower temperatures, as with the fluxes in iron extraction (Box 3.10). Although they save process energy by allowing processes to be carried out at lower temperatures, such reagents obviously add to costs, not least because they themselves require extraction and purification.

Not only are such processes grossly energy intensive, they are intrinsically polluting, not just from the combustion necessary

Box 3.11 Crystallization and Phase Changes

Phase changes are familiar in everyday life: a solid melting to a liquid, a liquid boiling to a gas, the freezing of a liquid, the condensation of a vapor, the crystallization of a precipitate out of an evaporating solution. They're so familiar, in fact, that it's often unappreciated how extraordinary some of those phase changes are. Sure, the changes with an *increase* in temperature seem reasonable. It seems intuitive that substances break up into more disordered forms as their average energy increases. A solid melts to a liquid when the molecular motions become strong enough to intermittently overwhelm the bonds between them; similarly, a liquid boils to a gas when the motions become so intense that the molecules (or atoms) separate completely.

In turn, it's intuitive that as the molecular motions decrease; i.e., as the substance is cooled, the molecules or atoms begin to stick together again. It's less obvious, however, that when cooled enough the molecules or atoms usually arrange themselves into highly symmetrical, indefinitely repeating patterns—that is, they crystallize. Frost forming on a winter night, as water molecules spontaneously leave the vapor phase to arrange themselves in the ordered arrays of ice crystals; feldspar crystals precipitating from a cooling magma; the examples can be multiplied indefinitely.

Of course, the initial condensed phase is typically a liquid, which is only poorly ordered. Yet on further cooling an ordered crystalline solid nearly always results. (A *glass* is a *dis*ordered solid phase, which is why glasses are sometimes called "supercooled liquids." Glasses usually result from rapid cooling, "rapid" being a relative term. Even glasses, however, tend to recrystallize ("devitrify") given enough time. Volcanic glass more than a few million years old is extremely rare, because most of what once was glass has devitrified. To be sure, this indicates that the recrystallization can be a very leisurely process indeed.)

The initial condensation results, of course, from the huge mutual forces acting between the atoms or molecules. The reason why *organized* phases ultimately result is that a crystal is a minimum energy configuration, so once again there is a tradeoff between disorder and the energy costs of that disorder. But why does an *ordered* arrangement have minimum energy? Roughly, because if a particular configuration of atoms has minimum energy, then for the *overall* energy to be minimum all atoms of the same kind must be in that same configuation. This in turn leads to an indefinitely repeating pattern. This also underscores, once again, that the belief that the progress of entropy must ultimately lead to fully disordered matter is a profound misconception.

The dissolution of crystals in a solvent, such as salt in water, is a similar process to melting. Indeed, "melt" can still mean "dissolve" in everyday English. Similarly, crystallization of a precipitate from a solution is somewhat analogous to the crystallization of a solid phase from a melt during freezing. Consider salt crystallizing from evaporating water, as sodium and chloride ions coalesce from the thickening brine to array themselves like bricks in a wall. In these cases, though, the phase changes are not driven by changes in temperature, and thus by flows of heat, but by changes in concentration.

Crystallization is not perfect; entropy does have an effect. Some "mismatched" "building blocks" (i.e., atoms or molecules) do get incorporated into a growing crystal, and the more similar the building blocks the more get incorporated. This leads to the phenomenon of solid solution (Box 3.3). Nonetheless, there is an energy cost to the mismatch, and this is reflected in a diffuse melting point. As an impure substance melts, the initial liquid formed does not have the same composition as the solid: it is enriched in the impurity component. By contrast, a sharp melting point is a routine criterion of a substance's purity. Hence, a traditional way to purify a substance is to subject it to repeated cycles of melting and recrystallization; cf. the "zone-refining" of silicon mentioned in Box 2.16. As with exploiting phase changes for separation, though, it is simple but highly energy intensive. However, crystallization is one example of so-called *self-organization*, a phenomenon that has enormous potential applications in nanofabrication, as discussed in Chapter 4.

to generate the heat, but also because the separation is never complete.

Thermal partitioning is also practical only with concentrated raw materials, for several reasons. Most obviously, the more concentrated the deposit, the fewer steps needed to deal with it. Less rock needs to be dug up, less energy is needed to heat it up, and less reagent is needed to react it with. There are also fewer side reactions, due to impurities in the rock, to consume those reagents. Obviously this has an extraordinary effect on the economics of the operation. (Even so, preprocessing— beneficiation (Box 3.2)—is usually needed to make a feedstock that can be smelted.) This need for preconcentrated feedstocks is why these processes require ores that *already* are geologically anomalous! That's why the travel to remote locales is so often necessary.

In addition, the partitioning is never complete. Some of the desired element is always lost into the waste phase, through those same "solid solution" phenomena that account for background concentrations of rare elements in ordinary rock-forming minerals (Box 3.3), and such losses can be significant in the economics of a mine or smelter. Some copper, for example, is lost into the slag.[8] Depending on the element(s) being extracted, such losses can be a pollution problem, too. Obviously, too, the problem of elements partitioning into undesired phases gets worse the lower grade the source. Such losses are also greater the higher the temperature at which processing is carried out, because the degree of solid solution becomes greater.

Thus element separation, despite conventional wisdom, is not intrinsically an energy-intensive process. The prodigious energy costs of conventional resource extraction, pyrometallurgy in particular, are due to clumsy technology. In particular, the costs result from the application and extraction of vast amounts of heat to force separation through phase changes. Hence the

---

[8]Rosenqvist, T. *Principles of Extractive Metallurgy*, 2nd ed., McGraw-Hill, 1983, pp. 331–2.

Promethean paradigm dominates extractive processes even more than it does energy-transformation processes.

The only virtue of conventional pyrometallurgy is simplicity. Humankind has been digging up and "cooking" anomalous geologic deposits for the last 5000 years or so. But even though smelting has not fundamentally changed since antiquity, it not only squanders energy but is extremely dirty. Moreover, the "footprint" left behind, consisting of large holes and huge piles of waste rock, also is a long-term environmental legacy that at best must be expensively reclaimed.

And it bears repeating: all the above are merely statements of current *engineering* limitations. They are *not* fundamental limits set by natural laws.

Non-thermal Extractive Processes

Although conventional pyrometallurgy dominates metals extraction, non-thermal processes are important even now, and this importance will only increase. These fall into two broad categories, electrolysis and solution-based extraction, the latter in turn subsuming several different approaches. They will be described briefly in turn.

Electrolysis

The use of electricity to split water into hydrogen and oxygen is a familiar school experiment, but at present it is not a competitive method of making hydrogen or oxygen. Electrolysis *is*, however, currently of commercial importance in extracting certain metals, in particular aluminum (Box 3.6) and magnesium. In these cases, what is electrolyzed is typically not an aqueous solution but a molten metal salt (traditionally, and perversely, referred to as a "fused" salt). Since electricity is a highly organized form of energy, the ultimate thermodynamic efficiency of electrolytic processes should be greater than for pyrometallurgical processes. "Real-world" efficiencies, however, are considerably less than the theoretical maximum. Electrical resistance leads to losses as waste heat, and in any case heat is often required to melt the substance so that it can be electrolyzed in the first place. More— sometimes considerably more—voltage must also be applied to drive the reaction than the thermodynamic minimum, and this "overvoltage" leads to additional losses. (The overvoltage is necessary to overcome the activation energy to reaction at the electrode; cf. Box 4.6.) Finally, the salts themselves must be extracted and purified before being processed into metal. Bauxite, for example, is mined and beneficiated like any ore mineral, whereas magnesium salts are extracted from brine solutions, such as seawater, before being electrolyzed.

Separation from fluid media

Undoubtedly the most important category of extraction processes, and the one that will be a major focus of this book, is the separation of solutes from fluid media (gases, liquids, or both). In theory, anyway, the engineering should also be easier, both because the material is free to flow, and because the problem of comminuting solid materials doesn't arise. A fluid is *already* broken up at the molecular level! Moreover, the fluid automatically remains well-mixed because of the thermal motion of the molecules making it up.

The "solution solution"

Aqueous solutions in particular are of enormous importance and also subsume purification and pollution control as well as extractive metallurgy. Most simply, elements can be extracted from seawater, as is already the case for magnesium and bromine salts. More concentrated natural solutions are also already practical for certain elements. Lithium and boron, for example, are extracted from saline-lake brines in which they are unusually concentrated. For example, currently the only US domestic source of lithium is brines in Clayton Valley, near Tonopah,

Nevada, and *salars* (saline lakes) in the Bolivian Andes are important world sources of lithium.

Such applications, however, merely hint at the possibilities. There are a great many solutions, many highly concentrated, of both natural and artificial origin that are potential feedstocks: oil-field brines, waste industrial "pickling" solutions from metal plating, acid-mine drainage (Box 3.7), to name a few. Indeed, many such solutions currently are important sources of pollution, and separating out the polluting solutes is already a technological driver. Of course, as such technologies improve they again will blur the distinction between a pollutant and a resource.

An important point is that solutes in aqueous solutions are often present as ions—that is, as species with a net electric charge (Boxes 3.3, 4.3). When ordinary salt (sodium chloride, NaCl) dissolves, for example, it is present in solution as positive sodium ions (cations), Na^+, and negative chloride ions (anions), Cl^-. Water, as a so-called polar solvent, is able to break up the NaCl crystal because of the asymmetric distribution of electric charge between the hydrogen and oxygen atoms in the water molecule. Their electric charge proves to be a useful "handle" in manipulating ionic solutes, as will become evident in the discussions below and in Chapter 6.

Phase-based extraction

Most simply, extraction from solution can involve a phase change, typically a precipitated solid containing the desired (or undesired) substance. Such phase changes are driven by changes in solution concentration, not directly by heat, in contrast to pyrometallurgical processes. Increasing concentration of a solute causes a solid phase containing it to become stable (Box 3.11), so that it then precipitates. Commonly, however, the solutions are concentrated by evaporation, which *does* require the application of heat. Solar evaporation can sometimes be used, which obviates the need for fuel, but it is not always practical.

The precipitation of salt (sodium chloride, NaCl) from brine is an example that goes back to antiquity. For a contemporary, more complicated example, the lithium-bearing brines at Clayton Valley, Nevada are first concentrated by solar evaporation. Then lime (calcium oxide) is added to precipitate magnesium, which otherwise would precipitate along with the lithium. Finally, lithium is precipitated as the carbonate by the addition of sodium carbonate.[9]

As this example shows, additional reagents are usually required in the processing, and represent an additional energy cost, especially since such those reagents themselves commonly were purified by thermal means. A further disadvantage is that the brine must be relatively concentrated to begin with, and finally, the extraction is not "quantitative"—that is, some of the desired material remains in solution (Box 3.12).

Liquid–liquid partitioning

A somewhat different twist is to exploit the partitioning of a solute between immiscible liquids—liquids, like oil and water, that do not mix. This obviously exploits the partitioning of a substance between coexisting phases, analogous to the differential partitioning between coexisting molten phases exploited in pyrometallurgy. It is thus akin to a phase separation, but the phase changes are not driven by heat.

A "hybrid" approach is to dissolve a substance—a "complexing agent"—having a strong affinity for the desired solute in one of the liquid phases, which strongly enhances the extraction into that phase. In practice, the agent is dissolved in the non-aqueous liquid, which is later separated from the aqueous phase by physical means. Under the name "solvent extraction," this technique has become important in a number of separation process, for copper in particular, and it will be discussed further below (p. 126). Moreover, complex-based

---

[9]Averill, WA; Olson, DL. A review of extractive processes for lithium from ores and brines. *Energy 3*, 305–13, 1978.

Box 3.12 The Solubility Product

The solubility of a substance is quantitatively described by the *solubility product*, which has the following form:

$$K_{sp} = \{a_1\}\{a_2\}\{a_3\} \ldots /\{s\}$$

where the $\{a_i\}$ are the *activities* of the dissolved consituents that make up the solid substance, and $\{s\}$ is the activity of the solid ($=1$ for a pure substance). In the case of ordinary salt (sodium chloride, $NaCl$), for example, the dissolved consituents are positive sodium ion (cation), Na^+, and the negative chloride ion (anion), Cl^-. There is a threshold value of K_{sp}, different for different substances, at which the solutes and the corresponding solid are in equilbium. If the product of the activities is greater than this value, more solid will tend to precipitate; if less, the solid will dissolve. (As is usual in chemical processes, activation barriers can exist to immediate reaction. So-called supersaturation, for example, is common because new solid phases do not precipitate immediately due to difficulty in nucleating the initial crystalline "seeds." Although these issues are of engineering importance and treated in the enormous literature on crystallization (Box 3.11), they are peripheral here.)

Note that the solutes need not all be derived from the phase that is dissolving. For example, adding sodium sulfate (Na_2SO_4) to a sodium chloride solution boosts the concentration of sodium cations, because the sodium sulfate dissociates into sodium cations and sulfate (SO_4^{2-}) anions, and the sodium cations don't "remember" which phase they originally came from. This phenomenon is the basis of separations by precipitation: a component is added to increase its activity such that a solid phase containing the desired component precipitates. In the case of lithium extraction from brine, for example, addition of sodium carbonate (Na_2CO_3) raises the activity of the carbonate anion such that the solubility product for lithium carbonate (Li_2CO_3)

is exceeded, and so it precipitates. Note, however, that *all* lithium does not precipitate; an amount remains in solution that is in equilibrium with the solid Li_2CO_3 phase.

The activity of solutes in an ideal solution is simply their concentration, and so for sufficiently dilute solutions the activities can be replaced by concentrations, as all solutions become ideal in the limit of infinite dilution. Real solutions, of course, are not ideal, and the nonideality is represented by an "activity coefficient", which when multiplied by the concentration gives the correct activity. For dilute solutions of ionic species, the activity coefficients can be approximately calculated by the so-called Debye–Hückel expression.[1] In general, however, they must be determined experimentally.

Fundamentally, dissolution and precipitation again reflect a tradeoff between the energy of different phases, so the solubility product is an equilibrium constant. Hence, if the free energies of the dissolved and solid species are known the solubility product can be calculated directly from the expression:

$$\Delta G = -RT \ln K,$$

where R is the gas constant, T is the absolute temperature, and ln is the natural logarithm.[2]

---

[1]Ref. 1, p. 83.
[2]Cf. discussion in Stumm, W; Morgan, JJ. *Aquatic Chemistry; an Introduction Emphasizing Chemical Equilibria in Natural Waters.* 583 pp. Wiley-Interscience, 1970, Ch. 2.

approaches to separation prove to be extremely important in general and will be treated at length in Chapter 6.

Non-phase separation techniques

At least in theory, these can avoid the problems noted above by direct, differential extraction of a particular component from the fluid, such as a solute out of a solution.

Semipermeable membranes

Semipermeable membranes are in essence "molecular filters" that strain out one or more components out of a fluid medium. They are currently attracting enormous interest, especially in pollution control and purification,[10] and exemplify yet another general approach to non-thermal separation.

One current set of applications is in the separation of gas mixtures,[11] often termed *vapor permeation*. Examples include the separation of nitrogen or oxygen from air, the separation of carbon dioxide, the purification of natural gas and other HCs,[12] and separating the components of *syngas* (Box 5.4), hydrogen in particular.[13]

Pervaporation (PV) is the evaporation of particular solutes from a liquid phase through a membrane, usually into a vacuum.[14] Obviously, it is best suited for volatile solutes, i.e., solutes with a lower boiling point than the solvent, and is already used for removal of volatile organic solvents from aqueous solution and in the manufacture of anhydrous ethanol. Although like reverse osmosis (see below) it requires the use of pressures different from

---

[10]E.g., Caetano, A; et al., eds. *Membrane Technology: Applications to Industrial Wastewater Treatment*. Kluwer Academic, 1995; Cardew, PT; Le, MS. *Membrane Processes: A Technology Guide*, RSC, 2006.

[11]E.g., Baker, RW. KOECT, 4th ed., 16, 60–90, 2001; Puri, PS. Chapter 8. Commercial Applications of Membranes in Gas Separations, *Membrane Engineering for the Treatment of Gases 1*, 215–44, 2011.

[12]E.g., Baudot, A. Gas/Vapor Permeation Applications in the Hydrocarbon-processing Industry, *Membrane Engineering for the Treatment of Gases 1*, 150–95, 2011.

[13]E.g., Di Donato, A. H_2 *Production and* CO_2 *Separation*. GE&T, 145–67, 2011.

[14]E.g., Ong, YK; Shi, GM; Le, NL; Tang, YP; Zuo, J; Nunes, SP; Chung, T-S. Recent membrane development for pervaporation processes. *Prog. Polym. Sci.* 57, 1, 2016; Wynn, N. Pervaporation comes of age. *CEP Magazine* October, 66–72, 2011; Jyoti, G; Keshav, A; Anandkumar, J. Review on Pervaporation: Theory, Membrane Performance, and Application to Intensification of Esterification Reaction. *J. Eng.* Article ID 927068, 2015; Bowen et al. (in Ref. 3, Chapter 6) also has some review.

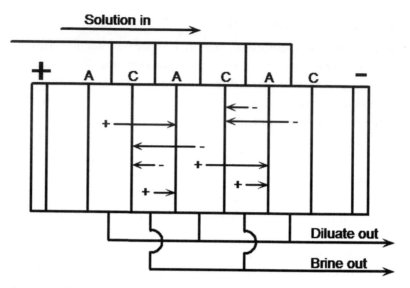

Figure 3.6 Electrodialysis.

Electrodialysis purifies water by means of the differential movement of ions, driven by an applied electric field, through permeable membranes. Alternating compartments, each bounded by an anion-permeable (A) and cation-permeable (C) membrane, thus become fresher and brinier.

ambient, it has promise as a low-energy alternative to distillation and similar processes.[15]

Several different membrane approaches are used with aqueous solutions. In *electrodialysis* (ED)[16] an electric potential is used to move ions differentially. ("Dialysis" in general is the purification of a solution by differential passage of solutes through a membrane or other "sieve.") A vessel containing the solution is partitioned by alternating membranes permeable to negative ions (anions) and positive ions (cations), respectively (Fig. 3.6). Applying a potential across the vessel causes cations

---

[15]Jonquieres, A; Clement, R; Lochon, P; Neel, J; Dresch, M; Chretien, B. Industrial state-of-the-art of pervaporation and vapour permeation in the western countries. *J. Membr. Sci. 206*, 87, 2002.

[16]E.g., Korngold, E. Electrodialysis—membranes and mass transport. In Belfort, G, ed., *Synthetic Membrane Processes*, 191–220, Academic, 1984.

and anions to migrate in opposite directions, such that the compartments walled by the membranes become alternately saltier and fresher. Electrodialysis has the advantage of ambient pressure operation, but the resistance of the solution becomes a serious source of efficiency loss for dilute solutions. Obviously, too, the solutes must be in the form of ions, but this is commonly the case. Indeed, the technique illustrates how ions' electric charge provides a "handle" for manipulating them.

In *reverse osmosis* (RO) particular dissolved species are "strained" out of a solution by a pressure gradient. Its big disadvantage is that it requires manipulation of pressures above ambient. This is a reasonably mature technology for such applications as desalination; indeed, both ED and RO are now used in desalination, RO for seawater, ED typically for brackish water.

Ultrafiltration (UF), as its name suggests, refers to the filtering out of extremely fine particles. These can be *colloidal* particles, particles so small that they remain suspended indefinitely (to be discussed more fully in Chapter 4); viruses; or even giant molecules (macromolecules). It thus lies between reverse osmosis and conventional filtration. An interesting recent application is in the recycling of organic solvents.[17]

Binding-based approaches

This is a very general description of several processes in which separation is effected by the binding of a component of the mixture to another species, such as a surface.

Adsorption: Most simply, the desired (or undesired, in the case of a pollutant) solute is not precipitated but *adsorbed* onto a substrate from which it can later be removed. (*Ad*sorption is specifically the process(es) by which a molecule or atom sticks to a surface. At the molecular level, most bulk *ab*sorption is

---

[17]Marchetti, P; Jimenez Solomon, MF; Szekely, G; Andrew G. Livingston, AG. Molecular separation with organic solvent nanofiltration: a critical review. *Chem. Rev., 114,* 10735–806, 2014.

really adsorption.) Such subtrates must have high surface area as well as some sort of affinity for the solute. Traditionally the mechanism of binding has not always clear, depending as it does on details of the surface structure, and the suitability of various adsorbents has been empirically determined. Indeed, a high surface area substance will typically exhibit significant adsorption due simply to van der Waals forces (Box 4.5). One of the oldest, and still widely used, adsorbers is "activated" carbon (often called activated "charcoal" in the older literature), which is a preparation of extremely high-surface-area carbon, but many other materials have been extensively investigated.[18]

Ion exchange: Ion *exchange* is an important subset of adsorption-based approaches that is another well-established non-thermal extraction technology.[19] As noted, many important solutes are present in solutions as ions, and such ions can be swapped out for others. In another example of molecular recognition, the dissolved ions displace ions already bound to the surface because they have slightly greater affinity for the binding site (hence ion "exchange.") Water softeners are a familiar example: hard water contains sufficient divalent cations (calcium (Ca^{++}) and/or

---

[18] E.g., Lee, SC; Kwon, Y-M; Park, YH; Lee, WS; Park, JJ; et al. Structure effects of potassium-based TiO_2 sorbents on the CO_2 capture capacity. *Top. Catal.* 53, 641–7, 2010; Wang, X; Ma, X; Xu, X; Sun, L; Song, C. Mesoporous-molecular-sieve-supported polymer sorbents for removing H_2S from hydrogen gas streams. *Top. Catal.* 49, 108–17, 2008; Zaspalis, V; Pagana, A; Sklari, S. Arsenic removal from contaminated water by iron oxide sorbents and porous ceramic membranes. *Desalination 217*, 167–80, 2007; Ho, KY; McKay, G; Yeung, KL. Selective adsorbents from ordered mesoporous silica. *Langmuir*, *19*, 3019–24, 2003; Babel, S; Kurniawan, TA. Low-cost adsorbents for heavy metals uptake from contaminated water: a review. *J. Hazard. Mater.* 97, 219–43, 2003.

[19] E.g., Helfferich, F. *Ion Exchange.* Dover, 1962; Slater, HJ. *Principles of Ion Exchange Technology.* Butterworth Heinemann, 1991; Sata, T. *Ion Exchange Membranes: Preparation, Characterization, Modification and Application.* RSC, 2004; Harland, CE. *Ion Exchange: Theory and Practice,* 2nd ed. RSC, 2006.

magnesium (Mg^{++})) to interfere with cleaning by soap, because these ions form insoluble salts with the stearate and other fatty-acid anions used in soaps, so they cannot act as surfactants. In a domestic water softener, Ca^{++} or Mg^{++} is exchanged for sodium ion (Na^+), which has no effect on soap action. The basis of the exchange is simply the greater electrostatic attraction of the doubly charged cations for the binding sites. Ion exchange is also currently used to extract heavy metals from wastewater effluents.

Ion exchange can also be employed to carry out difficult separations through what might be termed "dynamic" selectivity. Subtle differences exist in the dissociation constants between the active sites and the various solution species, so that on repeated exchanges over time a significant separation can build up. If a batch of mixed solution is inserted in the top of an ion-exchange column and allowed to flow through, for example, the solution that eventually exits will have varying composition depending on the time it exits. The solution that first exits will be enriched in the ion with the least affinity for the binding sites, and so forth. This can be used for quantitative separation if the ion-exchange column is sufficiently long. The same principle underlies the various kinds of chromatography, which are widely used on a laboratory scale for separation. Such a scheme is practical for separating solutes such as rare earth elements, whose separation is otherwise extremely difficult, but it is usually not economic otherwise.

Current ion-exchange is still polluting, though, because of that "exchange" of the ions. When (say) a lead ion is removed, it must be replaced with another ion to maintain charge balance. The replacement ion is more innocuous—obviously it's an improvement if (say) lead is replaced with calcium—but a solute still remains in solution.

Water softeners are currently even worse. Once the ion exchanger is filled with calcium, it must be regenerated by passing through a highly sodium-rich solution; i.e., an ordinary salt solution, or brine. This "recharges" the softener by replacing the calcium ions with sodium ions that can again be exchanged. The concentration of sodium ions in the brine overwhelms the preference of the exchange sites for calcium. Hence, a great deal of waste brine is generated, which traditionally just gets dumped into the wastewater stream. This issue foreshadows the "elution problem" discussed more fully in Chapter 6 (p. 292); once a solute is bound, it eventually must be unbound again.

Binding agents and complexation: Yet another approach is through specific molecules that can bind tightly with the desired component or solute to form highly stable "supramolecular" structures, so-called complexes. An example is in the so-called solvent extraction process for copper mentioned above. A copper-bearing leachate is mixed with a kerosene solution containing compounds (e.g., hydroxyoxime, Fig. 3.7) that bind to dissolved copper.[20] These compounds extract the copper essentially completely into the kerosene phase.

Hydrometallurgy

The solvent-extraction process for copper is an example of *hydrometallurgy*: not everything is already present in an aqueous solution, but raw materials can be dissolved and then the desired component(s) extracted from the solution. In fact, hydrometallurgy has a distinguished history. For example, the solvent-extraction process for copper has accounted for a growing percentage of copper production over the last few decades. Once again, too, biological systems have anticipated technology. Digestion, after all, may be viewed as solubilizing food so that particular compounds can be extracted from it.

---

[20]E.g., Kordosky, GA; Olafson, SM; Lewis, RG; Deffner, VL; House, JE. A state-of-the-art discussion on the solvent extraction reagents used for the recovery of copper from dilute sulfuric acid leach solutions. *Sep. Sci. Technol.*, 22, 215–33, 1987.

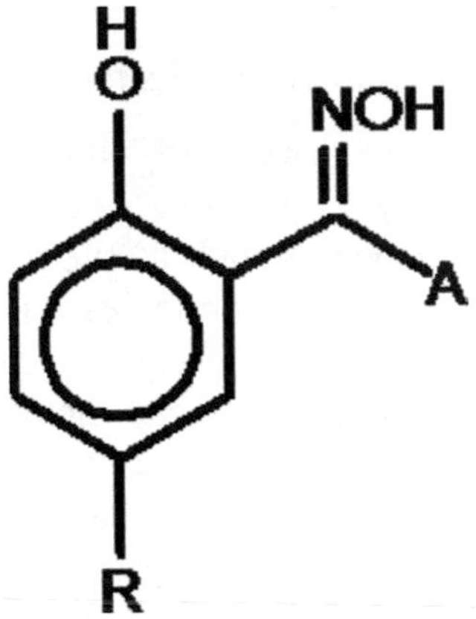

Figure 3.7 Hydroxyoximes.

These compounds are the basis of the solvent extraction of copper. They are hydrophobic and dissolved in a hydrocarbon phase, typically kerosene. They are highly specific complexing agents for Cu^{++} and extract it essentially completely out of an adjacent aqueous phase. By convention, such diagrams of molecular structure show only the carbon "skeleton." A carbon atom is present at every point where the lines come together at an angle, and it is implied that the bonding of the carbon atom is filled out with hydrogen atoms unless other atom(s) are shown. The hexagon enclosing the circle is a so-called "benzene ring," the circle showing that aromatic bonds (Box 2.8) link the carbons in the ring. Again, a hydrogen atom is implied at each angle unless another linkage is explicitly shown. A is H (hydrogen), methyl (CH_3), or ethyl (C_2H_5), R is a longer alkyl sidechain, typically C_9-C_{12}.

The most extraordinary current example of hydrometallurgy is that of gold. In the late 1800s, gold's extractive metallurgy was revolutionized by the discovery that cyanide, better known then and since as a virulent poison, has an extraordinary affinity for metallic gold (Box 3.13). Notwithstanding this example, however, the efficiencies of most current hydrometallurgical processes are still low. In part this is because most approaches still rely on phase changes, albeit isothermal changes, such as the precipitation or partitioning between immiscible liquids described above. Hence additional reagents, which probably *were* purified or synthesized thermally, are still required. The separation is also commonly not clean, a problem that plagues

Box 3.13 The Cyanide Process for Gold Extraction

"Revolutionize" is an overused word, but it applies to the effect of the cyanide process on gold extraction. Cyanidation, even more than gold discoveries such as the Witwatersrand and Klondike, was responsible for an increase in gold production that began in the 1890s and has continued off and on to the present. It proved practical—indeed lucrative—to reprocess old tailings and other mine waste for the gold that previous processes had left behind. Reprocessing old tailings wasn't nearly so glamorous as (say) prospectors slogging it up from Skagway, but it was considerably more effective in increasing gold output. Ongoing improvements in cyanidation in the 20th century, such as heap leaching, have now reached the point that, in favorable cases, gold can now be recovered from rocks containing less than 1 ppm (1 gram per tonne).

The overall reaction in the cyanide process is:

$$4Au + 8\ CN^- + O_2 + 2H_2O \Rightarrow 4Au(CN)^{2-} + 4OH^-$$

where Au is metallic gold, CN^- the cyanide anion, O_2 atmospheric oxygen, H_2O water, $Au(CN)^{2-}$ the dicyanoaurate anion, and OH^- the hydroxide ion. The reaction occurs in aqueous solution at remarkably low concentrations, a few hundred parts per million of cyanide[1] being enough. Note also that the presence of air (oxygen) is required.

After dissolution, the dicyanoaurate must be extracted from the solution. Usually it is adsorbed onto activated carbon, then eluted. The gold is then displaced from the dicyanoaurate by another metal, usually zinc.[2]

---

[1]Wadsworth, ME; Zhu, X; Thompson, JS; Pereira, CJ. Gold dissolution and activation in cyanide solution: kinetics and mechanism. *Hydrometallurgy* 57, 1–11, 2000; Kappes, DW. Evaluation & suitability of cyanide heap leach technology. [short course on heap leaching] Mackay School of Mines, University of Nevada, Reno, NV, USA, May 27–28, 1984.

[2]A great many details have been glossed over; see Marsden, J; House, I. *The Chemistry of Gold Extraction*, 2nd ed. Society for Mining, Metallurgy & Exploration (SME), 2006.

At ordinary temperatures gold is exceedingly unreactive, as is certainly indicated by its common use as a symbol of incorruptibility in cultures around the world, so its readiness to react with cyanide is more than a little surprising. Indeed, about the only other reagents that can dissolve gold under ambient conditions are aqua regia, an unstable and spectacularly corrosive mixture of hydrochloric and nitric acids, or free bromine or chlorine. Furthermore, the basis for gold's reactivity with cyanide is not understood in any detail; evidently by happenstance their electronic energy levels are highly "tuned" to each other. At present the detailed modeling of the electronic structure of complex atoms like gold lies at the very threshold of what is currently possible. It is probable, however, that with advances in computational chemistry (Box 4.5) the interaction of gold and cyanide can be understood from first principles. It may then be possible to design similar but less toxic extractive reagents for both gold and other rare metals, such as the platinum-group elements.

solvent extraction particularly: unmixing of the organic solvent from the aqueous solution is sufficiently incomplete that the aqueous phase becomes seriously contaminated and expensive reagents are also lost. Reaction rates also tend to be sluggish at low temperature, a problem that biological systems overcome with highly specific catalysts (enzymes).

Solute selectivity

Although some applications, such as desalination and purification, do not require selective extraction of solutes because the desired product is just water, *selective* extraction from solution is usually of paramount importance. Whether the application is pollution control or hydrometallurgical extraction, typically the toxic (lead, cadmium, etc.), or valuable (gold, lithium, etc.) solute is dispersed in a much more abundant background of innocuous solutes, and ideally only the particular solute is extracted. (Of course, this is why *ores* were important in the first place—they contain a desired substance already pre-concentrated.) Furthermore, many of those other solutes are chemically similar to the one of interest. Dissolved lithium, for example, is always much less abundant than sodium but is very similar chemically. Similarly, dissolved lead is chemically somewhat similar to innocuous but much more abundant calcium. In the case of the copper-bearing solutions resulting from hydrometallurgical leaching, iron is much more abundant than copper, and what makes the solvent extraction practical is that the oximes bind much more strongly with the copper.

Such considerations also apply to gaseous mixtures, as indeed indicated by the notion of "separating" gases in the first place. A current example of an adsorption-based approach is the use of molecular sieves (Box 6.3) to separate oxygen from air using one or another zeolite.[21] Nitrogen molecules are preferentially taken up in the voids in the zeolite, whereas oxygen molecules (which make up 21% of the atmosphere) are not.

Although much progress has been made with empirical "cut and try" approaches to selective extractants—cf. the accidental discovery of the cyanide process!—understanding the mechanisms of selectivity, not to mention to design new selective extraction agents, will require that the detailed chemical nature of the dissolved species be known. Such approaches will obviously rely on molecular recognition (Box 4.7) to an extraordinary degree, and will be the subject of extended discussion in Chapter 6.

---

[21]E.g., Sircar, S. Pressure swing adsorption technology. NATO-ASI, *158*, 285–321, 1989; Jasra, RV; Choudary, NV; Bhat, SGT. Separation of gases by pressure swing adsorption. *Sep. Sci. Tech.* 26, 885–930, 1991; Sherman, JD; Yon, CM. Adsorption, gas separation. KOECT 4th ed., *1*, 269 ff., 2001.

Oil as Non-fuel

Some 10% of oil production is used as a feedstock for the petrochemical industry, and it's been widely thought that petroleum is much too valuable as a source of chemically reduced carbon compounds to waste as fuel.[22] As mentioned above, however, an obvious alternative near-term source of reduced carbon is waste biomass. In fact, the petrochemical industry is fully aware that the days of abundant oil as a chemical feedstock are numbered, and they are already investigating biomass as an alternative.[23]

Use of biowaste as a chemical feedstock for reduced carbon is yet another example of the ongoing theme of "waste as resource," and it will be discussed in more detail in Chapters 5 (p. 242) and 6 (p. 331).

First, however, we turn to an overview of nanotechnology in the next chapter. How *can* we structure matter at near molecular scales? This will set the stage for nanotechnological applications to resource issues in the following chapters.

---

[22]E.g., Committee on Resources and Man, *Resources and Man* W.H. Freeman 1969, p. 7, 15.

[23]Spitz, P. *Petrochemicals: The Rise of an Industry.* John Wiley & Sons, 1988; R. Phair, pers. comm., 1997; Corma, A; Iborra, S; Velty; A. Chemical routes for the transformation of biomass into chemicals. *Chem. Rev.* 107, 2411–502, 2007; Lucia, LA; Argyropoulos, DS; Adamopoulos, L; Gaspar, AR. Chemicals, materials, and energy from biomass: a review, in Zhu et al., Ref. 98 (Chapter 5).

Chapter 4

Nanotechnology

The Biological Inspiration

The "biological inspiration" has been repeatedly emphasized in this book. Organisms have capabilities in separation, in energy gathering, in energy use, and indeed in the sheer organization of matter, that shame those of present technology. The staggering irony of growing biomass for fuel has been noted: generating an intricate nanostructured assembly, assembled molecule by molecule out of the environment using the diffuse and fitful energy of sunlight, so that it can be *burned*!

Biosystems are capable of these feats because they are organized at molecular scales and literally work molecule by molecule. Photosynthesis, for example, rips water molecules apart using the energy of sunlight absorbed with a molecular-scale arrangement of dye molecules, or pigments. As noted in Chapter 3, biosystems as diverse as diatoms, vertebrate kidneys, and plant roots extract specific molecules from a background of many other molecules, and they do this by means of molecular mechanisms that grab and move the desired (or undesired) molecules one by one. In using energy, biosystems must break chemical bonds; but they do not use anything so clumsy as a flow of heat from a hotter to a colder body. As little as possible

Nanotechnology and the Resource Fallacy
Stephen L. Gillett
Copyright © 2018 Pan Stanford Publishing Pte. Ltd.
ISBN 978-981-4303-87-3 (Hardcover), 978-0-203-73307-3 (eBook)
www.panstanford.com

of the energy released is thermalized, with most used directly in forming new chemical bonds, and all reactions are carried out at constant temperature or nearly so. Of course, warm-blooded creatures do use food to generate heat to maintain that isothermal environment, but they do not extract work by setting up a flow of heat from a hotter to a cooler body.

Nanotechnology and Resources

The above considerations, of course, motivate the relevance of nanotechnology for energy and resources. Not only could vast improvements in efficiency result, but new, low-grade resources will become practical—indeed, the distinction between "waste" and "resource" will become blurred. As will be shown in Chapter 6, even dirt and sewage ultimately can become resources. To foreshadow the discussion below, improvements in efficiency will come about from non-thermal energy usage, driven by such things as better catalysts, molecular separation, superstrong materials, more efficient energy collection, and so forth.

This leads to a consideration of fabrication issues—or rather, *nano*fabrication issues. Obviously, for such possibilities to become realities, nanostructured devices must be made somehow. Applications furthermore will depend on nanotechnological fabrication that's reasonably cheap and routine. Moreover, although biology provides ample proofs of concept—after all, organisms make molecular structures out of ambient materials, and do so with modest energy expenditures—organisms' capabilities aren't necessarily relevant for technological applications. Hence, this chapter will sketch a broad overview of current nanofabrication approaches. The literature on this subject continues to burgeon exponentially: indeed, in recent years entire new journals devoted to nanosystems have been founded. Many reviews and even treatises[1] have appeared, and at

---

[1]E.g., Cui, Z. *Nanofabrication: Principles, Capabilities and Limits*. Springer, 2009. Zhang, G; Manjooran, N. *Nanofabrication and its Application in Renewable Energy*, RSC, 2014; Liu, Q; Duan, X; Peng, C. *Novel Optical Technologies for Nanofabrication*. NS&T, 2014.

least one textbook.[2] Hence this summary can be no more than a sketch, and will already be outdated when you read it.

Nanotechnology

First, a more precise definition of "nanotechnology" is needed. Like any buzzword, "nanotechnology" has become a bit nebulous in its definition, referring to anything from structures at scales of a few nanometers (10–20 atom widths) to actual machines built out of individual atoms. Here, it will be defined as "design and organization of matter at near-molecular scales." The term when specifically referring to the atomic-level structuring of macroscopic objects, in which there's literally "a place for every atom, and every atom in its place," will be *molecular* nanotechnology, or "MNT."

As used in the literature, though, MNT often implies mechanical *machines* at a molecular level, with cams, rods, gears and so on that are themselves large molecules. Although such molecular devices are attracting much attention[3] and are even the subject of the 2016 Nobel Prize in Chemistry[4] their design and fabrication is a *much* more difficult problem, especially in the quantities required, and will not be a major focus of this book. Instead, it will be convenient to distinguish *nanostructured* materials from true molecular *machines*. Such materials are also (optimally) designed at a molecular level, but have no moving

---

[2]Ozin, GA; Arsenault, AC; Cademartiri, L. *Nanochemistry: A Chemical Approach to Nanomaterials*. RSC, 2009.

[3]E.g., Drexler, KE. *Nanosystems: Molecular Machinery, Manufacturing, and Computation*. Wiley Interscience, 1992; Balzani, V; Credi, A; Raymo, FM; Stoddart, JF. Artificial molecular machines. *Angew. Chem. Int. Ed. 39*, 3348–91, 2000; Ozin, G; Manners, I; Fournier-Bidoz, S; Arsenault, A. Dream Nanomachines. *Adv. Mater. 17*, 3011–8, 2005; Balzani, V; Credi, A; Venturi, M. *Molecular Devices and Machines-Concepts and Perspectives for the Nanoworld*. Wiley-VCH, 2008; Balzani, V; Credi, A; Venturi, M. Light powered molecular machines. *Chem. Soc. Rev. 38*, 1542–50, 2009.

[4]Awarded to Jean-Pierre Sauvage, J. Fraser Stoddart, and Ben L. Feringa. A summary is given in: Leigh, DA. Genesis of the Nanomachines: The 2016 Nobel Prize in Chemistry. *Angew. Chem. Int. Ed.* 10.1002/anie.201609841

parts. Not only are such structures much nearer term, being both easier to design and fabricate, but many of the "first-generation" resource applications treated in the following chapters are ideal for such passive systems. These include sieves, membranes, and films, some of which, unlike biosystems, could even be electrically powered from an external source, an example of the convergence of electronics with chemistry at the nanoscale level. Superstrength materials, which must be essentially defect-free at the molecular level, are another example. Structured *surfaces* are critical to catalysts and indeed to interfaces in general, and furnish a third example. More complicated 3D arrangements could be built up by adding successive layers, much as happens on the microscale with conventional microchip fabrication. Alternatively, larger scale 3D structures could be built using self-assembly approaches, as is more fully discussed below.

The routine fabrication of materials with unusual macroscopic properties, many of which are currently now curiosities, should also become practical. Some of these properties also have clear applications to resource issues, as with thermoelectric substances, which will be discussed more fully in Chapter 5.

Nanostructures also include materials with completely new macroscopic properties, largely due to having free electrons confined in dimensions of a few nanometers so that they exhibit quantum behavior.[5] Such confinement typically arises in nanoparticles that comprise the nanostructure. So-called quantum dot materials, for example, which are nanometer-sized semiconductor particles, have potential applications to catalysis, solar energy, sensing, and light generation, as discussed in Chapter 5.

Overall, nanotechnology is an obvious continuation of the technological trend to organization at ever-smaller scales. The most familiar example, of course, is electronics' trend to

---

[5] E.g., Kinge, S; Crego-Calama, M; Reinhoudt, DN. Self-assembling nano-particles at surfaces and interfaces. *ChemPhysChem 9*, 20–42, 2008.

miniaturization from the 1920s to the present. The replacement of vacuum tubes by transistors, and then of transistors by integrated circuits at ever-increasing density, has led not only to sweeping new capabilities, but vastly lower energy use with those capabilities. A "throwaway" digital wristwatch has more computing power than one of the room-filling behemoths of the 1960s. It's been commented that back in the 1960s the phrase "personal computer" would have sounded as preposterous as "personal nuclear submarine."[6] More recently, the wholesale replacement of wet-chemical photography by digital imaging illustrates how new capabilities, in this case also in electronics, can cause the obsolescence of long-established and completely unrelated technologies (cf. Box 4.1). It's noteworthy, too, that the ever-smaller and cheaper organization of matter, with profound economic effects, is not limited to electronics. The advent of cheap, abundant machine-made nails in the 19th century, for example, made so-called *balloon frame* construction practical, in which lots of small pieces of lumber are held together with lots of nails—a building style that dominates the present. Previously only very large timbers could be used, which were expensive and whose supply was dwindling.[7]

There is also a growing trend toward "designer" materials, which are optimized for various desirable properties, such as strength, biological inertness, uniform porosity, and a host of others. Of course, people have been seeking better materials ever since blacksmiths tinkered with quenching and tempering, or nameless artisans devised the composite bow by sheer trial and error, and in this sense this trend also is nothing new. However, even though there is now much more *understanding* of what's happening at the nanoscale to give those desired properties, control at the nanoscale has still been limited. For example,

---

[6]Wallace, J; Erickson, J. *Hard Drive: Bill Gates and the Making of the Microsoft Empire.* 1st HarperBusiness ed. 426 p., 1993, p. 25.
[7]Furnas, JC. *The Americans: A Social History of the United States.* 1587–1914. G. P. Putnam's Sons, 1969, pp. 622–3.

Box 4.1 "Replacement" vs. "Substitution"

"Infinite" substitutibility without loss of performance is fantasy.

—Preston Cloud[1]

Really? The "infinite" substitutibility of classical economics is commonly mocked by ecologists and non-traditional economists, as the quote above shows. But, as a generalization, this skepticism just isn't true. In fact, often the "substitute" is better as well as cheaper. Several millenia ago iron displaced bronze for just this reason. More recent examples include the replacement of vacuum tubes by transistors, and the subsequent replacement of individual transistors by integrated circuits. Indeed, if the "substitute" weren't often better, how could technologies ever become obsolete? People have been finding better ways of doing things for thousands of years.

To avoid these semantic quibbles, "replacement" is probably a better word than "substitute." Moreover, a useful generalization is that the replacement requires more highly organized matter. Ordinary copper or aluminum wire is just bulk matter. Fiber-optic cable, which is capable of vastly higher rates of data transmission than metal wires, is also much more organized at the nanoscale, albeit being made of more common elements. Similarly, the progression of vacuum tubes to transistors to microchips reflects ever higher degrees of small-scale organization. Magnetic tape is more highly organized than the film it replaced, and even offered new capabilities, being reusable in particular.

This last example further illustrates how often the replacement, although more organized, is made of cheaper materials. Unlike film, magnetic tape is not silver-based. Ironically, of course, magnetic media themselves may now be in the throes of replacement by silicon-based systems; cf. the growing popularity of all-solid-state "disk" drives.

---

[1]Ref. 6 (Chapter 3) in main text.

metal strengths depend upon such factors as the grain size of the metal and the distribution of alloying atoms, but those factors are manipulated only indirectly by traditional macroscopic procedures such as quenching, annealing, and so on. It is the prospect of increasing direct control at the nanoscale that promises to accelerate this trend beyond recognition.

Last, although the capabilities of living organisms are an important source of inspiration, it's important to emphasize that nanotechnology is *not* biotechnology (Box 4.2). There are vastly more possibilities for nanostructures than the subset that biosystems employ, and indeed they can be simpler and yet more robust. For one thing, they are not optimized for survival and reproduction! Because they are not biological, too, they are easier to keep separate from biosphere. There's no worry about (say) a metal-extracting bacterium mutating into something hazardous or pathological and then escaping into the environment.

Box 4.2 Biotechnology

Microorganisms, of course, are "natural nanomachines," and certainly the idea of exploiting their capabilities is hardly new. The fermentation of sugars into ethanol was carried out long before it was known that microorganisms were responsible for the conversion. Even in the modern world, microbiological transformations underlie many processes. Nonetheless, conventional biotechnology is usually a strange hybrid of natural molecular-scale machines and conventional bulk thermal technology. Examples include the distillation of alcohol from yeast fermentation, or the extraction and purification of antibiotics from mold cultures. It's rather like agriculture, in fact.

Some applications, to be sure, seem more fitted to bioprocessing than others. Biomass processing of biowastes seems particularly relevant, and indeed, biowastes are routinely digested by microbes for disposal. Until recently, however, little effort has been put into attempting to recover useful products from the digestion. The capabilities of certain chemosynthetic bacteria (cf. their role in acid-mine drainage, Box 3.7) has led to an onging interest in applying them to ore extraction.[1] There has also been research on using microorganisms for tertiary oil recovery[2] and even for breaking down oil shale.[3]

Nonetheless, biosystems as a rule are less promising as starting points for artificial molecular machines. They are extremely—indeed unnecessarily, as far as applications are concerned—complicated, which makes them both difficult to modify and delicate. For example, applications in biomining have been limited by the organisms' requirements for ongoing sources of carbon, water, and oxygen, which can easily be interrupted in the subsurface environment.

More subtly, organisms "have their own agendas." The human-designed modifications are not in the organism's interest, and any mutation that eliminates those modifications will tend to spread, because the host organism survives and reproduces better without them. Thus they tend to re-evolve spontaneously back into forms more suited for independent existence. Of course, this also fosters concerns about their possible effect on the rest of the biosphere should they "escape." Last, it's worth repeating that biosystems are only a small subset of possible nanosystems.

They remain a source of inspiration nonetheless.

[1]Brierley, JA; Brierley, CL. Reflections on and considerations for biotechnology in the metals extraction industry. In *Hydrometallurgy: Fundamentals, Technology and Innovation,* Hiskey, JB; Warren, GW, 4th International symposium, Aug 1993, Salt Lake City, UT, pp. 647–60, SME, 1993; Ehrlich, HL; Brierley, CL; eds. *Microbial Mineral Recovery.* McGraw-Hill, 1990.
[2]Ref. 21 (Chapter 2) in main text.
[3]Ref. 24 (Chapter 2) in main text.

Box 4.3 Atoms, the Building Blocks of Matter

Everything is made of atoms—people and plants, gasoline and exhaust, toxic waste and mountain tundras, rock and water and air. They're even largely the same kinds of atoms, just arranged in different ways. For example, the most common atom in Earth's crust is just oxygen! (Table 1.1).

This has been a "factoid" learned in elementary school, yet it has been of scant significance in engineering, much less in everyday life. But if atoms are the elementary building blocks of matter, then the ever-more precise structuring of matter inevitably means dealing directly with atoms. That, of course, is the basis of nanotechnology.

For chemical purposes, there are roughly 90 different "kinds" of atoms, the chemical elements. Some are vastly more common than others, and some are definitely more useful than others.

The basics of an atom are familiar: a tiny, positively charged nucleus roughly a hundred-trillionth of a meter across is surrounded by one or more "orbiting" electrons. (Why the quotes? Because the electrons don't really "orbit;" because of the wave-particle duality intrinsic to quantum mechanics, they're a lot fuzzier than that. They do occupy different energy states around the nucleus, however, which *are* traditionally termed "orbitals.") The nucleus is vastly smaller than its surrounding electron cloud: if the nucleus is the football on the 50-yard line, the whole atom is some ten times bigger than the Astrodome. Thus an atom is mostly empty space. Including the surrounding electrons, atoms are roughly a tenth of nanometer in diameter, a unit traditionally termed an Ångstrom. They don't vary much in size even with the changing nuclear charge, because higher charged nuclei hold the electrons in more tightly. That said, the atomic diameter is rather ill-defined because the electron orbitals are effectively "smeared out" due to the rules of quantum mechanics.

The nucleus, of course, is composed of protons and neutrons, with the nuclear strong force holding it together against the mutual repulsion of the protons (Box 2.12).

In an electrically neutral atom, the number of electrons cancels the positive charge on the nucleus. As mentioned in Chapter 3, an *ion* is an atom with a mismatch between the number of protons and the number of electrons, so that it has a net electical charge.

A "plasma" is a gas that consists of ions and free electrons. It is so hot that electrons have been knocked off the atoms by thermal collisions. Stars, including the Sun, are composed of plasma. However, ions are important in many everyday phenomena as well, as in "ionic solids" (Box 3.3). In particular, aqueous solutions commonly contain anions and cations in such proportions as to keep the solution as a whole electrically neutral. Ions therefore prove to be exceedingly important when separating solutes from solution, whether for pollution control or resource extraction, as is elaborated on in Chapter 6. They are not restricted to stellar or other high-temperature conditions.

Nanotechnological Fabrication

There are two fundamental issues in the assembly of nanotechnological materials: (i) the enormous number of atoms in a macroscopic object, and (ii) the tininess of atoms. The problem with treating atoms as Legos or Tinkertoys is putting all the atoms in their places. Snapping Legos together is easy, and there aren't so many Legos in an object that it can't be built in a reasonable time.

Neither is true of atoms. There are a *lot* of atoms in a typical object; roughly 10^{25} H_2O molecules in a glass of water, for example. Although technologies for manipulating small groups or even individual atoms exist, by means of so-called scanning probe microscopes (Box 4.4), and indeed will be critical in near-term

Box 4.4 Scanning Probe Microscopy

Even after the author was in graduate school in the late 1970s, it was common knowledge that one couldn't *see* individual atoms, because the wavelength of visible light so is much larger than a single atom.

That remains literally true. However, with the advent of the scanning-tunneling microscope (STM) in the mid 1980s,[1] individual atoms can certainly be *imaged*. The STM relies on the counter-intuitive laws of quantum mechanics, in particular the "tunneling" of electrons. Because the wave function of an electron extends in space, although dropping off very quickly with distance, an electron has a finite probability of "tunneling" through a barrier that by classical physics would be completely impenetrable. The STM moves an exceedingly thin, charged needle tip over a surface, and because of the extreme sensitivity to distance of the tunneling, can map the atomic-scale topography of that surface through variations in the tunneling current. These traverses can then be built up into images that look just like the space-filling models used in chemistry classes.

Since then, other, similar devices have been devised, generically termed scanning-probe microscopes (SPMs), of which the most important is the atomic-force microscope (AFM). All rely on exquisitely sensitive control and sensing at the nanoscale, based on so-called "piezoelectric" crystals, and have become standard tools for studies at the nanoscale.[2]

Piezoelectric substances develop an electric potential along particular crystal axes when deformed; or alternatively, deform mechanically when an electric field is applied. Such crystals must have a particular symmetry; or more precisely, a particular *lack* of symmetry. Only crystals without a center of symmetry can exhibit piezoelectric behavior. (A "center of

---

[1]Binnig, G; Rohrer, H. Scanning tunneling microscopy from birth to adolescence. *Angew. Chem. Int. Ed. 26*, 606–14, 1987.

[2]E.g., Bhushan, B. *Scanning Probe Microscopy in Nanoscience and Nanotechnology, NanoScience and Technology*, Springer, 2010.

symmetry" exists if for every atom at coordinates x, y, z in the crystal's unit cell there is also an atom at coordinates –x, –y, –z.) One classical piezoelectric material is simply quartz, the most common form of silica (silicon dioxide, SiO_2), and one of the most common minerals in the crust. Indeed, piezoelectrics have found application in electronics, particularly in sensing applications, for nearly a century. For example, piezoelectric crystals were the basis of the transducer that converted the minute variations in the grooves on traditional phonograph records into the varying electrical impulses that represented the original sound.

Besides their value in atomic-scale imaging, SPMs can be used for atomic or near-atomic manipulation as well, and indeed are "proto" molecular assemblers. Such "scanning probe lithography" is described more fully in the main text.

At a much larger scale, piezoelectric crystals may have potential for systems that convert mechanical motion into electricity without macroscopic moving parts such as turbines and armatures, as mentioned in Chapter 5.

nanofabrication, using only such devices to build a macroscopic device from scratch is utterly impractical. After all, if a million atoms a second are moved nonstop for a year, that's still less than 10^{14} atoms—which, in the case of water, is less than a trillionth of a gram. Therefore, either fabrication will somehow have to be "massively parallel;" that is, enormous number of atoms must be placed at the same time, or alternatively, the fabricated device must somehow have such a disproportionate "multiplier effect" in its operation that placing relatively few atoms suffices for it to be economically valuable.

Hence, approaches to nanofabrication are often described as "top-down" or "bottom-up," but in practice nanofabrication will partake of both approaches. A "top-down" approach is reminiscent of the sculptor who carves an elephant from a block of stone by "cutting away everything that doesn't look like an elephant." For nanofabrication, however, this begs the question of where the substrate to be modified—the "stone"—came from in the first place. In general it will need to have a degree of nanoscale organization present already.

Therefore, at least in the near term, nanofabrication will involve a hybrid approach, in which macroscopic objects themselves assembled in some sort of "bottom-up" fashion are further modified, most probably in a number of steps. The examples discussed below merely illustrate a few of the possibilities and are hardly exhaustive.

Bottom-Up Approaches

Conventional chemical synthesis: making molecules

The most obvious "bottom-up" strategy is simply to start with atoms and assemble them one by one. As noted, this straightforward approach founders on the sheer number of atoms in a macroscopic object. Alternatively, though, vast numbers of atoms could be moved all at once. This is exactly what's done in conventional chemical synthesis, and indeed, like Molière's character Monsieur Jourdain, who found out he'd been speaking prose all his life, synthetic chemists have now realized they've been carrying out nanofabrication all along.[8] (The goal in classic synthesis is usually, but not always, to link atoms via covalent bonds. A quick summary of chemical bonding is given in Box 4.5.)

In conventional synthesis, reagents are mixed together and then the statistics of colliding molecules take over. Usually, a mixture of possible products is obtained, only one of which is wanted, so then the desired product must be separated out. Then it is reacted with the next reagent (or set of reagents) and the process repeats. Those additional reagents are also an additional source of expense, having been synthesized themselves. Commonly heating or cooling of the reacting mixture also takes place during

---

[8]Cf. Roald Hoffman, quoted by Balzani, V. Nanoscience and nanotechnology: a personal view of a chemist. *Small* 1, 278–83, 2005, note 31.

Box 4.5 Chemical Bonding: Linking Atoms

Atoms are capable of linking ("bonding") together in various ways, and this, of course, forms the vast subject of chemistry. Bonding takes place through interaction between the outermost (i.e., highest-energy) electrons of different atoms, those in the so-called "bonding orbitals." Roughly speaking, the electrons arrange themselves into lower-energy configurations between the atoms, and chemical bonding thus is fundamentally a quantum-mechanical phenomenon. Nonetheless, there are lots of empirical rules for how atoms can combine, and chemists have distinguished several types of chemical bonding. All are fundamentally electrical in nature, as mediated by the laws of quantum mechanics, and the distinctions are gradational despite these hard-and-fast definitions. Ultimately, of course, they're all approximations to the ultimate quantum-mechanical reality, but in many practical cases they're much simpler and more convenient.

Covalent Bonding

In *covalent* bonding electrons are shared between two or more atoms. Covalent bonds are strong, directional (i.e., resistant to shear), and localized. They are the sort that leads to *molecules*, individual, discrete clusters of atoms tightly held together. Covalent bonds can make large structures as well, however; both diamond and graphite, crystalline forms of carbon, consist of indefinitely extended arrays held together with covalent bonds (Box 6.2). For many nanotechnological applications covalent bonds would be ideal, both because of their strength and because of their resistance to shear.

Ionic Bonding

Ionic (or electrostatic) bonding has already been mentioned in Box 3.6. In this case, one or more electrons is "stolen" from one atom by another, to leave the "stealing" atom with a net negative charge (i.e., an anion), while the atom with electron(s) removed has a positive charge (a cation). An atom that has high affinity for additional electrons is termed "electronegative," whereas "electropositive" atoms yield their outer electrons relatively readily. Ionic bonds typically form between highly electropositive and electronegative elements. Oxygen and the halogens (fluorine, chlorine, bromine, and iodine) are electronegative, whereas some (but not all) metals are highly electropositive. In the language of redox reactions (Box 2.15), electropositive elements are good reducing agents, while electronegative elements are good oxidizing agents.

There are intermediate cases as well, in which the electron hasn't been completely transferred from one atom to another. In such cases chemists speak of "partial ionic" or "partial covalent" character to the bond.

Metallic Bonding

Metallic bonding is typical of (yes!) metals, but is also found in some other compounds. Metals have only a few outer electrons, which are shared into an electron "sea," the so-called "conduction band" (Box 2.17), in which they are highly mobile because there are many unoccupied energy states available. This electron mobility is what makes metals good electrical and thermal conductors.

Weaker Bonds

Certain weaker bonds are also important. *van der Waals forces* lead to feeble attractions between molecules that can let them stick together if the environment isn't too energetic (i.e., too hot). The carbon dioxide molecules in a block of dry ice are held together with van der Waals bonds. *Hydrogen bonds* are important between hydrogen atoms and highly electronegative atoms such as oxygen. The relatively high boiling point and large liquid range of water, for example, are due to hydrogen bonds between a hydrogen atom in one molecule and the oxygen of another. Hydrogen bonds are particularly ubiquitous in biochemistry.

Computational Chemistry

Since chemistry fundamentally is nothing more than applied quantum mechanics, in theory the properties of any molecule—indeed, any chemical system—should be calculable from first principles. In fact, however, the mathematics rapidly becomes intractable even for extremely simple molecules. Due to the extraordinary improvements in computing power in recent years, however, direct calculation of chemical properties has become practical, at least in simple cases, and "computational chemistry" will become even more important in the future. It offers the prospect of addressing complicated real-world problems, such as catalytic mechanisms, from first principles, especially in cases where the empirical "rules of thumb" break down. It also holds out the prospect of understanding economically important chemical "quirks" and perhaps even designing analogs. For example, the cyanide process for gold extraction (Box 3.13) relies on an extraordinary affinity between the cyanide ion and the gold atom. If this affinity can be explained from first principles, it may be possible to design reagents with a similar degree of affinity for other valuable metal atoms, such as platinum or palladium.

one or more steps. Lots of the steps, moreover, merely involve putting on and taking off "molecular masking tape"—what an organic synthesist calls "protective groups," simply clusters of atoms stuck onto the molecular framework to keep parts of it from reacting so that other parts can react. It obviously would be better if just the desired parts of the molecule could be reacted without having to worry about shielding the other parts. Finally, if (say) the yield at each step averages 80%, then after 20 steps the total yield is $(0.8)^{20} \sim 1.15\%$. This dwindling yield also places practical limits on the number of steps, and hence the complexity of molecules that can be made. Little wonder that present synthesis techniques are often jocularly referred to as "shake and bake."

As an illustration, some years back a pair of papers[9] announced the synthesis of an intricate organic compound heretofore known only in an obscure fungus. The synthesis involved multiple steps with a low ratio of finished product to input raw materials. And even so, the synthesis still required other reagents as raw materials. It is unquestionably a synthetic tour de force, but after all, the fungus, and its plant precursors, started with only carbon dioxide, water, and sunlight!

Similarly, the traditional syntheses of organic polymers for organic membranes, ion exchange resins, and so forth result in tangles of macropolymers at the nanoscale with strands jumbled higgledy-piggledy. There's no direct control on the pore sizes between the strands, or the structure of the working surfaces in contact with the solution. The active binding sites for the solutes aren't even necessarily exposed! Just as in the conventional synthesis of isolated organic compounds, too, the synthesis is plagued by competing reactions. Lots of unwanted by-products always form, themselves to become a disposal and purification problem.

To be sure, the capabilities of present-day synthesis techniques *are* extraordinary. Synthetic chemists have a broad arsenal of techniques, the result of research going well back into the 19th century.[10] And for all its inefficiencies, it's unquestionably massively parallel. Nonetheless, there's lots of room for improvement. To return to a recurring theme, organisms are capable of chemical syntheses that put present technology to shame. And the way they do so is an inspiration to would-be nanotechnologists.

---

[9]Nicolaou, KC; Baran, PS; Zhong, Y-L; Choi, H-S; Yoon, WH; He, Y; Fong, KC. Total synthesis of the CP molecules CP-263,114 and CP-225,917—Part 1: Synthesis of key intermediates and intelligence gathering. *Angew. Chem. Int. Ed. 38*, 1669–75, 1999; Nicolaou, KC; Baran, PS; Zhong, Y-L; Fong, KC; He, Y; Yoon, WH; Choi, H-S. Total synthesis of the CP molecules CP-225,917 and CP-263,114—Part 2: Evolution of the final strategy. *Angew. Chem. Int. Ed. 38*, 1676–8.

[10]Cf. Hudlicky, T. Design constraints in practical syntheses of complex molecules: Current status, case studies with carbohydrates and alkaloids, and future perspectives. *Chem. Rev. 96*, 3–30, 1996.

As would be expected, in part biology gets much better yields by using highly specific catalysts, i.e., enzymes. If a catalyst is selective enough, it essentially excludes all the other products except the one desired, so that yields can approach 100%. By decreasing activation energy barriers (Box 4.6), too, the need for heat to drive a reaction step is also eliminated.

Nanotechnological approaches to synthesis will thus involve highly selective catalysts. Of course, there are important synergies here, because as is repeatedly emphasized in this book, catalysts have a host of applications related to energy and resources, and their development and (nano)fabrication are one of the best examples of proto-nanotechnology.

The capabilities of biosystems result from more than just highly specific catalysts, though. Biology uses highly specific synthetic assembly *systems*, in which the product of one reaction is handed off into the next. Consider the sequential molecular assembly of (say) a nucleic acid chain, or a spinnerets linking and spinning out strands of protein that are then woven into spider silk.[11] Another example is photosynthesis, in which the energy of an absorbed photon is used to synthesize simple sugars via an organized reaction chain. Yet another example is oxidative phosphorylation, in which the mitochondria in cells synthesize adenosine triphosphate (ATP), the energy-rich compound used to drive all "uphill" biochemical reactions.[12] It is often called the cellular "fuel," but—as per the discussion in Chapter 2—that is a misnomer because its energy is not turned into heat directly. Like biosystems, then, biomimetic ("life-mimicking") nanosynthesis will also require *systems*, structured molecular constructs that receive raw reagents and produce the finished molecular product.

---

[11]Cf. Cranford, SW; Buehler, MJ. *Biomateriomics*. SSMS, *165*, 2012, p. 148 ff.; cf. also Pugno, NM; Nanotribology of Spiderman. In *Physical Properties of Ceramic and Carbon Nanoscale Structures: The INFN Lectures, Vol. II.* Bellucci, S; ed. LNNS&T *11*, 111–36, 2011.

[12]von Ballmoos, C; Cook, GM; Dimroth, P. Unique rotary ATP synthase and its biological diversity. *Annu. Rev. Biophys.* 37, 43–64, 2008.

Such systems require a hierarchical nanoscale organization.[13] In biological systems this takes the form of compartmentalization, which controls the environment in which the intermediate products interact.[14] Indeed, nanoscale spatial constraints are critical in guiding reactions, and so-called templating approaches, in which surrounding molecular frameworks act like the nanoscale equivalents of workshop clamps, molds, and jigs, are currently attracting an enormous amount of attention, as discussed below. Such spatial constraints, by constraining the places at which molecules can react, obviate the need for the addition and removal of protective groups—the molecular "masking tape"—that so complicates conventional synthesis.[15] Indeed, certain catalysts, such as molecular sieves (Box 6.3), derive their selectivity from imposing spatial constraints: most simply, only molecules that can fit into the holes of the sieve can react. All this is but another aspect of *molecular recognition* (Box 4.7), which pervades approaches to molecular assembly and functionality, as will be elaborated in the rest of this chapter.

The ultimate goal is to produce finished molecules with as little energy expenditure as possible, with as few by-products as possible, and using the simplest starting materials possible, just as living things do. There will be further synergies here, too: molecules, of course, are building blocks for further syntheses, as in the polymers mentioned above. More broadly, however, intricate molecules are themselves essential parts of nanotechnological systems (cf. the crown ethers in Chapter 5),

[13]Cf. Bard, AJ. *Integrated Chemical Systems*. Wiley, 1994, Ch. 1; Adachi, M; David J. Lockwood, DJ; eds. *Self-Organized Nanoscale Materials*. NS&T, 2006, and papers therein; Cranford & Buehler, Ref. 11.

[14]Urban, PL. Compartmentalised chemistry: from studies on the origin of life to engineered biochemical systems. *New J. Chem. 38*, 5135–41, 2014.

[15]Cf. Ramanathan, M; Kilbey, MS, II; Ji, Q; Hill, JP; Ariga, K. Materials self-assembly and fabrication in confined spaces, *J. Mater. Chem. 22*, 10389–405, 2012; Yoshizawa, M; Klosterman, JK; Fujita, M. Functional molecular flasks: New properties and reactions within discrete, self-assembled hosts. *Angew. Chem. Int. Ed. 48*, 3418–38, 2009.

Box 4.6 Chemical Reaction: Change in
Atomic Arrangements

Not all atomic arrangements are created equal. Some arrangements are more stable (i.e., have lower energy) than others, and those are favored. Given the opportunity, therefore, a high-energy arrangement will rearrange into a lower-energy one, with the release of the stored energy— or potential energy, to use the language of physics. For an example familiar to any schoolchild, a mixture of hydrogen (H_2) and oxygen (O_2) gases will react spectacularly to make water (H_2O). Changing water back into hydrogen and oxygen, say by electrolysis, requires an input of energy. The physics is exactly analogous to releasing energy by letting a ball run downhill, and then storing the energy again by carrying the ball back uphill.

Of course, the ball will run back downhill again immediately unless it is restrained in some manner. Similarly, "downhill" chemical reactions don't always happen spontaneously. Chemists speak of an "activation energy," in effect a barrier like a hill or lip that must first be crossed before something can run downhill. Energy must be input to let the reaction occur. A match lighting a fire is an example: the fuel and the oxygen in the air are not in their lowest energy states, a condition described as "metastable," but they can remain without reacting indefinitely until enough energy is supplied to overcome the activation energy barriers. Once started, the fire continues as long as fuel and air are available, because the energy it releases keeps activating further reaction.

In general, reactions also happen more readily at high temperatures, simply because the thermal energy in the environment furnishes the activation energy. As previously described (Box 2.3), temperature describes the average energy of the particles (atoms or molecules) in a substances. Moreover, local fluctuations in energy occur, of magnitude roughly kT, where k is Boltzmann's constant, with a value of 1.38065 joules/kelvin, and T is absolute temperature. These fluctuations cause reactions to occur that would be precluded if all particles in the environment had exactly the average energy indicated by their temperature.

Obviously, the existence of activation energy barriers allows energy to be stored. But more than that; it allows the world we know to exist! At everyday temperatures, most of our environment, including all living things, exists metastably. A world in which reactions occurred automatically whenever energetically favorable would be boring indeed—although we couldn't be here to observe it.

Elsewhere in this book the importance of catalysts is repeatedly emphasized. Fundamentally, a catalyst lowers the activation energy barrier of a reaction so that can occur more readily. Often as an additional benefit undesirable "side" reactions are also excluded or minimized, because their activation energies remain unaffected.

including being the building blocks for other nanotechnological assemblies, such as the self-assembled structures described below.

In addition, as such nanoassembly systems improve, they will provide an evolutionary path to true "molecular assembly" systems. Such systems will initially supplement, and eventually supplant, the (relatively) large-scale atomic manipulators based on scanning probe microscopes described below. At first, such "molecular assemblers" will be used to fabricate nanostructured materials such as precisely nanostructured membranes, crystals, surfaces, and composites. The biomachinery assembling protein strains into hair in a follicle again furnishes a biological example. Ultimately, with the experience gained in designing, making, and handling structures at the molecular level, they should lead to assembly systems that are truly molecular.

Yet another example of the biological inspiration is the carrying out of syntheses at or near room temperature and usually in aqueous solution. Water, after all, is the "greenest" solvent possible, and the medium in which all biological reactions occur.

One example is a collection of techniques often termed *chimie douce* ("soft chemistry"). The motivation is not just greater energy efficiency by minimizing applied heat but the preparation of numerous materials not accessible by standard synthesis techniques.[16] So-called sol–gel syntheses are a particular focus, in which a suspension of nanoparticles (a "sol") is condensed into a gelatinous mass (a "gel") and thence into a low-temperature glass, often with a composition and/or structure that cannot be attained by conventional melting of the starting materials.

Self-organization

Crystallization (Box 3.11) was mentioned in the last chapter as an example of self-organization, and it was noted there that although it's simple, it's an extremely energy-intensive approach to element *separation*. However, self-organization is an extremely valuable approach to nanotechnological *assembly* that is likely to become of even greater importance in the future.

Crystallization itself has an enormous literature[17] that, however, is largely focused on the optimization of macroscopic

---

[16]E.g., Figlarz, M. Chimie douce. A new route for the preparation of new materials: some examples. *Chem. Scr. 28*, 3–7, 1988; Soft chemistry: Thermodynamic and structural aspects. *Mater. Sci. Forum 152–153*, 55–68, 1994; Gopalakrishnan, J. Chimie douce approaches to the synthesis of metastable oxide materials. *Chem. Mater. 7*, 1265–75, 1995; Levin, D; Soled, SL; Ying, JY. Chimie douce synthesis of nanostructured layered materials. In *Nanotechnology: Molecularly Designed Materials,* Chow, G-M; Gonsalves, KE; eds. ACSSS 622, 237–49, 1996; Rouxel, J; Tournoux, M. Chimie douce with solid precursors, past and present. *Solid State Ionics 84*, 141–9, 1996; Corriu, RJP. Chimie douce: wide perspectives for molecular chemistry. A challenge for chemists: control of the organization of matter. *New J. Chem. 25*, 2, 2001; Livage, J. Chimie douce: from shake-and-bake processing to wet chemistry. *New J. Chem. 25*, 1, 2001; Sanchez, C; Rozes, L; Ribot, F; Laberty-Robert, C; Grosso, D; Sassoye, C; Boissiere, C; Nicole, L. "Chimie douce": A land of opportunities for the designed construction of functional inorganic and hybrid organic-inorganic nanomaterials. *C. R. Chimie 13*, 3–39, 2010.

[17]E.g., Tiller, WA. *The Science of Crystallization* (2 Vols). Cambridge University Press, 1991; Mullin, JW. *Crystallization*, 4th ed. Butterworth-Heinemann, 2001.

environmental parameters (temperature, pressure, solute concentration, etc.) that favor crystal growth, and indeed that favor particular desired morphologies of the crystals. Empirically it also has been found that certain additives favor or inhibit the crystallization of one or another crystal structure. These effects can often be rationalized atomistically, such as by the inhibition of growth on particular crystal faces due to solute adsorption, and such effects have been a research focus for some years.[18] For example, the high-density aragonite form of calcium carbonate can easily be precipitated from seawater, whereas the calcite form, the one stable at ordinary temperatures and pressures, is not usually seen. This seems to result from inhibition of calcite crystallization nuclei by magnesium ions, which are abundant in seawater.[19] The small magnesium ions have a smaller effect on aragonite crystal nuclei, because its structure doesn't accomodate them very well.

Indeed, molecular sieves such as *zeolites* (Box 6.3) are routinely precipitated from empirical "cocktails" containing large organic molecules as well as metal ions and silica gel. The large molecules, so-called structure-directing agents, evidently act like templates around which particular silicate "building blocks" can crystallize, and so determine the crystal structure

---

[18]E.g., Clydesdale, G; Roberts, KJ; Docherty, R. Modelling the morphology of molecular crystals in the presence of disruptive tailor-made additives, *J. Cryst. Growth 135*, 331–40, 1994; Clydesdale, G; Roberts, KJ; Lewtas, K; Docherty, R. Modelling the morphology of molecular crystals in the presence of blocking tailor-made additives. *J. Cryst. Growth 141*, 443, 1994.

[19]E.g., Falini, G; Gazzano, M; Ripamonti, A. Crystallization of calcium carbonate in presence of magnesium and polyelectrolytes. *J. Cryst. Growth 137*, 577, 1994; House, WA; Howson, MP; Pethybridge, AP. Crystallisation kinetics of calcite in the presence of magnesium ions, *Faraday 1 84*, 2723–34, 1988; Mucci, A; Morse, JW. The solubility of calcite in seawater solutions of various magnesium concentration, $I_t = 0.697$ m at 25°C and one atmosphere total pressure, *Geochim. Cosmochim. Acta 48*, 815–22, 1984; Meldrum, FC; Hyde, ST, Morphological influence of magnesium and organic additives on the precipitation of calcite, *J. Cryst. Growth 231*, 544, 2001.

obtained.[20] Of course, such templating underscores once again the key role of molecular recognition (Box 4.7) in crystallization.

Once formed, crystals can sometimes be further modified by variation of the environmental parameters. For example, solid phases formed at high temperature can become thermodynamically unstable as they cool, and separate into two interleaved phases often having the form of alternating layers of different composition. Such "spinodal decomposition" is one approach to crystal nanostructuring.[21] More usually, however, crystallizing phases remain homogeneous. They are nonetheless valuable as substrates for further modification, as discussed below. Indeed, the "pre-organized" substrates for top-down modification are usually crystalline.

Although our atomistic understanding of why substances crystallize as they do has improved considerably, we certainly have nothing like direct nanoscale control of the crystallization. Moreover, crystal sizes are typically small in conventional syntheses, however modified with templating agents and other additives, because the energy driving the crystallization is proportional to the surface area of the crystal, and the larger the crystal, the smaller proportionately the surface area is. To be sure, as discussed further below, nanocrystals and other nanoparticles have a host of applications themselves. In many cases, however, macroscopic perfect or near-perfect crystals would be considerably more useful. In the case of molecular sieves (Box 6.3), for example, whose conventional synthesis yields micrometer-sized crystals, the microcrystals are added to a reaction vessel just like reagents. They end up used not so much

---

[20]E.g., Anthony, JL; Davis, ME. Assembly of zeolites and crystalline molecular sieves. In *Self-Organized Nanoscale Materials*, Adachi, M; Lockwood, DJ; eds. NS&T, 159–85, 2006.

[21]E.g., Lee, JS; Hirao, A; Nakahama, S. Polymerization of monomers containing functional silyl groups. 7. Porous membranes with controlled microstructures *Macromolecules* 22, 2602–6, 1989; Brambilla, A; Calloni, A; Aluicio-Sardui, E; Berti, G; Kan, Z; Beaupré, S; Leclerc, M; Butt, HJ; Floudas, G; Keivanidis, PE; Duò, L. X-ray photoemission spectroscopy study of vertical phase separation in F8BT:PDI/ITO films for photovoltaic applications, *Proc. SPIE* 9165, 2014.

Box 4.7 Molecular Recognition

Molecular recognition[1] is fundamental to understanding structuring at the molecular scale. Most simply, molecules have shapes and sizes, and, just like pieces of a three-dimensional jigsaw puzzle, can fit together to varying degrees depending on how well their shapes match. Much more is at work, however, than simply complementary shapes. Molecules and individual atoms also have differing electrical properties, and electrostatic attraction is commonly a major factor in recognition. Even if species have no net charge (i.e., are not ionized), more subtle electrical properties can play key roles. These latter include dipole moment and "polarization"—the susceptibility of the electron clouds forming the chemical bonds to distortion by an external electric field. Other weak forces such as van der Waals interactions and hydrogen bonding (Box 4.5), although feeble, commonly are important as well. Indeed, the reasons why a complex is favored can get extremely subtle indeed, and are not always understood. The extraordinary affinity of the simple cyanide ion (CN^-) for gold has been mentioned (Box 3.13).

Such recognition is basic to many phenomena of great interest in nanotechnology, and some have already been mentioned in passing. One example is the binding of particular solutes by macrocycles, which underlies nanotechnological approaches to separation (p. 288). Recognition is also basic to selective catalysis, one of the most spectacular examples being the "lock and key" fit of enzymes with the substrates they act on. A technological example is the selectivity of zeolite catalysts (p. 186).

---

[1]E.g., Rebek, J, Jr. Model studies in molecular recognition. *Science 235*, 1478–84, 1987; Gellman, SH. Introduction: Molecular recognition. *Chem. Rev. 97*, 1231–2, 1997, and papers in this issue; Mallouk, TE; Gavin, JA. Molecular recognition in lamellar solids and thin films. *Acc. Chem. Res. 31*, 209–17, 1998; Ariga, K; Ito, H; Hill, JP; Tsukube, H. Molecular recognition: From solution science to nano/materials technology. *Chem. Soc. Rev. 41*, 5800–35, 2012.

The burgeoning field of supramolecular chemistry,[2] which involves the synthesis and study of entities bigger than individual molecules (hence "supra" molecular) that are held together by forces weaker than ordinary covalent bonds, furnishes another spectacular example. Similar recognition and assembly due to subtle intermolecular forces directly underlies much biochemical interaction, as in the folding-up of protein strands. Predicting how a strand will fold is a notoriously difficult problem, largely because of the subtle interactions among the weak forces involved. Molecular recognition also underlies molecular templating, an approach to constraining how molecules can react and thereby favor desired products. Indeed, the phenomena of self-assembly, ranging from the crystallization of simple ionic crystals to the intricate structure of something like the tobacco mosaic virus, fundamentally are all consequences of molecular recognition.

[2]E.g., Vögtle, F. *Supramolecular Chemistry: An Introduction.* Wiley, 1991. Lehn, JM. *Supramolecular Chemistry: Concepts and Perspectives.* Wiley-VCH, 1995. Various editors, *Comprehensive Supramolecular Chem.* Pergamon Press, 1996; Atwood, JL; Steed, JW, eds. *Encyclopedia of Supramolecular Chemistry,* Vol. 1&2, 2004.

as "sieves" but as selective absorbers. Even though a macroscopic, monocrystalline zeolite film or membrane would be exceeding useful, as discussed in Chapter 6, their direct fabrication remains impossible with current techniques; current "membranes" are actually aggregates of tiny crystals. Indeed, in some applications macroscopic near-perfect crystals will be critical, as in the superstrength materials also discussed in Chapter 6.

Parenthetically, some large monocrystals occur in extraordinary natural settings, in which they have grown despite exceedingly slow crystallization rates, because geologic timescales are available. One example is the celebrated enormous gypsum crystals in Mexico's Cave of Crystals.[22] An

---

[22]E.g., Wysession, ME. *The World's Greatest Geological Wonders—36 Spectacular Sites,* Lect. 13. The Great Courses, Chantilly, VA, USA. Because

example that remains of commercial importance is so-called Iceland spar, which is a large, optically perfect crystal of pure calcite (rhombohedral calcium carbonate, $CaCO_3$). It still finds use in high-end optics. As might be expected, many gemstones represent similar leisurely crystallization as well.

Biomimetic crystallization

Of course, in general achieving such macroscopic crystallization, and indeed long-range nanoscale ordering in general, will require direct nanoscale control as geologic timescales are not available. Indeed, such control is will be critical in many applications, even at much smaller dimensions. How can it be achieved? Here again, a look at biology is illuminating. Organisms control the placement, orientation, shape, and very structure of the crystals making up tissues such as bone or shell by a matrix of organic macromolecules—by "templates" writ large, in effect. One example is the use of successive layers of β-protein sheets to localize the growth of inorganic ionic crystals.[23] Indeed, the formation of mineralized tissue by organisms has been an enormous focus of research for decades.[24]

Additionally, the bioproduct is typically a composite, with oriented inorganic crystals embedded in a matrix of polymer, typically protein, strands. This yields an extraordinary improvement in strength over the brittle crystals or protein fibers by themselves. Abalone shell, for example, consists of

---

of their size and well-developed crystal faces, these crystals are often called "selenite," which refers to highly perfect, almost "gem" grade gypsum crystals.

[23]Lowenstam, HA; Weiner, S; *On Biomineralization*, Oxford University Press, 1989.

[24]E.g., Falini, G; Albeck, S; Weiner, S; Addadi, L. Control of aragonite or calcite polymorphism by mollusk shell macromolecules. *Science 271*, 67–9, 1996; Falini, G; Fermani, S; Gazzano, M; Ripamonti, A. Biomimetic crystallization of calcium carbonate polymorphs by means of collagenous matrixes, *Chem.– Eur. J. 3*, 1807–14, 1997; Mann, S. Biomineralization: the form(id)able part of bioinorganic chemistry!, *Dalton* 3953–62, 1997.

aragonite plates (orthorhombic $CaCO_3$, a brittle ionic solid) and protein strands, but the composite has much greater strength than aragonite alone, in particular because of its nanoscale organization.[25] Such materials further underscore the point that "the structure is as important as the material."[26]

Unsurprisingly, therefore, approaches to "biomimetic" structuring are attracting an enormous amount of interest from nanotechnology researchers.[27] The templating materials are not just macromolecules roughly analogous to proteins, but also include such things as nanoporous materials of various sorts, themselves commonly nanofabricated by one or another technique.[28] In particular, many templating materials are made

---

[25]Gao, H; Ji, B; Jger, IL; Arzt, E; Fratzl, P. Materials become insensitive to flaws at nanoscale: lessons from nature, *Proc. Natl. Acad. Sci. USA 100*, 5597–600, 2003.

[26]Cranford & Buehler, in Ref. 11, p. 8.

[27]E.g., Archibald, DD; Mann, S. Template mineralization of self-assembled anisotropic lipid microstructures. *Nature 364*, 430–33, 1993; Yang, J; Meldrum, FC; Fendler, JH. Epitaxial growth of size-quantized cadmium sulfide crystals under arachidic acid monolayers. *J. Phys. Chem. 99*, 5500–4, 1995; Mann, S. Biomineralization and biomimetic materials chemistry, biomineralization and biomimetic materials chemistry. *J. Mater. Chem. 5*, 935–46, 1995; Yang, P; Zhao, D; Margolese, DI; Chmelka, BF; Stucky, GD. Block copolymer templating syntheses of mesoporous metal oxides with large ordering lengths and semicrystalline framework. *Chem. Mater. 11*, 2813–26, 1999; Mann, S; Davis, SA; Hall, SR; Li, M; Rhodes, KH; Shenton, W; Vaucher, S; Zhang, B. Crystal tectonics: Chemical construction and self-organization. *Dalton* 3753–63, 2000; Cölfen, H; Mann, S. Higher-order organization by mesoscale self-assembly and transformation of hybrid nanostructures. *Angew. Chem. Int. Ed. 42*, 2350–65, 2003; Bae, C; Yoo, H; Kim, S; Lee, K; Kim, J; Sung, MM; Shin, H. Template-directed synthesis of oxide nanotubes: Fabrication, characterization, and applications. *Chem. Mater. 20*, 756–67, 2008; Li, M; Mann, S; Emergent hybrid nanostructures based on non-equilibrium block copolymer self-assembly. *Angew. Chem. Int. Ed. 47*, 9476–9, 2008; Meldrum, FC; Cölfen, H. Controlling mineral morphologies and structures in biological and synthetic systems. *Chem. Rev. 108*, 4332–432, 2008; Docampo, P; Guldin, S; Stefik, M; Tiwana, P; Orilall, MC; Hüttner, S; Sai, H; Wiesner, U; Steiner, U; Snaith, HJ. Control of solid-state dye-sensitized solar cell performance by block-copolymer-directed TiO_2 synthesis. *Adv. Funct. Mater. 20*, 1787–96, 2010; Ref. 26.

[28]E.g., Sides, CR; Martin, CR. Deposition into templates. In Schmuki & Virtanen, in Ref. 62, pp. 9–320.

by the techniques of molecular self-assembly described below, such as with *surfactants* (Box 4.8),[29] colloidal crystallization, and even viruses.[30] In such cases, especially where the template is held together only with weak noncovalent bonds, it is commonly later removed, as by heat or etching, such that the template has effectively acted as a mold. In such cases it commonly is termed *nanocasting*.[31]

Nanoparticles, nanocrystals, and colloids

As mentioned above, conventional syntheses typically yield tiny particles, which themselves are finding a host of new applications. Indeed, they form a branch of nanotechnology all unto themselves, as evidenced by their burgeoning literature.[32] A

---

[29]E.g., Nakanishi, K; Amatani, T; Yano, S; Kodaira, T. Multiscale templating of siloxane gels via polymerization-induced phase separation. *Chem. Mater. 20,* 1108–15, 2008, and references therein.

[30]Mao, C; Flynn, CE; Hayhurst, A; Sweeney, R; Qi, J; Georgiou, G; Iverson, B; Belcher, AM. Viral assembly of oriented quantum dot nanowires, *Proc. Natl. Acad. Sci. USA 100,* 6946–51, 2003; Mao, C; Solis, DJ; Reiss, BD: Kottmann, ST; Sweeney, RY; Hayhurst, A; Georgiou, G; Iverson, B; Belcher, AM. Virus-based toolkit for the directed synthesis of magnetic and semiconducting nanowires. *Science 303,* 213–7, 2004; Nam, KT; Kim, DW; Yoo, PJ; Chiang, CY; Meethong, N; Hammond, PT; Chiang, YM; Belcher, AM. Virus-enabled synthesis and assembly of nanowires for lithium ion battery electrodes. *Science 312,* 885, 2006.

[31]E.g., Caruso, RA. Nanocasting and nanocoating. *Top. Curr. Chem. 226,* 91–118, 2003; Sogo, K; Nakajima, M; Kawata, H; Hirai, Y., Reproduction of fine structures by nanocasting lithography. *Microelectron. Eng. 84,* 909–11, 2007.

[32]E.g., Henglein, A. Small-particle research: physicochemical properties of extremely small colloidal metal and semiconductor particles. *Chem. Rev. 89,* 1861–73, 1989; Caruso, F, ed. *Colloids and Colloid Assemblies.* Wiley, 2004; Zrínyi, M; Hórvölgyi, ZD. From colloids to nanotechnology. *Prog. Colloid Polym. Sci. 125,* 2004; Nagarajan, R; Hatton, TA; eds. *Nanoparticles: Synthesis, Stabilization, Passivation, and Functionalization.* ACSSS, 996, 2008; Hórvölgyi, ZD; Kiss, E; eds., Colloids for Nano- and Biotechnology, *Prog. Colloid Polym. Sci. 135,* 2008; Schmid, G. *Nanoparticles: From Theory to Application,* 2nd ed. Wiley-VCH, 2010; Cargnello, M; Gordon, TG; Murray, CB. Solution-phase synthesis of titanium dioxide nanoparticles and nanocrystals. *Chem. Rev. 114,* 9319–45, 2014; *Nanoparticle Synthesis and Assembly, Faraday Discuss., 181,* 2015 (themed issue); Corr, SA. Metal

Box 4.8 Surfactants

Surfactants were already mentioned in the context of flotation agents for separating ore minerals, but they have much broader importance. A surfactant is a molecule, usually linear, whose ends differ greatly in their affinity for different solvents. Typically one end can dissolve in a polar solvent such as water. As mentioned elsewhere, the molecules of polar solvents have an asymmetric distribution of electric charge, and this lets them form complexes with charged solute species; that is, with ions. Thus the "hydrophilic" ("water-loving") end of a surfactant molecule is typically ionic. The other end, straightforwardly enough, has an affinity with non-polar solvents, such as hydrocarbons, and is termed "hydrophobic."

Surfactants are most familiar in everyday life as soap and detergents. Common surfactants have a long hydrocarbon "tail" at the hydrophobic end, and a hydrophilic "head" consisting of an ionic molecular group. In ordinary soaps this is a carboxylic group, COO^-. Hence, soap molecules in a water environment tend to arrange themselves so that their ionic heads are in the water and their hydrocarbon tails away from it to form a film on the water surface. Indeed, it's not hard to make surfactant films a molecule thick, the basis for the "Langmuir-Blodgett" fabrication technique described in the main text.

Alternatively, surfactants can form tiny spheres— "micelles"—within the solvent whose outer surface is formed by the hydrophilic heads, while all the hydrocarbon tails poke into the interior of the ball. This, in fact, is how soaps and detergents work as cleaning agents—oily matter on the object to be cleaned ends up encapsulated in micelles and so can be washed away. A colloidal dispersion (see the main text) of micelles in an aqueous (or similar) solution is called an "emulsion." They're vastly important both biologically and technologically.

great deal of research has focused on optimizing their synthesis to yield particles as uniform as possible in both size and shape, both by optimization of conventional synthesis,[33] as well as by more exotic methods.[34]

The science of small particles, however, goes back well into the 19th century. Traditionally a "colloidal suspension" is a suspension that shows little or no tendency to settle under gravity. Indeed, colloidal particles and colloidal suspensions are already of great importance both technologically and biologically, including such familiar substances as milk[35] and paint. A "colloidal particle" may be defined as a particle with such a large ratio of surface area to volume that surface effects dominate its properties, and such particles have dimensions from roughly 1 to 1000 nm. (Usually the term *colloid* refers to the colloidal suspension, but sometimes it's applied to the colloidal particles themselves. For clarity the term will not be used here.) A classic literature exists on agents, so-called flocculants, that causes colloidal particles to clump and settle by changing their surface properties. Another criterion for a colloidal particle is that such particles show Brownian motion; that is, the random jiggling in the fluid suspension due to collisions with molecules in the fluid. The particles are small enough that the random collisions don't

---

oxide nanoparticles. *Nanoscience 3*, 31–56, 2016; Malik, MA; Ramasamy, K; Revaprasadu, N. The recent developments in nanoparticle synthesis. *Nanoscience 3*, 57–153, 2016.

[33]E.g., Matijevic, E. Preparation and properties of uniform size colloids. *Chem. Mater. 5*, 412–26, 1993; Monodispersed inorganic colloids: achievements and problems. *Pure Appl. Chem. 64*, 1703–7, 1992; Ribeiro, C; Leite, ER. Assembly and properties of nanoparticles. In *Nanostructured Materials for Electrochemical Energy Production and Storage*, Leite, ER, ed. NS&T, 2009, 33–79; cf. also Ref. 32.

[34]E.g., Douglas, T; Strable, E; Willits, D; Aitouchen, A; Libera, M; Young, M. Protein engineering of a viral cage for constrained nanomaterials synthesis. *Adv. Mater. 14*, 415–8, 2002; Ossi, PM; Nisha R; Agarwal, NR; Fazio, E; Neri, F; Trusso, S. Laser-mediated nanoparticle synthesis and self-assembling. In *Laser Physics for Materials Scientists: A Primer*, Haglund, RF; ed. SSMS, *191*, 175–212, 2014.

[35]Milk is strictly an *emulsion*, a colloidal suspension of *micelles* (Box 4.8).

average out. Colloidal particles have an enormous literature, including a number of dedicated journals.[36]

Recently the term *nanoparticle* has come into vogue, being defined as a particle in the size range 1–100 nm; that is, at the lower end of the colloidal size range. The distinction is useful because such small particles are also "quantum dots," particles small enough to exhibit quantum confinement of the electrons. In other words, the highest-energy electrons within the particle start to exhibit discrete energy levels, like the orbitals in isolated atoms (Box 4.3). This means that not only is the electronic structure quite different from the bulk material, but that it can be varied with the particle dimensions. Thus they can have very different, and useful, properties from the bulk material. (Ironically, quantum dots also antedate modern technology: gold nanoparticles are responsible for the colors of some stained glass. The particles are small enough they absorb different light wavelengths than bulk gold.[37])

As might be expected from their high surface area, colloidal particles are also important in many catalysts, as synthetic precursors[38] and also directly, being commonly immobilized on

[36]Current colloid journals and book series include: *Journal of Nanoparticle Research* (Springer), *Progress in Colloid & Polymer Science* (Springer); *Colloid & Polymer Science* (Springer); *Advances in Colloid and Interface Science* (ScienceDirect); *Colloids & Surfaces A & B* (ScienceDirect); *Current Opinion in Colloid and Interface Science; Journal of Colloid and Interface Science.* Many now have a strong nanotechnology focus.

[37]Daniel, M-C; Astruc, D. Gold nanoparticles: Assembly, supramolecular chemistry, quantum-size-related properties, and applications toward biology, catalysis, and nanotechnology. *Chem. Rev. 104*, 293–346, 2004.

[38]E.g., Burton, PD; Lavenson, D; Johnson, M; Gorm, D; Karim, AM; Conant, T; Datye, AK; Hernandez-Sanchez, BA; Boyle, TJ. Synthesis and activity of heterogeneous Pd/Al$_2$O$_3$ and Pd/ZnO catalysts prepared from colloidal palladium nanoparticles. *Top. Catal. 49*, 227–32, 2008; Zhong, Z; Teo, J; Lin, M; Ho, J. Synthesis of porous α-Fe$_2$O$_3$ nanorods as catalyst support and a novel method to deposit small gold colloids on them. *Top. Catal. 49*, 216–26, 2008; Didillon, B; Pagès, T; Verdier, S; Uzio, D. Synthesis of palladium-based supported catalysts by colloidal oxide chemistry. In Zhou et al., in Ref. 8 (Chapter 5), pp. 255–87; Kariuki, NN; Wang, X; Mawdsley, JR; Ferrandon, MS; Niyogi, SG; Vaughey, JT; Myers, DJ. Colloidal synthesis and

a substrate, or "support" to form the catalytically active site.[39] Sometimes such a substrate is referred to as "decorated" with the catalytic particles. Quantum dots are of intense interest in catalysts at present because of the prospect of tuning the electronic levels for highly specific catalysis, as will be described in the next chapter.

Among the most interesting emerging applications is the use of colloidal particles themselves as nanoscale building blocks, another approach to nanofabrication that is now attracting enormous attention. Under the right conditions colloidal suspensions can "crystallize" to yield ordered arrays of bound nanoparticles,[40] sometimes termed *supracrystals*.[41] Indeed, colloidal particles have been used as (relatively) macroscopic

---

characterization of carbon-supported Pd-Cu nanoparticle oxygen reduction electrocatalysts. *Chem. Mater.* 22, 4144–52, 2010.

[39]E.g., Zhong et al., in Ref. 38; Baturina et al., Lim et al., Kortlever et al., all in Ref. 78 (Chapter 5); Patel et al., in Ref. 104 (Chapter 5); Boahene, PE; Soni, K; Dalai, AK; Adjaye, J. Influence of different supports on hydrosulfurization and hydrodenitrogenation of heavy gas oils using FeW catalysts. In *Nanocatalysis for Fuels and Chemicals.* Dalai, AK; ed. ACSSS 1092, 2012; Refs. 19, 27 (Chapter 5); Didillon et al., in Ref. 38; Eppler et al., in Ref. 8 (Chapter 5).

[40]E.g., Gates, B; Qin, D; Xia, Y. Assembly of nanoparticles into opaline structures over large areas. *Adv. Mater.* 11, 466–69, 1999; McLellan, Joe; Yu Lu, Xuchuan Jiang, Younan Xia, Self-Assembly of Colloidal Building Blocks into Complex and Controllable Structures, in *Nanoscale Assembly: Chemical Techniques*, WTS Huck, ed., NS&T, 2005; Caruntu, G; Caruntu, D; O'Connor, CJ. Nanoparticles and colloidal self-assembly. In *Scanning Microscopy for Nanotechnology: Techniques and Applications*, Zhou, W; Wang, ZL; eds. 306–56, 2007, Springer; Pileni, MP. Self-Assemblies of Organic and Inorganic Materials. SSMS 99, 47–66, 2008; Self-assembly of inorganic nanocrystals: Fabrication and collective intrinsic properties. *Acc. Chem. Res.* 40, 685–93, 2007; Vogel, N; Retsch, M; Fustin, C-A; del Campo, A; Jonas, U. Advances in colloidal assembly: The design of structure and hierarchy in two and three dimensions. *Chem. Rev.* 115, 6265–311, 2015.

[41]Pileni, M. P. Supracrystals of inorganic nanocrystals: An open challenge for new physical properties. *Acc. Chem. Res.* 41, 1799–809, 2008; Goubet, N; Pileni, MP. Analogy between atoms in a nanocrystal and nanocrystals in a supracrystal: Is it real or just a highly probable speculation?. *J. Phys. Chem. Lett.* 2, 1024–31, 2011.

models for crystallization, but they can achieve a much wider variety of structures because their interactions are both not quantum-mechanical and can be tailored.[42] Surprisingly, natural supracrystals exist: the mineral opal results from crystallization of colloidal silica particles, and the "fire" of gem-quality opal results when the particles are of uniform size.

Molecular self-assembly (SA)

Considerably more complex structures than typical crystals are possible with adroit choice of the self-assembling entities. As usual, biology furnishes a host of examples, of which the most extraordinary may be the tobacco mosaic virus.[43] Like all viruses, it consists of a sheath of proteins around a core of nucleic acids, and it can be disaggregated by certain mild reagents or by heat. But if the constituent macromolecules themselves haven't been disrupted, they will spontaneously reassemble into a virus when conditions change. Obviously, such self-assembly relies on molecular recognition to an extraordinary degree.

A more humdrum example is a soap film. Soaps are *surfactants* (Box 4.8), which have already been mentioned above in conjunction with templated assembly. The soap molecules in the film minimize the forces on them by arraying themselves at the surface of the water with their hydrophilic heads in the water and their hydrophilic tails pointing away, to yield a unimolecular layer. This almost laughably simple approach to nanoscale structuring is the basis of the decades-old Langmuir–Blodgett technique touched on below.

Unsurprisingly, molecular SA is yet another pathway to nanoscale organization that has been receiving an enormous

---

[42]Manoharan, VN. Colloidal matter: Packing, geometry, and entropy. *Science 349*, 1253751–1, 2015; Day, C. Colloidal particles crystallize in an increasingly wide range of structures. *Phys. Today*. 113–15, June 2006; van Blaaderen, A. Colloids get complex. *Nature 439*, 545–46, 2006.

[43]Klug, A. From From macromolecules to biological assemblies (Nobel lecture). *Angew. Chem. Int. Ed. 22*, 565–636, 1983.

amount of attention over the last couple of decades.[44] The structures range from modest aggregates of molecules held together by non-covalent forces (so-called supramolecular clusters; Box 4.7) to full three-dimensional structures. A further motivation for SA is the potential of much greater energy efficiency in nanofabrication.[45]

Syntheses of the molecular building blocks needed for self-assembly are also receiving attention and have obvious synergies with the molecular synthetic techniques described above.[46] Block copolymers, in which the polymerizing units are themselves made up of smaller molecular blocks, are a particular focus of interest.[47] They can undergo a great deal of self-organization due to nanophase separation,[48] including the formation of 3D

---

[44]E.g., Bard, in Ref. 13, pp. 54–68; Lehn, J-M. Perspectives in supramolecular chemistry: from molecular recognition towards self-organisation. *Pure Appl. Chem. 66*, 1961–66, 1994; Bell, TW. Molecular trees: A new branch of chemistry. *Science, 271*, 1077–8, 1996; Huck, Wilhelm TS. *Nanoscale Assembly: Chemical Techniques*. NS&T, 2005; Adachi & Lockwood, in Ref. 13; Chapter 8 in Cui, Ref. 1; Chapters 3, 5, 6, 8, 9, 11 in Ref. 2; Wang, Y; Lin, H-X; Chen, L; Ding, S-Y; Lei, Z-C; Liu, D-Y; Cao, X-Y; Liang, H-J; Jiang, Y-B; Tian, Z-Q. What molecular assembly can learn from catalytic chemistry. *Chem. Soc. Rev. 43*, 399–411, 2014; Müller, SC; Parisi, J. *Bottom Up Self-Organization in Supramolecular Soft Matter: Principles and Prototypical Examples of Recent Advances*. SSMS 217, 2015.

[45]Palma, C-A; Cecchini, M; Samori, P. Predicting self-assembly: From empirism to determinism. *Chem. Soc. Rev. 41*, 3713–30, 2012.

[46]E.g., Shenhar, R; Rotello, VM. Nanoparticles: Scaffolds and building blocks. *Acc. Chem. Res. 36*, 549–61, 2003; Mansoori, GA; George, TF; Assoufid, L.; Zhang, G. *Molecular Building Blocks for Nanotechnology: From Diamondoids to Nanoscale Materials and Applications*. TAP, 2007; Sang, L; Zhao, Y; Burda, C. TiO$_2$ Nanoparticles as functional building blocks. *Chem. Rev. 114*, 9283–318, 2014; Tatum, LA; Su, X; Aprahamian, I. Simple hydrazone building blocks for complicated functional materials. *Acc. Chem. Res. 47*, 2141–49, 2014.

[47]E.g., Drzal PL, Barnes JD, Kofinas P. Path dependent microstructure orientation during strain compression of semicrystalline block copolymers. *Polymer 42*, 5633–42, 2001; Abetz, Volker, ed. *Block Copolymers I. Adv. Polym. Sci. 189*, 2005; *Block Copolymers II. 190*, 2005; Theato, P; Kilbinger, AFM; Brya, E; eds. *Non-Conventional Functional Block Copolymers*, ACSSS 1066, 2011.

[48]E.g., Park, M; Harrison, C; Chaikin, PM; Register, RA; Adamson, DH. Block copolymer lithography: Periodic arrays of ~10^{11} holes in 1 square

ordered units,[49] and have also found application as templates.[50] They provide one approach toward making polymeric materials with ordered nanoscale structures, thus avoiding the nanoscale disorganization, mentioned above, resulting from conventional syntheses.

So-called polyoxometalates are also attracting attention as building blocks.[51] These are large anions, representing an extraordinary degree of self-organization, which commonly can be nucleated directly from aqueous solution. An example is the Keggin structure (Fig. 4.1), $XW_{12}O_{40}^{-3}$, which contains 4 sets of 3 linked WO_6 octahedra arranged around an atom in tetrahedral coordination ("X" in the formula), which can be silicon, phosphorus, iron, cerium, titanium, and others. Other structures can also be nucleated depending on the conditions. Polytungstate anions, such as "paratungstate Z," $W_{12}O_{41}^{-10}$, were the first known, but other so-called early transition metals (those with four or fewer d-electrons) are now known to make similar structures. As with the tungsten compounds, many are stable in aqueous solution over a wide range of conditions.

centimeter. *Science 276*, 1401, 1997; Park, C; Yoon, J; Thomas, EL. Enabling nanotechnology with self assembled block copolymer patterns. *Polymer 44*, 6725–60, 2003; O'Reilly, RK; Hawker, CJ; Wooley, KJ. Cross-linked block copolymer micelles: functional nanostructures of great potential and versatility. *Chem. Soc. Rev. 35*, 1068, 2006; Park, S; Lee, DH; Xu, J; Kim, B; Hong, SW; Jeong, U; Xu, T; Russell, TR. Macroscopic 10-terabit-per-square-inch arrays from block copolymers with lateral order. *Science 323*, 1030, 2009.

[49]E.g., Hillmyer, MA. Nanoporous materials from block copolymer precursors. *Adv. Polym. Sci. 190*, 137–81, 2005.

[50]Templin, M; Franck, A; Du Chesne, A; Leist, H; Zhang, Y; Ulrich, R; Schädler, V; Wiesner, U. Organically modified aluminosilicate mesostructures from block copolymer phases. *Science 278*, 1795–98, 1997.

[51]E.g., Borras-Almenar, JJ; Coronado, E; Mueller, A; Pope, M. *Polyoxometalate Molecular Science*. *NATO Science Series 98*, 2003; Yamase, T; Pope, M. *Polyoxometalate Chemistry for Nano-Composite Design*. NS&T, 2004; D'Cruz, B; Samuel, J; Sreedhar, MK; George, L. Green synthesis of novel polyoxoanions of tungsten containing phosphorus as a heteroatom: characterization, non-isothermal decomposition kinetics and photocatalytic activity. *New J. Chem. 38*, 5436–44, 2014.

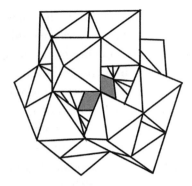

Figure 4.1 The Keggin ion.

Formula $XW_{12}O_{40}^{-3}$, with X = Si (silicon), P (phosphorus), or another element in tetrahedral coordination, and where W = tungsten (wolfram). Four sets of three linked WO_3 octahedra surround the tetrahedral ion, shown by the shading in the middle. The octahedra each have a tungsten atom in the center and an oxygen atom at each vertex. The oxygen atom is shared where vertices are linked.

Polyoxometalates are also of note as nanotechnological building blocks that contain no carbon.

Two-Dimensional Structures

A major difficulty with nanofabrication by self-assembly is that all the design information is implicit in the shapes of the fitting molecules.[52] Although complex 3D structures are certainly possible, as is shown by the tobacco mosaic virus itself, they are extremely difficult to design a priori, depending as they do on the details of the weak forces that determine molecular recognition.[53] Indeed, the issues are akin to the notoriously difficult protein-folding problem.

Although progress is being made with prediction of 3D structures,[54] for now it is considerably simpler to focus on two-

---

[52] E.g., Chapter 8 in Cui, Ref. 1.

[53] Bishop, KJM; Wilmer, CE; Soh, S; Grzybowski, BA. Nanoscale forces and their uses in self-assembly. *Small* 5, 1600–30, 2009.

[54] Ref. 45.

dimensional (2D) structures, and the self-assembly of molecules in two dimensions—so-called self-assembled monolayers (SAMs)—has received an enormous amount of attention.[55] Indeed, the Langmuir–Blodgett technique, which involves construction and transfer of surfactant monolayers, dates back to the 1930s but remains an active field of research.[56]

SAMs adsorbed to gold surfaces by a thiol (–S–H) link have been an ongoing focus.[57] To achieve planar structures, such SAMs have been structured on vapor-deposited gold layers (made by one of the techniques mentioned below), including cleaved mica.

---

[55]E.g., Ulman, A. Formation and structure of self-assembled monolayers. *Chem. Rev. 96*, 1533–54, 1996; Ref. 2, Ch. 2; Kondo, T; Yamada, R; Uosaki, K. Self-assembled monolayer (SAM). In *Organized Organic Ultrathin Films: Fundamentals and Applications* (1st ed.), Ariga, K; ed. Wiley, pp. 7–42, 2013; Borges, J; Mano, JF. Molecular interactions driving the layer-by-layer assembly of multilayers. *Chem. Rev. 114*, 8883–942, 2014, esp. Refs. 30–47 therein; 1–14 (Chapter 3); Wen, J; Li, W; Chen, S; Ma, J. Simulations of molecular self-assembled monolayers on surfaces: packing structures, formation processes and functions tuned by intermolecular and interfacial interactions. *Phys. Chem. Chem. Phys. 18*, 22757–71, 2016.

[56]Ariga, K; Yamauchi, Y; Mori, T; Hill, JP. 25th anniversary article: What can be done with the Langmuir-Blodgett method? Recent developments and its critical role in materials science. *Adv. Mater. 25*, 6477, 2013; Acharya, S; Hill, JP; Ariga, K. Soft Langmuir-Blodgett technique for hard nanomaterials. *Adv. Mater. 21*, 2959–81, 2009; Tao, AR; Huang, J; Yang, P. Langmuir-Blodgettry of nanocrystals and nanowires, *Acc. Chem. Res. 41*, 1662–73, 2008; Moridi, N; Wäckerlin, C; Rullaud, V; Schelldorfer, R; Jung, TA; Shahgaldian, P. Langmuir–Blodgett monolayer stabilization using supramolecular clips. *Chem. Commun. 49*, 367–9, 2013.

[57]E.g., Love, JC; Estroff, LA; Kriebel, JK; Nuzzo, RG; Whitesides, GM. Self-assembled monolayers of thiolates on metals as a form of nanotechnology. *Chem. Rev. 105*, 1103–69, 2005; Vericat, C; Vela, ME; Benitez, G; Carro, P; Salvarezza, RC. Self-assembled monolayers of thiols and dithiols on gold: new challenges for a well-known system. *Chem. Soc. Rev. 39*, 1805–34, 2010; Park, JH; Hwang, S; Kwak, J. Nanosieving of anions and cavity-size-dependent association of cyclodextrins on a 1-adamantanethiol self-assembled monolayer. *ACS Nano 4*, 3949–58, 2010; Volkert, AA; Subramaniam, V; Ivanov, MR; Goodman, AM; Haes, AJ. Salt-mediated self-assembly of thioctic acid on gold nanoparticles. *ACS Nano 5*, 4570–80, 2011.

This is also an example of exploiting the extraordinary flatness, at literally atomic scales, of cleaved mica (Box 3.5).

Although a great degree of useful organization can be achieved, such SAMs are weak due to the absence of covalent binding to the substrate. Covalent binding leads to much stronger monolayers, and covalently bound layers, typically bound to oxide surfaces, have also been a subject of a great deal of research.[58]

Two-dimensional surfaces obviously have valuable uses, for example in catalysis and specific adsorption, but most applications will require some sort of three-dimensional structure. One set of approaches, building on the by now vast number of approaches to forming 2D layers, is "layer-by-layer assembly," which now has a burgeoning literature.[59] Layers consisting of a wide variety of constituents (e.g., surfactants, nanoparticles, macromolecules), and bound together by a variety of mechanisms, are laid down in successive steps. Such stepwise assembly obviously provides a great degree of interactive control over the building of complex structures, especially if intermediate layers are in addition modified by one or more of the techniques described below. The disadvantage is that the construction of multilayered structures can become very time-consuming.

---

[58]E.g., Wang, Y; Supothina, S; De Guire, MR; Heuer, AH; Collins, R; Sukenik, CN. Deposition of compact hydrous aluminum sulfate thin films on titania particles coated with organic self-assembled monolayers. *Chem. Mater.* 10, 2135–44, 1998. Onclin, S; Ravoo, BJ; Reinhoudt, DN. Engineering silicon oxide surfaces using self-assembled monolayers. *Angew. Chem. Int. Ed.* 44, 6282–304, 2005; Wen, K; Maoz, R; Cohen, H; Sagiv, J; Gibaud, A; Desert, A; Ocko, BM. Postassembly chemical modification of a highly ordered organosilane multilayer: New insights into the structure, bonding, and dynamics of self-assembling silane monolayers, *ACS Nano* 2, 579–99, 2008.

[59]Chapter 3 in Ref. 2; Ariga, K; Yamauchi, Y; Rydzek, G; Ji, Q; Yonamine, Y; Wu, KC-W; Hill, JP. Layer-by-layer nanoarchitectonics: Invention, innovation, and evolution. *Chem. Lett.* 43, 36, 2014; Ma, R; Sasaki, T. Organization of artificial superlattices utilizing nanosheets as a building block and exploration of their advanced functions. *Acc. Chem. Res.* 48, 136–43, 2015; Borges & Mano, Ref. 55.

Ordering by Electric Fields

An interesting synergy of nanofabrication with nanoelectronics seems to be emerging. Using electric fields to direct nanoparticle or even macromolecule assembly has been the object of extensive research,[60] and higher degrees of control can be expected as (nano)electronics becomes ever smaller. Indeed, so-called electrocrystallization—crystallization under the influence of an electric field—has been known since the 19th century.[61] Of course, such reactions have long been known in the context of metal plating, in which metal ions in solution are reduced to metal on a metal substrate acting as a cathode. However, the degree of nanoscale control in traditional electroplating is limited. Recently attempts have been to increase the degree of control, and with application to more complicated species than metal ions.[62] Anodic oxidation of aluminum has been used to prepare

[60]E.g., O'Riordan, A; Delaney, P; Gareth Redmond, G. Field configured assembly: Programmed manipulation and self-assembly at the mesoscale. *Nano Lett.* 4, 761–5, 2004; Dickerson, JH; Boccaccini, AR; eds. *Electrophoretic Deposition of Nanomaterials.* NS&T, 2012; esp. Sides, PJ; Christopher L. Wirth, CL; Dennis C. Prieve, DC. Mechanisms of Directed Assembly of Colloidal Particles in Two Dimensions by Application of Electric Fields, therein; Smith, BD; Mayer, TS; Keating, CD. Deterministic assembly of functional nanostructures using nonuniform electric fields. *Annu. Rev. Phys. Chem.* 63, 241–63, 2012; Pescaglini, A; O'Riordan, A; Quinn, AJ; Iacopino, D. Controlled assembly of Au nanorods into 1D architectures by electric field assisted deposition. *J. Mater. Chem. C* 2, 6810–16, 2014; Marmisollé, WA; Azzaroni, O. Recent developments in the layer-by-layer assembly of polyaniline and carbon nanomaterials for energy storage and sensing applications. From synthetic aspects to structural and functional characterization. *Nanoscale 8*, 9890–918, 2016.
[61]E.g., Winand, R. Electrocrystallization—theory and applications. *Hydrometallurgy 29*, 567, 1992.
[62]E.g., Batail, P. Electrocrystallization, an invaluable tool for the construction of ordered, electroactive molecular solids. *Chem. Mater. 10*, 3005–15, 1998; Abe, M; Yamamoto, A; Orita, M; Ohkubo, T; Sakai, H; Momozawa, Z. Control of particle alignment in water by an alternating electric field. *Langmuir 20*, 7021–26, 2004; O'Riordan, A; Delaney, P; Redmond, G. Field configured assembly: Programmed manipulation and self-assembly at the mesoscale. *Nano Lett. 4*, 761–65, 2004; Staikov, DG; ed. *Electrocrystallization in*

microporous alumina, which can then be used as templates for futher nanofabrication.[63] Magnetic field-based ordering and the self-ordering of magnetic nanoparticles are also subjects of study.[64]

Nanoscale electrochemical methods also provide approaches for imposing covalent binding on an assembled structure. In particular, electropolymerization, the linking up of particular monomers under the influence of an electric current, has been an object of study for decades[65] and is receiving major attention as another way to structure layers at surfaces.[66]

Nanotechnology. Wiley, 2007; Schmuki P; Virtanen, S; eds. *Electrochemistry at the Nanoscale.* NS&T, 2009, esp. Roy, S. Electrochemical Fabrication of Nanostructured, Compositionally Modulated Metal Multilayers (CMMMs), therein, pp. 349–76; Kondo, K; Akolkar, RN; Barkey, DP; Yokoi, M. *Copper Electrodeposition for Nanofabrication of Electronics Devices.* NS&T, 2014; Liu, L; Choi, BG; Tung, SO; Hu, T; Liu, Y; Li, T; Zhao, T; Kotov, NA. Low-current field-assisted assembly of copper nanoparticles for current collectors. *Faraday Disc. 181,* 383–401, 2015.

[63]E.g., Ref. 28.

[64]Raman, V; Bose, A; Olsen, BD; Hatton, TA. Long-range ordering of symmetric block copolymer domains by chaining of superparamagnetic nanoparticles in external magnetic fields. *Macromolecules 45,* 9373–82, 2012; Singh, G; Chan, H; Udayabhaskararao, T; Gelman, E; Peddis, D; Baskin, A; Leitus, G; Král, P; Klajn, R. Magnetic field-induced self-assembly of iron oxide nanocubes. *Faraday Disc. 181,* 403–21, 2015. Ghosha, S; Puri, IK. Changing the magnetic properties of microstructure by directing the self-assembly of superparamagnetic nanoparticles. pp. 423–35; Hoffelner, D; Kundt, M; Schmidt, AM; Kentzinger, E; Bender, P; Disch, S. Directing the orientational alignment of anisotropic magnetic nanoparticles using dynamic magnetic fields. pp. 449–61.

[65]Waltman, RJ; Bargon, J. Electrically conducting polymers: a review of the electropolymerization reaction, of the effects of chemical structure on polymer film properties, and of applications towards technology. *Can. J. Chem. 4,* 76, 1986.

[66]Schab-Balcerzak, E. *Electropolymerization.* InTech, 2011; Pernites, RB; Foster, EL; Felipe, MJL; Robinson, M; Advincula, RC. Patterned surfaces combining polymer brushes and conducting polymer via colloidal template electropolymerization. *Adv. Mater. 23,* 1287–92, 2011. Rydzek, G; Terentyeva, TG; Pakdel, A; Golberg, D; Hill, JP; Ariga, K. Simultaneous electropolymerization and electro-click functionalization for highly versatile surface platforms. *ACS Nano 8,* 5240–48, 2014.

Modification Processes

It was noted above that "top-down" vs. "bottom-up" nano-fabrication is an oversimplification, because the substrate modified in a top-down approach already possesses useful nanoscale ordering; commonly it's a crystal, in fact. But modification, typically in successive steps,[67] is often a critical element of current approaches to nanofabrication, as has already been mentioned above in the context of layer-by-layer fabrication.

Modification techniques can broadly be divided into "chemical" and "physical" methods, and in practice a combination is usually used. Both subsume a host of different approaches, only some of which can only be highlighted; for a not extreme example, in one study surface-functionalized gold nanoparticles were then self-assembled and deposited via the Langmuir–Blodgett method onto a silicon surface.[68]

Chemical modification

Much falls under the rubric of "surface engineering," which has already been encountered above in the context of applying and modifying self-assembled molecular layers. Surface engineering, however, also includes a broad array of techniques for the replacement or alteration of surface-exposed chemical groups (often termed *functionalization*) that do not necessarily lead to complete coverage, but yield desired properties, such as for sensors, active sites for catalysts, binding sites, and so on.[69]

---

[67]Gooding, JJ; Ciampi, S. The molecular level modification of surfaces: from self-assembled monolayers to complex molecular assemblies. *Chem. Soc. Rev.* 40, 2704–18, 2011.

[68]Kundu, S. Layer-by-layer assembly of thiol-capped Au nanoparticles on a water surface and their deposition on H-terminated Si(001) by the Langmuir-Blodgett method. *Langmuir* 27, 3930–6, 2011.

[69]E.g., Willenbockel, M; Maurer, RJ; Bronner, C; Schulze, M; Stadtmüller, B; Soubatch, S; Tegeder, P; Reuter, K; Tautz, FS. Coverage-driven dissociation of azobenzene on Cu(111): a route towards defined surface functionalization. *Chem. Commun.* 51, 15324–27, 2015; Dong, A; Ye, X; Chen, J; Kang, Y;

So-called "click chemistry," for example, has recently become a research focus because it can be used for highly specific reactions.[70]

Chemical modification is not restricted to surfaces, however. For example, under certain circumstances low-temperature modification of host phases can be carried out that leaves the original crystal framework unaffected. Such "topotactic reaction" is another aspect of *chimie douce*, and has been known for decades: it is used to activate molecular sieve (Box 6.3) catalysts by exchanging the cations in the as-synthesized crystal for catalytically effective species, for example.

Vapor deposition

Deposition from a vapor, by a variety of mechanisms, is a well-established set of techniques that straddles both "chemical" and "physical" processes. In "physical" vapor deposition (PVD) the vapor condenses directly onto the substrate,[71] whereas in "chemical" vapor deposition (CVD) chemical reaction of vapor-phase precursors causes the deposition.[72] A variant of CVD, so-called atomic layer deposition (ALD), has attracted much

Gordon, T; Kikkawa, JM; Murray, CB. A generalized ligand-exchange strategy enabling sequential surface functionalization of colloidal nanocrystals. *J. Am. Chem. Soc. 133*, 998–1006, 2011.

[70]E.g., Zheng, J; Chen, Y; Karim, A; Becker, ML. Dopamine-based copper-free click kit for efficient surface functionalization. *ACS Macro Lett. 3*, 1084 87, 2014; Selvanathan, A; Popik, VV. Attach, remove, or replace: Reversible surface functionalization using thiol-quinone methide photoclick chemistry. *J. Am. Chem. Soc. 134*, 8408–11, 2012; Ziarani, GM; Hassanzadeh, Z; Gholamzadeh, P; Asadi S; Badiei, A. Advances in click chemistry for silica-based material construction. *RSC Adv. 6*, 21979–2006, 2016.

[71]Mattox, DM. *Handbook of Physical Vapor Deposition*, 2nd ed. Elsevier, 2010; Maitani, MM; Allara, DL. Issues and challenges in vapor-deposited top metal contacts for molecule-based electronic devices. *Top. Curr. Chem. 312*, 239–, 2012.

[72]Jones, AC; Hitchman, ML; eds. *Chemical Vapour Deposition: Precursors, Processes and Applications*. RSC, 2009.

interest recently. As its name suggests, it offers the prospect of improved control at the nanoscale.[73]

Physical modification

Lithography

A great many physical modification techniques are lumped under the rubric "lithography." The most familiar "micro" lithographic technique is the optical lithography used to fabricate semiconductor chips, in which a chemically bound surface layer (the "resist") is modified by illumination in the desired pattern. The light induces chemical changes in the resist so that it can be etched away, and new layers can then be applied by various chemical or physical means. Then the next step can be carried out. This interplay between the chemically deposited resist in conventional lithography and its subsequent illumination also illustrates how practical nanofabrication, at least in the near term, is likely to involve a combination of chemical and physical procedures in a series of steps, as already encountered above in the context of layer-by-layer assembly. The patterning is also relatively fast because it occurs in parallel, all at once across the illuminated surface.

As suggested by the quoted modifier "micro," though, conventional semiconductor lithography is not at molecular scales, due to the limitations imposed by the wavelength of the light used. Hence, the quest for fabrication at ever-smaller scales has led to a great deal of effort on alternative forms of lithography, including shorter wavelengths (e.g., X-rays,[74]

---

[73]E.g., Jiang, X; Bent, SF. Area-selective ALD with soft lithographic methods: Using self-assembled monolayers to direct film deposition. *J. Phys. Chem. C 113*, 17613–25, 2009; Ritala, M; Niinisto, J. Atomic Layer Deposition, in Ref. 72, Ch. 4; Foong, TRB; Shen, YD; Hu, X; Sellinger, A. Template-directed liquid ALD growth of TiO_2 nanotube arrays: Properties and potential in photovoltaic devices. *Adv. Funct. Mater. 20*, 1390–96, 2010.

[74]E.g., Jackson, MJ. Microfabrication using X-ray lithography. In *Micro and Nanomanufacturing*, 55–98, 2007.

charged-particle lithography with ions or electron beams,[75] and a growing inventory of more exotic techniques[76]). In general, all such lithographic techniques, although in favorable circumstances capable of resolutions <10 nm, are slow both because of limitations of the techniques themselves and because they are serial in nature. They also typically require expensive equipment; e.g., charged particle beams require high vacuum.

SPM-based lithography

Since the invention of scanning probe microscopes three decades ago (Box 4.4), they have attracted much attention as tools for the modification of surfaces. Subnanometer patterns have been produced by a variety of mechanisms.[77] Most are "subtractive," in that atoms already present are removed or rearranged, which limits the possible structures that can be achieved.[78] "Additive" lithography places new material on the surface, which leads

---

[75]Chapters 2 & 3 in Cui, Ref. 1; Kant, Krishna; Dusan Losic, eds. *Focused Ion Beam (FIB) Technology for Micro- and Nanoscale Fabrications.* LNNS&T *20,* 2013; Shan, J; Chakradhar, A; Anderson, K; Schmidt, J; Dhuey, S; Burghaus, U. Butane adsorption on silica supported MoOx clusters nanofabricated by electron beam lithography. In *Nanotechnology for Sustainable Energy*, Hu, YH; Burghaus, U; Qiao, S; eds. ACSSS *1140*, 295–310, 2013.

[76]E.g., Gates, BD; Xu, Q; Love, JC; Wolfe, DB; Whitesides, GM. Unconventional nanofabrication. *Annu. Rev. Mater. Res. 34*, 339–72, 2004; Castillejo, M; Ezquerra, TA; Oujja, M; Rebollar, E. Laser nanofabrication of soft matter, SSMS, *191*, 2014; Liu, Q; Duan, X; Peng, C. Laser Interference Nanofabrication, in Liu et al., Ref. 1, pp. 153–78.

[77]E.g., Kramer, S; Fuierer, RR; Gorman, CB. Scanning probe lithography using self-assembled monolayers, *Chem. Rev. 103*, 4367–418, 2003; Tseng, AA; Notargiacomo, A; and Chen, TP. Nanofabrication by scanning probe microscope lithography: A review. *J. Vac. Sci. Technol. B 23*, 877–93, 2005; Chapter 4 in Cui, Ref. 1. Moriarty, P. Atom-technology and beyond: manipulating matter using scanning probes. *Nanoscience: Volume 1: Nanostructures through Chemistry 1*, 116–44, 2012, RSC.

[78]Mirkin, CA. The power of the pen: Development of massively parallel dip-pen nanolithography. *ACS Nano 1*, 79–83, 2007; Chapter 4 in Cui, Ref. 1.

potentially to a much wider variety of structures.[79] The most mature additive technique is probably "dip-pen lithography," in which the SPM tip acts as a "pen" from which molecules in aqueous suspension are deposited as "ink."[80]

Like most of the other lithographic techniques mentioned above, SPM-based litholography is slow and serial. There has been much interest, therefore, in "ganged" SPMs; that is, an array operating in parallel. Such arrays have been fabricated by conventional lithographic techniques and indeed are now available commercially (e.g., the "Millipede", which, however, is directed at high-density data storage[81]), although they are have to be optimized for the particular substrate.[82] Dip-pen arrays have been a particular focus.[82,83] It has even been proposed that the individual elements of a dip-pen array could be independently addressed to build specific macromolecules;[82] this would be a true, albeit primitive, molecular assembler. As fabrication scales continue to decrease, such arrays may provide one approach to a true molecular assembler.

[79]E.g., Cohen, SR; Maoz, R; Sagiv, J. Constructive Nanolithography, in Vilarinho, PM, Rosenwaks, Y, Kingon, A. *Scanning Probe Microscopy: Characterization, Nanofabrication and Device Application of Functional Materials, NATO Science Series II: Mathematics, Physics and Chemistry, 186*, 2005; Ref. 78.

[80]E.g., Brown, KA; Eichelsdoerfer, DJ; Liao, X; He, S; Mirkin, CA. Material transport in dip-pen nanolithography. *Frontiers of Physics 9*, 385–397, 2014; Salaita, K; Wang, Y; Mirkin, CA. Applications of dip-pen nanolithography. *Nat. Nanotechnol. 2*, 145–55, 2007.

[81]Binnig, G; Cherubini, G; Despont, M; Dürig, U; Eleftheriou, E; Pozidis, H; Vettiger, P. The "Millipede" - A Nanotechnology-Based AFM Data-Storage System. *Springer Handbook of Nanotechnology*, 1457–1486, Springer, 2007.
[82]Ref. 78.

[83]E.g., Salaita, K; Wang, Y; Fragala, J; Vega, RA; C; Mirkin, CA. Massively parallel dip-pen nanolithography with 55000-pen two-dimensional arrays. *Angew. Chem. Int. Ed. 45*, 7220–23, 2006; Park, S; Wang, WM; Bao, Z. Parallel fabrication of electrode arrays on single-walled carbon nanotubes using dip-pen-nanolithography-patterned etch masks. *Langmuir 26*, 6853–59, 2010; Wang, S; Hosford, J; Heath, WP; Wong, LS. Large-area scanning probe nanolithography facilitated by automated alignment of probe arrays. *RSC Adv. 5*, 61402–409, 2015.

As emphasized above, building macroscopic objects atom-by-atom founders on the sheer number of atoms required. The structuring of catalyst surfaces, however, might be one application in which the individual placement of atoms or clusters, say by ganged SPMs, could yield a useful product. Practical catalysts must have a strong "multiplier effect" because to be economic each catalytic site must be used repeatedly a vast number of times. Hence even a modest amount of nanostructuring could yield a useful amount of product (Box 4.9).

Pattern transfer

Another lithographic approach is nanoscale pattern transfer, by various techniques under various names (e.g., nanoprinting, microcontact printing, molecular printing).[84] It is also potentially much faster—cf. printing a page vs. copying it serially! Although direct molecular resolution cannot be achieved, surprisingly small-scale features can be duplicated. Indeed, nanopattern transfer has been suggested as a lower-cost alternative for conventional semiconductor fabrication.[85] Relatively cheap

---

[84] E.g., Nealey, PF; Black, AJ; Wilbur, JL; Whitesides, GM. Micro- and nanofabrication techniques based on self-assembled monolayers. In *Molecular Electronics*, Jortner, J. and Ratner, M; eds., Blackwell Science, pp. 343–67, 1997; Gates, BD; Xu, QB; Stewart, M; Ryan, D; Willson, CG; Whitesides, GM. New approaches to nanofabrication: Molding, printing, and other techniques. *Chem. Rev.* 105, 1171–96, 2005; Zeira, A; Chowdhury, D; Maoz, R; Sagiv, J. Contact electrochemical replication of hydrophilic-hydrophobic monolayer patterns. *ACS Nano* 2, 2554–68, 2008; Braunschweig, AB; Huo, F; Mirkin, CA. Molecular printing. *Nat. Chem.* 1, 353–8, 2009; Mele, E; Pisignano, D. Nanobiotechnology: Soft Lithography, *Progress in Molecular and Subcellular Biology* 47, 341–58, 2009; Jeong, S; Hu, L; Lee, HR; Garnett, E; Choi, JW; Cui, Y. Fast and scalable printing of large area monolayer nanoparticles for nanotexturing applications. *Nano Lett.* 10, 2989–94, 2010; Li, J-R; Yin, N-N; Liu, G-Y. Hierarchical Micro- and Nanoscale Structures on Surfaces Produced Using a One-Step Pattern Transfer Process. *J. Phys. Chem. Lett.* 2, 289–94, 2011, & refs therein.

[85] Costner, EA; Lin, MW; Jen, WL; Willson, CG. Nanoimprint lithography: Materials development for semiconductor device fabrication. *Annu. Rev. Mater. Res.* 39, 155–80, 2009; Voorthuijzen, WP; Yilmaz, MD; Naber, WJM;

> **Box 4.9** The "Molecular Loom"
>
> To determine whether a "ganged" SPM could synthesize interesting amounts of material, assume it modifies 10^6 sites a second on a "die". Assume that this die can be stamped onto a surface once a second to generate the actual catalytic sites. Assume further that each site generates 10^6 product molecules per second (i.e., at atmospheric pressure, only ~1% of the molecular collisions with a site lead to reaction). The number of die sites in existence is then $10^6 t$, the number of catalyst sites $\sim(1/2)10^6 t^2$, and
>
> $$N = (1/6)\ 10^{12} t^3$$
>
> is the number of product molecules in existence, where the coefficient accounts roughly for the fact that a newly fabricated die or surface can't be put into production immediately.
>
> The first mole of product is produced after $\sim 1.5 \times 10^4$ seconds (~0.2 days), and nearly 70,000 moles have been produced by the end of the first week, as obviously unit production increases the longer the "fab" is run. Of course, this model is unrealistic in that it assumes that both dies and catalytic surfaces work indefinitely once made, whereas obviously they will wear out or otherwise become unusable (e.g., due to fouling) eventually. Nonetheless, it appears a significant amount of product could be made within the several-year timeframe of venture capital if the product were valuable enough.

fabrication of catalytic surfaces might be another application of nanoprinting, especially because occasional defects in the surface would be tolerable.

Huskens, J; van der Wiel, WG. Local doping of silicon using nanoimprint lithography and molecular monolayers. *Adv. Mater. 23*, 1346–50, 2011.

Of course, even when a nanofabrication capability exists, we need to know *what* to build with it. We now turn, therefore, to some specifics on resource applications, starting with nanotechnological applications to energy in the next chapter.

Chapter 5

Nanotechnology and Energy

In this chapter, we consider nanotechnogical approaches to the extraction and use of energy, with a special focus on some of the alternative energy sources sketched in Chapter 2. Although this review aspires to give the flavor of the possibilities, there is no way it can be comprehensive, particularly with research moving as fast as it is. Furthermore, there are no doubt applications that have been completely overlooked.

Before sketching specific aspects of nanotechnological applications to energy, some generalizations on the technological issues are worth highlighting:

- nanoscale structuring at an *interface* is fundamental to a great many applications, including
    - catalyst surfaces (p. 184);
    - large-area semiconductor p–n junctions, critical to thermoelectric (p. 235) and photovoltaic (p. 244) materials;
    - nanolayered structures, such as for high-performance capacitors (p. 223), thermoelectric materials (p. 235), multijunction photovoltaics (p. 248), piezoelectric stacks (p. 238), and sensitized photovoltaic materials (p. 249).

Nanotechnology and the Resource Fallacy
Stephen L. Gillett
Copyright © 2018 Pan Stanford Publishing Pte. Ltd.
ISBN 978-981-4303-87-3 (Hardcover), 978-0-203-73307-3 (eBook)
www.panstanford.com

- *solid electrolytes* prove to be widely applicable:
  - critical in fuel cells (p. 190) and supercapacitors (p. 226),
  - relevant to intercalation batteries (p. 221), smart windows (p. 216), electrosynthesis (p. 228), and also to ionic separation (p. 302).

Such applications have no moving parts other than electrons and (in some cases) individual ions. They therefore seem particularly accessible to the relatively primitive approaches to molecular structuring sketched in the previous chapter. However, in many cases, as for bulk energy generation, large quantities of material will be required, and this remains a significant challenge to make cheaply.[1] Nonetheless, as also noted in the previous chapter, synergies will exist, as better nanofabrication techniques lead to better syntheses, which in turn help improve nanofabrication.

Nanotechnology vs. Prometheus: Efficiency

It was commented in Chapter 2 that the cheapest joules are the ones that are never used. Hence a section on "efficiency" is certainly appropriate. The efficient use of energy is not so glamorous as new sources, perhaps, but it is extremely cost-effective, not least because undesirable consequences due to wasted energy, such as thermal pollution, are also minimized or avoided.

Catalysis

As repeatedly mentioned, catalysts cause a thermodynamically favorable reaction to occur under circumstances where it otherwise would not. They do so by decreasing the activation

---

[1]For a recent review see Lin, Z; Wang, J. *Low-cost Nanomaterials: Toward Greener and More Efficient Energy Applications*. GE&T, 2014.

energy barrier to reaction, and *that* occurs via molecular-scale interactions between the catalyst and the reactants. As already noted, catalysts are fundamental to energy savings and usage in a host of different applications. Indeed, enormous savings are possible due both to greatly minimized process energy, and to much greater selectivity for the desired products so that far less waste material is generated. A further incentive is that the alternative energy sources described later also require better catalysts.

Three parameters, which are commonly independent, describe catalysts: their *activity*, or how effective they are at promoting a desired reaction; their *selectivity*, or the degree to which they promote the desired reaction only; and their *stability*, or resistance to degradation while operating. All are important for energy efficiency. Activity is important for minimizing the amount of catalyst required. Selectivity is even more important, because the greater the degree to which the desired product is formed, the fewer the unwanted by-products, and thus the less the purification and disposal costs. Finally, a practical catalyst must remain active over a vast number of catalytic cycles before requiring replacement or reactivation.

All catalysis fundamentally requires that the reacting molecule(s) bind to the catalyst, at least briefly. Thus catalysis must involve some degree of molecular recognition (Box 4.7)—although, as will be seen, molecular recognition is only a part of the story.

Homogeneous catalysts are mixed in with the reactants. Because they are molecular, they can be highly selective, but they have the serious disadvantage of being difficult to separate from the reacting medium. *Heterogeneous* catalysts are solids, with catalysis taking place at the interface between the solid and the fluid phase containing the reactants. Potentially they are considerably easier to use because the catalytic surface (which commonly is a high-surface area solid fixed in place) remains separate from the reactants and products. The distinction, however, is not always significant in practice; e.g., heterogeneous catalysts mixed into the reacting fluid in the form of fine powders can also be difficult to separate afterward.

Heterogeneous catalysis is also intrinsically more complicated because the selectivity and activity are intimately bound up with atomistic details of the surface. In some cases, as for biological enzymes and catalysts based on molecular sieves (Box 6.3), the *selectivity* is reasonably straightforward to rationalize on the basis of molecular recognition. Only reactants that can "fit" have access to the reacting site.

Even so, however, the very nature of the active sites is usually uncertain, commonly depending on minute details of the (very complicated) electronic structure due to such things as adsorbed atoms, surface dislocations or defects, and so on.[2] Even in the case of obvious molecular recognition, too, it can remain unclear as to what makes an active site "active." Mere molecular recognition in general need not lead to reaction; the active site must also somehow change the local electronic environment so that the species don't just bind but react. Furthermore, the site generally is small enough, a few nanometers or less, that quantum confinement effects are likely to be important, as indicated by the increasing interest in quantum-dot-based catalysts,[3] and extreme electric fields may be important as well.[4]

Even if the catalytic mechanisms are understood, with conventional fabrication techniques there is little nanoscale

---

[2]E.g., Nørskov, JK. From quantum physics to heterogeneous catalysis. *Top. Catal. 1*, 385–403, 1994, and refs therein.

[3]E.g., Kiyonaga, T; Akita, T; Tada, H. Au nanoparticle electrocatalysis in a photoelectrochemical solar cell using CdS quantum dot-sensitized TiO_2 photoelectrodes. *Chem. Commun.* 2011–3, 2009; Frame, FA; Osterloh, FE. CdSe-MoS_2: A quantum size-confined photocatalyst for hydrogen evolution from water under visible light. *J. Phys. Chem. C 114*, 10628–33, 2010; Shen, J; Zhu, Y; Yang, X; Li, C. Graphene quantum dots: Emergent nanolights for bioimaging, sensors, catalysis and photovoltaic devices. *Chem. Commun. 48*, 3686–99, 2012.

[4]E.g., Ref. 2; Fried, SD; Bagchi, S; Boxer, SG. Extreme electric fields power catalysis in the active site of ketosteroid isomerase, *Science 346*, 1510–4, 2014.

control over the structures that direct those mechanisms. Hence conventional syntheses are empirical recipes that commonly are poorly reproducible. In many cases one batch of catalyst differs enough from the next that the chemical processes have to be adjusted for each new batch.[5] Traditionally, therefore, catalysis is an "art" that relies on a host of empirical observations with limited understanding of the actual molecular mechanisms. Some years back, practical catalysis was called[6] "still magic," which occasioned a heated discussion.[7]

Overall, therefore, there are two issues: the first is controlled nanofabrication of the catalyst; but even if nanofabrication capability were available, there's then the question of *what* to fabricate!

For this reason an enormous amount of catalysis research in recent years has focused on the nanofabrication of experimental and model catalysts, using techniques such as those sketched in Chapter 4.[8] The idea is to isolate the variables responsible for

---

[5]R. Phair, pers. comm., 1997.

[6]Schlögl, R. Heterogeneous catalysis—still magic or already science? *Angew. Chem. Int. Ed. 32*, 381–83, 1993.

[7]Thomas, JM; Zamaraev, KI. Rationally designed inorganic catalysts for environmentally compatible technologies. *Angew. Chem. Int. Ed. 33*, 308–11, 1994.

[8]This literature is huge; some representative references are: Eppler, AS; Zhu, J; Anderson, EA; Somorjai, GA. Model catalysts fabricated by electron beam lithography: AFM and TPD surface studies and hydrogenation/ dehydrogenation of cyclohexene + H_2 on a Pt nanoparticle array supported by silica. *Top. Catal. 13*, 33–41, 2000; Thomas, JM; Raja, R; Lewis, DW. Single-site heterogeneous catalysts. *Angew. Chem. Int. Ed. 44*, 6456–82, 2005; Contreras, AM; Grunes, J; Yan, X-M; A. Liddle, A; Somorjai, GA. Fabrication of 2-dimensional platinum nanocatalyst arrays by electron beam lithography: ethylene hydrogenation and CO-poisoning reaction studies. *Top. Catal. 39*, 123–29, 2006; Anpo, M; Thomas, JM. Single-site photocatalytic solids for the decomposition of undesirable molecules. *Chem. Commun.* 3273–8, 2006; Österlund, L; Grant, AW; Kasemo, B. Lithographic techniques in nanocatalysis. In *Nanocatalysis*. Heiz, U; Landman, U; eds. NanoScience and Technology, 269–341, 2007; Zhou, B; Han, S; Raja, R; Somorjai, GA. *Nanotechnology in Catalysis: Volume 3*. NS&T, 2007; Kung, H; Kung, M. Nanotechnology and heterogeneous catalysis. In Österlund et al. above, 1–11; Somorjai, GA; Tao, F; Park, JY. The nanoscience revolution: Merging

the catalytic activity, with, of course, the ultimate goal being the fabrication of "designer catalysts."[9] In addition, a "designer" approach to molecular sieves and their molecular recognition capabilities is also occurring.[10]

A further motivation is to devise "greener" (i.e., more environmentally benign) syntheses using alternative catalysts.[11] This is another example of the biological inspiration: after all, the solvent used in all biological syntheses is just water! By contrast, conventional chemical processes typically require reagents, such as concentrated acids or bases, or volatile organic solvents, whose handling is at best difficult because of their toxicity and/ or corrosiveness, and which therefore also are environmental hazards.

Additionally, as understanding of the catalytic mechanisms grows, there is the prospect of replacing expensive catalytic materials, such as the platinum-group metals basic to many conventional catalysts, with cheaper alternatives.[12] In particular,

---

of colloid science, catalysis and nanoelectronics. *Top. Catal. 47*, 1–14, 2008; Kung, MC; Davis, RJ. Understanding Catalysis Through Characterization and Synthesis of Catalysts: Gabor A. Somorjai Award and Symposium for Creative Research 2011. *Top. Catal. 55*, 108–15, 2012; Shan et al., in Ref. 75 (Chapter 4).

[9]E.g., Scott, SL; Crudden, CM; Jones, CW; eds. *Nanostructured Catalysts.* NS&T, 2003; Rolison, DR. Catalytic Nanoarchitectures-the Importance of Nothing and the Unimportance of Periodicity, *Science 299*, 1698–701, 2003; Musselwhite, N; Somorjai, GA. Atomic scale foundation of covalent and acid-base catalysis in reaction selectivity and turnover rate. *Top. Catal. 58*, 184–89, 2015.

[10]E.g., Thomas, JM; Klinowski, J. Systematic enumeration of microporous solids: Towards designer catalysts. *Angew. Chem. Int. Ed. 46*, 7160–63, 2007.

[11]E.g., Sharma, VK; Chang, S-M; Doong, R-A; eds. *Green Catalysts for Energy Transformation and Emission Control.* ACSSS, *1184*, 2014, and refs. therein.

[12]E.g., Royer, S; Duprez, D; Can, F; Courtois, X; Batiot-Dupeyrat, C; Laassiri, S; Alamdari, H. Perovskites as substitutes of noble metals for heterogeneous catalysis: Dream or reality. *Chem. Rev. 114*, 10292–368, 2014; Melchionna, M; Marchesan, S; Prato, M; Fornasiero, P. Carbon nanotubes and catalysis: the many facets of a successful marriage. *Catal. Sci. Technol. 5*, 3859–75, 2015.

the importance of quantum confinement effects in catalysis suggests that controlled nanostructuring will be important, as it will allow tuning of the electronic levels.[13]

A catalyst that expedites redox reactions is often termed an *electrocatalyst*,[14] and as will become abundantly clear, such catalysts are critical in many energy-related technologies such as fuel cells, electrolysis, and electrosynthesis. Nanostructuring such catalysts also ties in with a long-standing electrochemical literature on so-called functionalized electrodes, in which an electrode surface has been chemically modified, often at the nanoscale, to direct reactions at the electrode surface.[15] Additionally, another possibility inspired by the continued convergence of nanoelectronics and nanochemistry would be the ability to directly modulate a catalytic site by applying an electrical potential to it.[16] In other words, to some degree the catalytic activity would be "switchable."

---

[13]E.g., Peles, A. Nanostructured electrocatalysts for oxygen reduction reaction: First-principles computational insights. In Shao, in Ref. 21, pp. 613–35.

[14]E.g., Cui, C-H; Yu, S-H. Electrocatalysis at restructured metal and alloy surfaces. In *Nanotechnology for Sustainable Energy*, Hu, YH; Burghaus, U; Qiao, S; eds. ACSSS *1140*, 265–94, 2013; Lavacchi, A; Miller, H; Vizza, F. *Nanotechnology in Electrocatalysis for Energy*. NS&T, 2013.

[15]E.g., Murray, RW. Chemically modified electrodes. *Acc. Chem. Res. 13*, 135–41, 1980; Rolison, DR. Zeolite-modified electrodes and electrode-modified zeolites. *Chem. Rev. 90*, 867–78, 1990; Merz, A. Chemically modified electrodes. *Top. Curr. Chem. 152*, 49–90, 1990; Šljukic, B; Banks, CE; Mentus, S; Compton, RG. Modification of carbon electrodes for oxygen reduction and hydrogen peroxide formation: The search for stable and efficient sonoelectrocatalysts. *Phys. Chem. Chem. Phys. 6*, 992–7, 2004; Nelson, GW, Foord, JS. Nanoparticle-based diamond electrodes. In Ref. 77 (Chapter 6), pp. 165–204; da Costa, PRF; dos Santos, EV; Peralta-Hernández, JM; Salazar-Banda, GR; da Silva, DR; Martínez-Huitle, CA. Modified diamond electrodes for electrochemical systems for energy conversion and storage. In Ref. 77 (Chapter 6), pp. 205–35; Duan, T; Chen, Y; Wen, Q; Duan, Y. Different mechanisms and electrocatalytic activities of Ce ion or CeO_2 modified Ti/Sb-SnO_2 electrodes fabricated by one-step pulse electro-codeposition. *RSC Adv. 5*, 19601–12, 2015.

[16]E.g., Somorjai, GA. The 13th international symposium on relations between homogeneous and heterogeneous catalysis–an introduction. *Top. Catal. 48*, 1–7, 2008.

Non-thermal engines

The inefficiency of conventional, Carnot-limited heat engines has been repeatedly emphasized. Most of the energy released by the burning fuel is merely ejected as heat to the environment, with the result that the supposedly "irreplaceable" high energy density of conventional fuels is merely a brute-force compensation for the inefficiency with which they are used. As might be expected, however, transforming chemical energy without thermalizing it requires controlling the reactions at the nanoscale. (Nearly the only advantage of fire is that it's simple.) Nanotechnology promises to make such transformations considerably more practical.

As usual, biosystems are far ahead of technology. As reviewed in Chapter 2, even though organisms are loosely spoken of as "burning" food for energy, biological systems are not heat engines and so are not Carnot-limited. Rather than converting the energy of chemical fuel (i.e., food) to heat, and converting only some of that heat to work as it flows to a cooler body, living things oxidize their fuel non-thermally, via a cascade of molecular-scale mechanisms that approach the reversible limit set by the difference in free energy. Since they carry out their chemical processes isothermally in the range 25–40°C or so, the irreversible losses are also much lower.

Fuel cells

Probably the best example of a technological non-thermal conversion device is a fuel cell, which transforms chemical energy directly into electricity via the controlled reaction of fuel and oxidizer (Box 5.1).[17] Fuel cells first attracted attention as long ago as the 19th century, but although practical fuel cells

---

[17]E.g., Kartha, S; Grimes, P. Fuel cells: Energy conversion for the next century. *Phys. Today 47*, 54 ff, Nov 1994; Winter, M; Brodd, RJ. What are batteries, fuel cells, and supercapacitors? *Chem. Rev. 104*, 4245–70, 2004.

Box 5.1 Fuel Cells

A fuel cell can be thought of as a battery (Box 5.2) in which the fuel that is oxidized is continually replenished, and where the oxidizer is just atmospheric oxygen (O_2). Alternatively, it can be viewed as the inverse of electrolysis: rather than applying an external electric potential to drive a redox reaction "uphill", as in the splitting of water into hydrogen and oxygen, those products are instead combined to yield an electric current directly. The concept is simple, but as noted in the main text, the realization is difficult.

Fuel molecules in a fuel cell must be ionized to react. This follows because, as in any battery, it is the electrons transferred during the oxidation reaction that do useful work in the external circuit. Hence the fuel must lose electrons at the anode. To use hydrogen as an example, hydrogen gas (H_2) must yield two hydrogen ions (H^+) and two electrons (e^-):

$$H_2 \rightarrow 2\,H^+ + 2\,e^-.$$

Conversely, at the cathode atmospheric O_2 must be reduced by electrons with as low an overvoltage as possible:

$$O_2 + 2\,e^- \rightarrow 2\,O^{2-}.$$

As discussed in the main text, finding and fabricating catalysts that can make these ionization reactions occur is one of the biggest challenges in designing practical fuel cells.

While the fuel cell is operating, electrons are flowing from the anode to the cathode through the external circuit. To maintain charge balance, therefore, there must be a compensating flow of ions through the electrolyte within the fuel cell (Fig. 5.1). Either anions must flow toward the anode, or cations must flow toward the cathode. Both sorts of flow happen in different types of fuel cells. In any case, this ionic flow must occur as freely as possible. As discussed in the text, several design approaches are possible, but it is another challenge for practical fuel cells.

Box 5.2 Batteries

Batteries are certainly familiar from everyday life. They exploit a "downhill" redox reaction (Box 2.15) in which the positive electrode Pos is reduced by the transferred electrons:

$$Pos + e^- \rightarrow Pos^-$$

In turn, the negative electrode Neg is oxidized by losing electrons:

$$Neg \rightarrow Neg^+ + e^-.$$

The overall electron transfer is responsible for the generated current. If the battery is reversible, application of current in the opposite direction regenerates the original species Pos and Neg from Pos$^-$ and Neg$^+$, respectively.

In the reversible limit, the energy obtained is the Gibbs free energy for the reaction:

$$Pos + Neg \rightarrow Pos^- + Neg^+$$

which is also equal to the charge Q transferred times the change ΔV in potential, in volts:

(E3) $$\Delta G = Q\Delta V.$$

For real substances $\Delta V < {\sim}4$ V.[1]

---

[1] E.g., Ref. 66 in main text.

have been around since the 1960s, only recently have they begun attracting serious attention outside high-value niche markets such as spacecraft. Indeed, interest has exploded in the last decade or so, as evidenced by the enormous and growing literature devoted

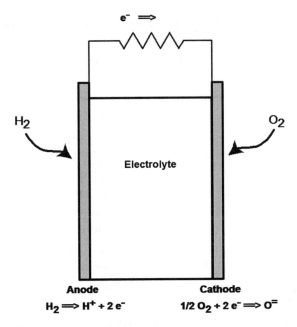

Figure 5.1 Fuel cell.

Fuel cells convert the energy of a redox reaction directly into electricity, in essence by intercepting the flow of electrons from the oxidant to the reductant to drive an electrical circuit. For illustration, in this diagram the reductant (fuel) is H_2 while the oxidant is atmospheric O_2, but conceptually any fuel would work similarly. The practical issue lies in the electrodes, which must catalyze the ionization of oxidizer and fuel without significant degradation. In any fuel cell, an ionic flow in the electrolyte must also compensate the external electron flow.

to them,[18] motivated by their potential as a clean and efficient technology.

---

[18]E.g., Jensen, JO; Li, Q. Fuel cells. In *Hydrogen Technology: Mobile and Portable Applications*. Léon, A; ed. GE&T, 151–84, 2008; Herring, AM; Zawodzinski, TA, Jr; Hamrock, SJ; eds. *Fuel Cell Chemistry and Operation*. ACCSS *1040*, 2010; Corbo, P; Migliardini, F; Veneri, O. *Hydrogen Fuel Cells for Road Vehicles*. GE&T, 2011; Basualdo, MS; Feroldi, D; Outbib, R. *PEM Fuel Cells with Bio-Ethanol Processor Systems: A Multidisciplinary Study of Modelling, Simulation, Fault Diagnosis and Advanced Control*. GE&T, 2012; McPhail, SJ; Cigolotti, V; Moreno, A. *Fuel Cells in the Waste-to-Energy Chain: Distributed Generation Through Non-Conventional Fuels and Fuel Cells*. GE&T, 2012; Guerrero-Lemus, R; Martínez-Duart, JM. Fuel cells. In *Renewable Energies and CO₂: Cost Analysis, Environmental Impacts and Technological Trends- 2012 Edition*. LNE *3*, 289–306, 2013.

Traditionally, working fuel cells have been beset with several problems: high cost due to fabrication issues, including the use of precious-metal catalysts; limited lifetimes; and restriction to hydrogen as the fuel. These problems are due both to the limited current capabilities in nanofabrication and to inadequate catalysts. (For convenience, the term *catalyst* will be used here instead of *electrocatalyst*.) As might be expected, converting chemical energy to electricity without thermalizing it requires nanostructuring, as the fuel and oxidizer must combine in a highly controlled manner. In other words, highly organized matter yields highly organized (i.e., non-thermal) energy. In fact, the failure of 19th-century attempts to convert coal electrochemically into electricity, as well as the long delay in commercializing fuel cells, stem from the need for better catalysts, and probably, even if working catalysts existed, the further need for their cheap nanofabrication.

Another motivation, besides efficiency, for alternatives to thermal engines is the global push to minimize vehicle emissions. In effect, "zero-emission" rules mandate electric vehicles, and that is giving a big boost to fuel cells.

The two main issues are first, better catalysts, which ideally would be

- tolerant of common contaminants such as carbon monoxide (CO) and sulfur species;
- work at room temperature;
- work with carbon-containing fuels, not just hydrogen;
- cheap!

Unsurprisingly, no current catalysts fit these desiderata, most especially the last.

The second issue is better internal electrolytes. Fuel cells require that ions be conducted between anode and cathode to compensate for the electron flow in the external circuit, and ideally this flow would occur in the solid state.

These two issues will be elaborated below.

Fuel cell catalysts

The nanotechnological prospects for catalysts were overviewed at the beginning of this chapter. Here just the issues relevant to fuel cells are highlighted. First, the lower the operating temperature of the cell, the greater the theoretical thermodynamic efficiency possible, not to mention collateral improvements in safety and engineering convenience. Unfortunately, the lower the temperature, the more difficult the catalysis. Moreover, *two* different catalysts are required: one at the anode, where the fuel is oxidized, and the other at the cathode, where atmospheric oxygen (O_2) is reduced. Both must be optimized for good performance. Of course, the need for two different catalysts adds complexity, although since all practical fuel cells are likely to use oxygen at the cathode, that catalyst system should have wider applicability, depending on such factors as the operating temperature of the cell.

Unfortunately, conventional catalysts for low-temperature fuel cells, both at the anode and cathode, use platinum-group metals (PGMs), and indeed these are often a major part of the fuel cell cost.[19] The oxygen-reduction reaction is particularly difficult and typically requires a high overvoltage to achieve adequate reaction rates, again especially at low temperatures. One approach to minimizing platinum-group metal usage is nanostructuring of the interface so that nanoscale precious-metal particles are deployed most effectively, such that their effective surface area is maximized.[20] Even better, of course,

---

[19]Heinzel, A; König, U. Nanotechnology for fuel cells. In *Nanostructured Materials for Electrochemical Energy Production and Storage*, Leite, ER, ed. NS&T, 151–83, 2009; Ref. 77.

[20]E.g., Yazici, MS. Nanoscale materials for hydrogen and fuel cell systems. In *Functionalized Nanoscale Materials, Devices and Systems. NATO Science for Peace and Security Series B*, 283–90, 2008; Lu, Y; Chen, W. 1D Pd-based nanomaterials as efficient electrocatalysts for fuel cells. In Ref. 1, pp. 321–57.

would be alternative catalysts that *don't* use rare elements, and a great deal of research is being directed along this line.[21] Such catalysts will unquestionably require nanostructuring.

Because they use specific molecular mechanisms, non-thermal converters like fuel cells are also much more "picky" about their reactants than is combustion. Hence different catalysts are needed for different fuels, which is an additional source of complexity. Indeed, the reason that present-day low-temperature commercial fuel cells are restricted to hydrogen is that currently it is the only fuel for which practical catalysts exist. Hydrogen is often touted as a "next generation" fuel, and it definitely has advantages, but it has serious disadvantages as well (Box 5.3).

Liquid fuels would be more compatible with current infrastructure, being also considerably easier to store and handle. Hence a great deal of research is currently taking place on non-hydrogen fuel cells. Conventional liquid fuels contain carbon, however, which proves difficult to react in a fuel cell. Fuels containing a carbon–carbon (C–C) bond are especially recalcitrant; indeed, no low-temperature catalyst for cleaving C–C bonds has been found. Carbon monoxide (CO) is also an inevitable intermediate product of carbon oxidation, and it tends to "poison" the catalysts; i.e., deactivate the catalytic action by binding to and blocking the active sites. Sulfur compounds, which are common impurities in petroleum-derived fuels, also tend to be poisons.

One way around these issues is by "reforming" the fuel: i.e., processing it first to extract and purify the hydrogen. This obviously adds complexity and decreases efficiency, especially if done on-site with the fuel cell, rather than as part of the manufacturing process for the hydrogen. Such reactions, of

---

[21]E.g., Martin, KE; Kopasz, JP; McMurphy, KW. Status of fuel cells and the challenges facing fuel cell technology today. In Herring et al., in Ref. 18, pp. 1–13; Ferrell, JR., III; Herring, AM. Metal oxides and heteropoly acids as anodic electrocatalysts in direct proton exchange membrane fuel cells. In Herring et al., in Ref. 18, pp. 153–77; Shao, M; ed. *Electrocatalysis in Fuel Cells: A Non- and Low- Platinum Approach* LNE 9, 2013.

Box 5.3   A Note on the "Hydrogen Economy"

The "hydrogen economy" has been a subject of discussion since the early 1970s, and hydrogen (H_2) does have some clear advantages as a fuel. It is easy to react in fuel cells; in fact, nearly all current commercial fuel cells are hydrogen-fueled. It can also be used in conventional internal combustion engines with minimal modification, which would help ease the transition away from petroleum fuels. The combustion product is also just water, so that the exhaust is nonpolluting. It may also provide a way to store electricity, via electrolysis of water, although despite the tabletop demonstrations of water electrolysis, the process is still too inefficient to be economic, as discussed in the main text.

Hydrogen does have, however, serious disadvantages. First, like electricity, hydrogen (H_2) is not an energy *source* but an energy *carrier*. Naïve politicians notwithstanding, it is not a replacement for oil. It must be made with an input of energy, and the source of that energy is problematic. (To be sure, this is true also of other next-generation fuels such as methanol.) Indeed, most hydrogen currently is made from natural gas. Alternative sources from biomass are being pursued, but are as yet minor.

Second, because it's the lightest gas, hydrogen is extremely bulky. Even though it has roughly 3 times the energy of a hydrocarbon fuel per *kilogram*, a kilogram of H_2 gas at standard temperature and pressure (STP) occupies about 11 cubic meters, or a cube some 2¼ meters on a side. This yields a very low energy density, roughly 30% the value of methane, volume for volume. Its extremely low boiling point ($\sim -253°C$), the second-lowest of any substance after helium, also means that transporting it in liquid form is impractical. By comparison, the temperature of LNG (p. 49) is downright balmy. The Space Shuttle could be run on liquid hydrogen, but it is not reasonable for an automobile!

These transportation and storage issues are exacerbated by the fact that hydrogen cannot be stored in conventional tanks nor moved in conventional pipelines. Not only does the small hydrogen molecule readily leak through gaps that are impervious to larger molecules, but many conventional structural metals take up hydrogen, which leads to embrittlement, cracking, and ultimately failure. As is discussed in the main text, there are a number of nanotechnological approaches to hydrogen storage, and—as emphasized elsewhere—one must beware of mistaking the limitations of current engineering approaches for laws of nature. Nonetheless, having to replace the entire fuel transportation and distribution infrastructure with hydrogen-proof materials would not help ease the transition to non-petroleum fuels.

course, also need catalysts, and indeed the issues are much the same as with syngas (Box 5.4).

Because it has only a single carbon atom, methanol (Box 5.5) has been a particular focus of attention, since the issue of cleaving carbon–carbon bonds does not arise.[22] Indeed, direct methanol fuel cells (DMFCs) are commercially available, although their efficiency is still low. Much effort is now focused on finding better (and cheaper) catalysts.[23] Indeed, some current DMFCs operate on a methanol–water mixture, in which the methanol reacts directly with the H_2O at the anode to yield hydrogen and CO_2.[24] Thus reforming of the mixture takes place at the anode, while the hydrogen alone reacts in the cell at the cathode to form H_2O.

[22]E.g., Wasmus S; Kuver A; Methanol oxidation and direct methanol fuel cells: a selective review. *J. Electroanal. Chem.* 461, 14–31, 1999; Olah, GA; Goeppert, A; Surya Prakash, GK. *Beyond oil and gas: The Methanol Economy*, 2nd. ed. Wiley, 2009, 207–12.

[23]E.g., Yu, J-S; Kim, M-S; Kim, M-S. Combinatorial discovery of new methanol-tolerant non-noble metal cathode electrocatalysts for direct methanol fuel cells. *Phys. Chem. Chem. Phys.* 12, 15274–281, 2010; Shao, in Ref. 21; Kakati, N; Maiti, J; Lee, SH; Jee, SH; Viswanathan, B; Yoon, YS. catalysts for direct methanol fuel cells in acidic media: Do we have any alternative for Pt or Pt-Ru? *Chem. Rev.* 114, 12397–429, 2014.

[24]Olah et al., Ref. 22.

Box 5.4 Synthesis Gas

Synthesis gas ("syngas") is a mixture of carbon monoxide (CO), carbon dioxide (CO_2) and hydrogen (H_2), and a host of simple carbon-containing compounds can be made from this mixture. Indeed, most simple feedstock organic chemicals, such as methanol (CH_3OH; Box 5.5), acetic acid (CH_3COOH), formic acid (HCOOH), and so on are now made by catalytic processing of syngas.[1] Even higher hydrocarbons can be synthesized, via the so-called "Fischer–Tropsch" process (Box 5.6).

The products obtained depend on the composition of the syngas (the H_2/CO and CO/CO_2 ratios), the temperature, the pressure—and the catalysts used. Although syngas synthesis is a reasonably mature technology, it depends critically on the effectiveness of the catalysts, both as to their selectivity as well as to their efficiency in generating the final products. However, the search for new catalysts is still largely empirical, and their fabrication even more so. This is an outstanding example of where better catalysts could have a large effect on energy economics. The conventional processes for making syngas are also thermally driven and thus intrinsically inefficient, even though some of the applied energy is recovered as the stored energy of the products.

Natural Gas

Called "steam reforming" or "steam cracking," the reaction of methane (from natural gas) and steam is the most common way of making synthesis gas:

$$CH_4 + H_2O \rightarrow CO + 3\ H_2.$$

Obviously this process yields a hydrogen-rich mixture, and in fact this is also the current source of most hydrogen gas.

---

[1]E.g., Klier, K. Methanol synthesis. *Adv. Catal. 31*, 243–313, 1982; Sneeden, RPA. Preparation and purification of carbon dioxide and carbon monoxide. *Compreh. Organomet. Chem. 8*, 1–17, 1982; Reactions of carbon dioxide. *Compreh. Organomet. Chem. 8*, 225–83, 1982; Olah et al., in Ref. 22 main text, pp. 239–42.

An alternative approach is CO_2 or "dry" reforming:

$$CH_4 + CO_2 \Rightarrow 2CO + 2H_2$$

which has the advantages of both consuming carbon dioxide and yielding a high carbon to hydrogen ratio in the gas. Unfortunately, current catalysts are not economic, although research is continuing.[2]

Coal gasification

Synthesis gas can also be made from coal via the so-called "water-gas" reaction:

$$C + H_2O \rightarrow CO + H_2. \text{ ("water gas")}$$

This is the basis of approaches to coal gasification. It was also much used for fuel gas generation a century or so ago, but now has been largely supplanted by natural gas. Obviously it is quite toxic because of the carbon monoxide content, and this was how Victorian heroines could kill themselves by "turning on the gas."

Biomass

Pyrolysis (baking in absence of air, formerly called "destructive distillation") of organic matter, including waste biomass, is another route to syngas that continues to attract attention.[3] As another thermal technology, however, it is unlikely to be long-term solution. As is described elsewhere, non-thermal approaches to processing biomass seem more promising in the long term.

---

[2]E.g., Hu, YH. Advances in catalysts for CO_2 reforming of methane, 155–74; Liu, C; Shi, P; Jiang, J; Kuai, P; Zhu, X; Pan, Y; Zhang, Y. Development of coke resistant Ni catalysts for CO_2 reforming of methane via glow discharge plasma treatment, 175–80; Guo, J; Zhang, A-J; Zhu, A-M; Xu, Y; Au, CT; Shi, C. A carbide catalyst effective for the dry reforming of methane at atmospheric pressure, 181–96, all in ACSSS *1056*, 2010.

[3]E.g., Jones, JL; Radding, SL; eds. *Thermal Conversion of Solid Wastes and Biomass*. ACSSS, 130, 1980; Puigjaner, L; ed. *Syngas from Waste: Emerging Technologies*. GE&T, 2011.

Finally, whatever its source, the CO/CO_2 ratio in the syngas is typically varied by the water-gas shift reaction:

$$CO_2 + H_2 \leftrightarrow CO + H_2O.$$

to optimize the composition for different products. It obviously requires the addition of carbon dioxide. As noted in Chapter 2, though, the use of permselective membranes to separate the gases, particularly the hydrogen, is also of growing importance.

This obviously leads to efficiency losses, as well as to placing new demands on the anode catalyst. Catalysts for methanol reaction should also have application in methanol synthesis.

Because ethanol is expected to be an abundant product of next-generation waste biomass processing (Box 5.8), another "wish list" item would be a direct-ethanol fuel cell. It would also have the advantage of lower toxicity than methanol, or liquid hydrocarbons for that matter, and such cells continue to be the subject of much research.[25] Moreover, purifying ethanol derived from fermentation is a major expense for ethanol to be used as fuel. Indeed, if purification is carried out thermally(!), by distillation, there may be a net energy cost.[26] This issue would vanish if dilute ethanol could be used directly in a fuel cell. Indeed, an ethanol–water mixture could reform at the anode, as in a DMFC. As with higher hydrocarbon fuels, however, the carbon–carbon bond is difficult to cleave,[27] even when temperatures $>100°C$ are

[25]Antolini, E. Catalysts for direct ethanol fuel cells. *J. Power Sources 170*, 1, 2007; Antolini, E; Perez, J. Anode catalysts for alkaline direct alcohol fuel cells and characteristics of the catalyst layer. In Shao, Ref. 21, pp. 89–127.

[26]E.g., Giampietro, M; Ulgiati, S; Pimentel, D. Feasibility of large-scale biofuel production. *BioScience 47*, 587, 1997.

[27]E.g., Shen, S; Zhao, TS; Xu, J; Li, Y. High performance of a carbon supported ternary PdIrNi catalyst for ethanol electro-oxidation in anion-exchange membrane direct ethanol fuel cells. *Energy Environ. Sci. 4*, 1428–33, 2011.

Box 5.5 Methanol

Methanol (a.k.a. methyl alcohol or wood alcohol), the simplest alcohol, is an attractive next-generation fuel. To a chemist an "alcohol" is a hydrocarbon where one, or more, of the hydrogen atoms has been replaced with an oxygen-hydrogen, or "hydroxyl" group. Thus methanol, with formula CH_3OH, can be thought of as derived from the simplest hydrocarbon methane (CH_4) by replacing one hydrogen with an OH group.

Methanol is toxic to humans, but not at low concentrations; indeed, it is present in normal blood at low levels, as well as in a great many foodstuffs. It's also widely present in the environment at low levels and can be metabolized by many organisms. Hence it's much less of a biological hazard than pure hydrocarbon fuels such as gasoline.[1]

It is usable in present-day internal combustion engines with minimal modification; in fact, methanol is the main component of additives that rid fuel of water. Because one end of the molecule looks like a water molecule, whereas the other end looks like a hydrocarbon, methanol, as well as other simple alcohols, is miscible in all proportions with both hydrocarbons and with water. This allows a small amount of water to "dissolve" into a hydrocarbon-methanol mixture. It also means that spills are less ecologically hazardous, because the methanol disperses freely into water where it can be metabolized by microbes. The chemical resemblance to water at the "OH" end of the molecule, however, means that the methanol is more corrosive to metals than are pure hydrocarbons. It is particularly corrosive to aluminum. Methanol is also potentially useful in next generation fuel cells, as described in the main text.

Historically, methanol was obtained by the pyrolysis ("destructive distillation") of wood, which accounts for the name "wood" alcohol. Now, however, methanol is largely made from synthesis gas (Box 5.4). Methanol synthesis

---

[1]E.g., Olah et al., in Ref. 22 main text, pp. 220–5.

directly from H_2 and CO_2, as a product of CO_2 hydrogenation (Box 5.7), is another possibility.

Perhaps the most interesting "wish list" item, however, would be a catalyst or catalytic system that could synthesize methanol directly from methane by its controlled partial oxidation with atmospheric oxygen:

$$2\ CH_4 + O_2 \Rightarrow 2\ CH_3OH,\ \Delta G = -231.8\ kJ/mol.$$

Approaches to such synthesis has been the subject of considerable research.[2] The reaction proves to be difficult to control because even this partial oxidation is significantly exothermic, as shown by the large ΔG value, so that it tends to "run away" to full oxidation. Once again, however, this problem has been solved by biosystems, in particular methane-consuming bacteria.[3]

---

[2]E.g., Olah et al., in Ref. 22 main text, pp. 244–5.
[3]Olah et al., in Ref. 22 main text, pp. 251–2.

used.[28] As biosystems routinely cleave carbon–carbon bonds at ordinary temperatures, no fundamental barrier to such catalysts exists. There may also be a "cultural" reason that contributes to the relative lack of research on cleaving C–C bonds, as organic chemists are typically interested in building up larger molecules, not breaking them apart,[29] although research along this line is increasing.[30]

Finally, a low-temperature fuel cell that could use methane (CH_4), the main component of natural gas, would find application

---

[28]E.g., Otomo J, Nishida S, Takahashi H, Nagamoto H. Electro-oxidation of methanol and ethanol on carbon-supported Pt catalyst at intermediate temperature. *J. Electroanal. Chem.* 615, 84–90, 2008.

[29]Cf. Dong, G. *C-C Bond Activation*. TCC 346, 2014.

[30]E.g., Zheng, Q-Z; Jiao, N. Ag-catalyzed C-H/C-C bond functionalization. *Chem. Soc. Rev.* 45, 4590–627, 2016, and refs. therein.

Box 5.6 Fischer-Tropsch Process

The Fischer-Tropsch (F-T) process provides the pathway to converting syngas (Box 5.4) into hydrocarbons more complicated than methane. The overall reactions ideally are:

$$n \, CO + (2n + 1) \, H_2 = C_n H_{2n+2} + n \, H_2O.$$

It is already finding use as a gas-to-liquids technology for stranded gas.[1] In addition, F-T is basis of converting coal into gasoline, which has been used *in extremis* when petroleum is unavailable, e.g., by Nazi Germany during the latter days of World War 2.

The overall conversion involves a host of sub-reactions, and unsurprisingly, good catalysts are again vital to the viability of the process. It's also worth noting that coal conversion ultimately yields as much CO_2 as if the coal were burned directly, so it does not directly yield a decrease in CO_2 emissions. Moreover, some energy is consumed in the conversion. However, CO_2 generated during the gasification process may prove easier to sequester than that from direct combustion of the coal.

[1]E.g., Böhringer, W; Kotsiopoulos, A; de Boer, M; Knottenbelt, C; Fletcher, JCQ. Selective Fischer-Tropsch wax hydrocracking - opportunity for improvement of overall gas-to-liquids processing. *Stud. Surf. Sci. Catal.* *163*, 345–65, 2007.

in the direct use of natural gas. At present, it is confined to solid-oxide fuel cells, as described below.

Fuel cell electrolyte

As noted above, fuel cells also require an electrolyte between the cathode and anode to conduct ionic species so that they can react, and fuel cells are commonly classified by the nature of this electrolyte (Fig. 5.1). Liquid electrolytes have been used in

Box 5.7 CO_2 Hydrogenation

The direct hydrogenation of carbon dioxide to make reduced compounds would have the dual advantage of storing hydrogen in a denser and more convenient form, while at the same removing, or "fixing" CO_2. Such syntheses are obviously closely akin to those involving syngas (Box 5.4) and Fischer-Tropsch reactions (Box 5.6); indeed, at least formally a H_2-CO_2 mixture can be regarded as a CO-free syngas.

One possible product is methanol[1]:

$$CO_2 + 3\ H_2 \rightarrow CH_3OH + H_2O,\ \Delta G\ (STP) = -9.1\ kJ/mol.$$

This reaction has the advantage of being slightly "downhill" (exothermic), as shown by the negative Gibbs free energy of formation. Other products, including formic acid ($HCOOH$)[2] and higher hydrocarbons[3] have also been investigated. Obviously, good catalysts will be again the key to the practicality of such processes.

---

[1]Jessop, PG; Ikariya, T; Noyori, R. Homogeneous hydrogenation of carbon dioxide. *Chem. Rev. 95*, 259–72, 1995; Olah et al., in Ref. 22 main text, pp. 264–78; Olah, GA; Goeppert, A; Surya Prakash, GK. Chemical recycling of carbon dioxide to methanol and dimethyl ether: from greenhouse gas to renewable, environmentally carbon neutral fuels and synthetic hydrocarbons. *J. Org. Chem. 74*, 487–98, 2009.

[2]Dorner, RW; Hardy, DR; Williams, FW; Willauer, HD. Catalytic CO_2 hydrogenation to feedstock chemicals for jet fuel synthesis using multi-walled carbon nanotubes as support. In Hu, YH; ed., in Ref. 80 (Chapter 6), main text, pp. 125–39.

[3]Himeda, Y. Utilization of carbon dioxide as a hydrogen storage material: Hydrogenation of carbon dioxide and decomposition of formic acid using iridium complex catalysts. In Hu, YH; ed., in Ref. 80 (Chapter 6), main text, pp. 141–53.

Box 5.8 Ethanol

Ethanol (a.k.a. ethyl alcohol or grain alcohol) is the second-simplest alcohol after methanol (Box 5.5). Just as methanol can be thought of as methane with one hydrogen replaced by a hydroxyl (OH) group, ethanol can be regarded as ethane (C_2H_6) with one hydrogen replaced. Ethanol is also the "alcohol" of everyday use, the stuff that makes alcoholic beverages "alcoholic". Hence its use is complicated by legal factors due to the restrictions on, and taxation of, beverage alcohol. As shown by the current prevalence of ethanol-based fuel blends, ethanol, like methanol, can be used directly in internal combustion engines with little modification.

Whereas methanol at present is nearly all made synthetically, from syngas (Box 5.4), ethanol is still largely made by fermentation of sugars through essentially the same processes by which alcoholic beverages are made. This means that competition with foodstuffs is potentially an issue in conventional ethanol production. Ethanol is also less suitable for fuel cells, because reacting the carbon–carbon bond leads to the same difficulties as with hydrocarbon fuels. Conversely, however, it is less toxic.

A promising and so-far embryonic approach to ethanol manufacture, however, is the fermentation of sugars derived from cellulosic material (Box 5.9), which constitutes a large percentage of waste biomass. This, of course, fits into the ongoing theme of "waste into resource," and better catalysts will no doubt be key.

"traditional" fuel cells. Examples include the "classical" alkaline cell, used on the Apollo missions, in which the electrolyte is aqueous potassium hydroxide (KOH) operating at ~70°C.[31] Phosphoric acid and molten carbonate fuel cells are not poisoned

---

[31]Hamnett, A. Fuel cells and their development. *Philos. Trans. Roy. Soc. Lond. A 354*, 1653–69, 1996.

Box 5.9 Lignocellulosic Materials

This constitutes the bulk of biomass, and is "woody matter," broadly speaking. It includes a swath of biomass ranging from waste paper to cornstalks to sewage sludge to grass clippings to scrap lumber. The details vary, but it's largely composed of *cellulose* (and related compounds such as hemicellulose), and *lignins*.[1] Such materials, of course, have been used as fuel since the discovery of fire, but the raw fuel is also not high-quality in terms of its energy content. It also must be dried, as anyone who's built a fire knows.

Lignins have a higher ratio of carbon and hydrogen to oxygen than cellulose, enough so that they are hydrophobic (i.e., water-repelling).[2] Indeed, 6-membered aromatic hydrocarbon rings (benzene rings) are a common structural group. They are complex polymers whose biosynthesis is still not fully understood,[3] and have historically not attracted as much attention as potential raw materials, although that is changing.

Cellulose, by contrast, has received a great deal of attention since it was first characterized in 1838. It is abundant, being the most common biopolymer on the planet. It is a polysaccharide, that is, a macromolecule made up of repeating sugar units, in this case glucose.[4] Glucose is a simple sugar that is much more familiar as a basic foodstuff, and it's unexpected to find that sugar is a major component of wood. Indeed, starch, an important foodstuff, is also a glucose polymer. In starch, however, the bonds between the glucose units, so-called "alpha linkages," are easily broken to yield

---

[1]Zhu, H; Luo, W; Ciesielsk, PN; Fang, Z; Zhu, JY; Henriksson, G; Himmel, ME; Hu, L. Wood-derived materials for green electronics, biological devices, and energy applications. *Chem. Rev. 116*, 9305–74, 2016.

[2]E.g., Yin, SY; Lebo, SE, Jr. Lignin. KOECT, 4th ed. *15*, 137–47, 2001.

[3]E.g., Boerjan, W; Ralph, J; Baucher, M. Lignin biosynthesis. *Annu. Rev. Plant Biol. 54*, 519–46, 2003.

[4]E.g., French, AD; Bertoniere, NR; Battista, OA; Cuculo, JA; Gray, DG. Cellulose. KOECT, 4th ed. *5*, 236–45, 2001.

individual glucose molecules. The "beta-linked" glucose units in cellolose, by contrast, are much more tightly packed and require more extreme chemical measures. As usual, biosystems are ahead of technology, but even the biosystems are more limited. Cellulases, enzymes capable of cleaving cellulose, are largely restricted to microbes and fungi, and organisms that digest cellulose, such as ruminants, have microbial symbionts in the gut that carry out the actual breakdown of the cellulose. In any case, cellulases are attracting a great deal of research attention[5] and furnish another example of the biological inspiration.

Technological cellulose solvents have been more extreme, including such things as strong alkalis or acids, or highly concentrated solutions of salts such as zinc chloride. Cellulose-based products from such processes have been important for decades—e.g., Cellophane, which is one of the first plastics and dates back to the early 20th century, is reconstituted cellulose, and the importance of paper and paper products hardly needs emphasis. Nonetheless, the complete saccharification (breakdown into glucose), of cellulosic material into a form that could (say) be fermented into ethanol has so far not been economic, not least because the corrosive reagents used to break up the cellulose would need to be separated out afterward.

---

[5] E.g., Golan, AE, ed. *Cellulase: Types and Action, Mechanism, and Uses.* Nova Science Publications (Series on Biotechnology in Agriculture, Industry and Medicine), 2011; Gilbert, HJ; Ed. *Cellulases* Academic Press (Methods in Enzymology *510*), 2012.

by carbon dioxide as is the alkaline cell, but they also operate at considerably higher temperatures.

Although they have the virtue of simplicity, liquid electrolytes have serious deficiencies. Corrosion of the electrodes places major constraints on their possible compositions, and high-temperature operation also decreases thermodynamic efficiency. Containing hot, corrosive liquid also does not decrease the engineering difficulties.

All-solid-state fuel cells are thus attractive. They require a *solid electrolyte*, a substance that conducts ions while remaining in the solid state.[32] For example, conventional hydrogen fuel cells operating below 100°C incorporate a proton-permeable membrane, which conducts protons from anode to cathode. The material used in commercial units is Nafion, a solid fluorocarbon polymer. It is hardly ideal, being expensive and subject to degradation through dehydration. As with any conventional polymer, too, there is hardly nanoscale control in its synthesis (p. 149). The membrane pores are not uniform in size or distribution, and the surface active sites involved in proton binding and release are haphazardly exposed. Thus there is much effort directed toward finding alternative proton-conducting materials.[33] Leakage of fuel across the membrane—"fuel crossover"—is also a problem, particularly with DMFCs.[34]

Recently, alkaline fuel cells using *anion*-exchange membranes have received attention.[35] These are selectively permeable to the

---

[32]Rickert, H. *Electrochemistry of Solids: An Introduction.* Springer-Verlag, 1982.

[33]E.g., Peled, E; Duvdevani, T; Melman, A. A novel proton-conducting membrane. *Electrochem. Solid-State Lett.* 1, 210, 1998; Wang, X; Wang, S. Nanomaterials for proton exchange membrane fuel cells. In *Energy Efficiency and Renewable Energy Through Nanotechnology*, Zang, L; ed. GE&T, 393–424, 2011; Tang, H; Li, J; Wang, Z; Zhang, H; Pan, M; Jiang, SP. Self-assembly of nanostructured proton exchange membranes for fuel cells. In *Nanotechnology for Sustainable Energy*, Hu, YH; Burghaus, U; Qiao, S; eds. ACSSS *1140*, 243–63, 2013; Kumar, SMS; Pillai, VK. Low-Cost Nanomaterials for High-Performance Polymer Electrolyte Fuel Cells (PEMFCs), 359–94, in Ref. 1; Sambasivarao, SV. Thermal stability and ionic conductivity of high-temperature proton conducting ionic liquid-polymer composite electrolyte membranes for fuel cell applications. In *Polymer Composites for Energy Harvesting, Conversion, and Storage*, Li, L; Wong-Ng, W; Sharp, J; eds. ACSSS *1161*, 111–26, 2014.

[34]Olah et al., Ref. 22, p. 208; Ref. 35.

[35]E.g., Pawar, MS; Zha, Y; Disabb-Miller, ML; Johnson, ZD; Hickner, MA; Tew, GN. Metal-ligand based anion exchange membranes. In *Polymer Composites for Energy Harvesting, Conversion, and Storage*, Li, L; Wong-Ng, W; Sharp, J; eds. ACSSS *1161*, 2014, and refs therein; Varcoe, JR; Atanassov, P; Dekel,

hydroxyl ion (OH⁻), which is what makes alkaline solutions "alkaline," and are both more robust and less prone to carbon dioxide poisoning than conventional alkaline cells. Catalysis in an alkaline environment is also less demanding.[36] In any case, the issues for membrane-based solid electrolytes are much as for permeable membranes in general.

Other solid electrolytes are 3D crystal structures. These work by having many ionic site vacancies: under the influence of an electric field an adjacent ion can hop into the vacancy, leaving a new vacancy behind. Thus in effect the vacancy moves, analogous to the motion of a "hole" in a semiconductor. Such a mechanism also minimizes or eliminates fuel crossover. Obviously (and unfortunately), however, this hopping works best at high temperature, where the thermal motion of the ions makes them easier to shift.

For example, one of the most practical solid electrolytes is at present doped zirconium dioxide (ZrO_2), which is the basis of conventional "solid-oxide" fuel cells (SOFCs). The oxide conduction is only effective at temperatures >600°C,[32] and practical SOFCs typically operate at temperatures >700°C.[37] The crystal consists of close-packed oxide (O^{2-}) ions, with the 4-positive zirconium cations (Zr^{4+}) in the interstices. Substitution of (e.g.) ~10% 3-positive yttrium cations (Y^{3+}) leaves some oxide sites vacant to preserve charge balance. Oxygen anions can therefore migrate toward the anode by shifting into these vacancies. In this case, too, the oxygen anions are obviously the mobile species.

DR; Herring, AM; Hickner, MA; Kohl, PA; Kucernak, AR; Mustain, WE; Nijmeijer, K; Scott, K; Xu, T; Zhuang, L. Anion-exchange membranes in electrochemical energy systems. *Energy Environ. Sci.* 7, 3135–91, 2014.

[36]E.g., Chen, R; Guo, J; Hsu, A. Non-Pt cathode electrocatalysts for anion-exchange-membrane fuel cells. In Shao, in Ref. 21, pp. 437–81; Chen, X; Xia, D; Shi, Z; Zhang, J. Theoretical study of oxygen reduction reaction catalysts: From Pt to non-precious metal catalysts. In Shao, in Ref. 21, pp. 339–73.

[37]Suzuki, T; Yamaguchi, T; Hamamoto, K; Fujishiro, Y; Awano, M; Sammes, N. A functional layer for direct use of hydrocarbon fuel in low temperature solid-oxide fuel cells. *Energy Environ. Sci.* 4, 940–43, 2011.

Such high-temperature operation obviously severely limits potential applications, increases engineering difficulties, and also decreases the ultimate thermodynamic efficiency. Nonetheless, commercial SOFCs already exist.[38] In addition, high-temperature operation has the advantage of less demanding catalysis. Not only is this because catalysis is easier through activation energies being lower, but also because catalytic poisons such as carbon monoxide and sulfur compounds tend to desorb more readily. Moreover, reforming carbon-bearing fuels becomes easier; indeed, a wide variety of carbon-bearing fuels can be handled.[39] Overall, current SOFCs seem to be a sort of a "halfway house": more efficient than combustion, and more tolerant of different fuel mixtures, but overall still close to a thermal technology.

The desire for lower temperature operation continues to motivate searches for alternative catalysts, in particular ones not requiring noble metals,[40] and inevitably such approaches require nanostructuring.[41] Alternative oxide electrolytes have also been investigated, such as the *perovskite* $(La,Sr)(Mg,Ga)O_3$ (LSMG) (Box 5.10) and cerium oxide-based compositions,[39] but so far they are still experimental.

Another possibility may be solid electrolytes based on molecular sieves (Box 6.3), which have already mentioned in connection with catalysis. The uniform molecular or supramolecular-scale channels traversing these crystals may be amenable to cation transport, especially for small cations such as

[38]Milewski, J; Swirski, K; Santarelli, M; Leone, P. *Advanced Methods of Solid Oxide Fuel Cell Modeling.* GE&T, 2011, esp. review, pp. 1–16; Irvine, JTS; Connor, P. *Solid Oxide Fuels Cells: Facts and Figures: Past Present and Future Perspectives for SOFC Technologies.* GE&T, 2013.

[39]Singhal, SC. Solid oxide fuel cells: Past, present and future. In Irvine & Connor, Ref. 38, pp. 1–23.

[40]E.g., Ref. 37; Ormerod, RM. Solid oxide fuel cells. *Chem. Soc. Rev. 32*, 17–28, 2003.

[41]E.g., Ruiz-Morales, JC; Canales-Vázquez, J; Savaniu, C; Marrero-López, D; Núñez, P; Zhou, W; Irvine, JTS. A new anode for solid oxide fuel cells with enhanced OCV under methane operation. *Phys. Chem. Chem. Phys. 9*, 1821–30, 2007.

> ## Box 5.10 The Perovskite Structure
>
> Perovskite (cubic calcium titanium oxide, $CaTiO_6$) is a rare mineral with an important structure. Each titanium atom is surrounded by six oxygen atoms at the corners of an octahedron, and each vertex of an octahedron is shared with another octahedron (Fig. 5.2). The calcium atoms fit in the large voids left between the octahedra. "Perovskite" is also used generically for a substance having this structure.
>
> One reason for the (nano)technological importance of the perovskites lies in their very open structure. In pure tungsten trioxide (WO_3), for example, the large cavities are unoccupied, the so-called "rhenium trioxide (ReO_3)" structure, although in such cases the cubic symmetry of the ideal structure is often distorted. In any case, other ions can be intercalated into the framework; this is the basis for electrochromism in tungsten bronzes, as described in the main text. Alternatively, such open frameworks could serve as solid electrolytes, whose importance has also been described in the main text. They also potentially have a host of applications in catalysis, and perovskite structures are attracting attention as potential replacements for precious-metal catalysts.
>
> In other cases, the small metal cations in the octahedral sites can be displaced by an applied electric field. This in turn leads to high polarizability and thus a high dielectric constant, which is desirable for high-performance capacitors.

hydrogen ion (H^+). Unfortunately, currently accessible structures are impractical, as they have conductivities several orders of magnitude lower than alternative ionic conductors.[42] Part of the problem seems to be that the sieves (zeolites) tested were polycrystalline aggregates consisting of tiny (~1 μm) unoriented crystals, so the bulk conduction properties probably reflect a

---

[42]E.g., Kelemen, G; Lortz, W; Schoen, G. Ionic conductivity of synthetic analcime, sodalite and offretite. *J. Mater. Sci.* 24, 333–8, 1989.

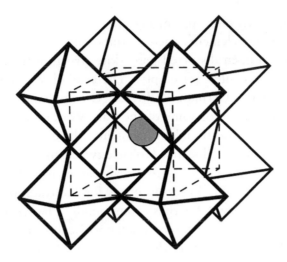

Figure 5.2 Perovskite structure.

An oxygen anion lies at each vertex shared between the octahedra. In tungsten trioxide (WO_3), a tungsten atom lies at the center of each octahedron (at the positions connected by the dashed cube), while the large site (the shaded circle) between the octahedra is empty. In tungsten bronzes, though, some of the large sites are occupied by the additional metal atoms necessary to maintain charge balance. In the case of strontium titanium oxide ($SrTiO_3$, "strontium titanate"), the titanium atoms occupy the octahedral centers, while the strontium cation occupies the large site. The titanium cations can "rattle around" to some degree in the octahedral sites, which leads to $SrTiO_3$'s high polarizability and hence its utility as a capacitor dielectric. In WO_3 and many other perovskites, the perfect cubic symmetry is distorted. This occurs especially when the large site is unoccupied.

great deal of grain-boundary and surface effects. Hydration of the cations can also increase their effective size and thus decrease their mobility significantly.[43] In any case, nanofabrication of larger crystals, as is already desirable for other purposes, may motivate a new look at these materials.

---

[43]Rees, LVC. Zeolites. In *Solid Electrolytes: General Principles, Characterization, Materials, Applications*. Hagenmuller, P; Van Gool, W; eds. 417–28. Academic, 1978.

Perhaps even more promising are open frameworks formed by redox-active elements such as tungsten. Tungsten trioxide, WO_3, has an open distorted perovskite structure (Box 5.10) in which the large sites are empty. (The symbol W for tungsten comes from the German Wolfram.) On applying a negative potential, reduction of some of the framework atoms from W^{6+} to W^{5+} puts the additional electrons into the conduction band, where they become delocalized to yield typical metallic behavior. The material thus changes from transparent to reflective, and the resulting metallic luster accounts for the traditional name of "tungsten bronze" for these compounds. Charge balance is maintained by intercalation of small positive ions. Reversing the applied potential reoxidizes the tungsten to W^{6+}, with expulsion of the cations and loss of reflectivity. Such an intercalation mechanism seems potentially relevant to solid electrolytes. It is also the basis of electrochromism (p. 217); furthermore, lithium batteries (p. 221) work on a similar principle, as do potential approaches to separation of solutes from aqueous solution (p. 302).

Alternatively, in an inversion of perspective, perhaps the interfaces themselves can instead provide the conductive channels.[44] In any case, these is little doubt that nanoscale fabrication of solid electrolytes will be critical for practical low-temperature fuel cells.

Superstrength Materials

As will be discussed in Chapter 6 (Box 6.5), a result from elementary materials science is that macroscopic materials are far weaker, typically by 1–2 orders of magnitude, than would be inferred from the strength of the chemical bonds making them up. One of the most promising applications of nanofabrication

---

[44]Skinner, SJ; Cook, S; Kilner, JA. Materials for next generation SOFCs. In Irvine & Connor, Ref. 38, pp. 181–201.

is the construction of materials that are essentially defect-free at the atomic level, such that their strength approaches the limit set by chemical bonds. (Such capabilities will also have profound implications for the mix of desired raw materials, as also discussed at length in Chapter 6, p. 323). Because strong materials involve no molecular moving parts, they are also likely to be accessible in the nearer term as their fabrication need not be carried out by full-blown molecular machines.

The stronger a material, the less required for a given application, and the less energy required to move it. This energy savings is likely to have its greatest effect on transportation by decreasing the weight of vehicles. Even though the sheer volume of transportation, particularly for bulk commodities, is likely to dwindle as nanotechnology matures (cf. p. 233), travel is unlikely to vanish completely.

Superstrength materials have many obvious applications in resource extraction. OTEC, for example, would profit from long cables that must hang under their own weight, to reach the sea floor, and yet still be strong enough to withstand the storm risk. At least in the nearer term, such materials also have obvious applications in turbine and windmill blades. Besides the obvious improvements in efficiencies of transport, drilling will benefit particularly. The "drill string," the column of rotating pipe tipped by a diamond-studded bit for "making hole," could be considerably lighter, with consequent energy savings in its operations. Additionally, much greater depths could become routinely accessible, although the destruction of permeability at great depth makes the value of this problematic, at least for hydrocarbon exploration. Strong materials are even more important for geothermal drilling because of the high temperatures encountered. Of course, the fabrication of cheap, megascopic diamond, which is commonly envisioned as a result of nanotechnology, is also going to make drill bits much cheaper.

Passive Energy Handling

As noted in Chapter 2, passive energy handling is also not so glamorous as the exploitation of raw energy sources. Nonetheless, it could make a significant contribution to the energy mix. At present, for example, some 6% of the electricity consumption in the US is for air conditioning.[45]

Smart Materials

A major part of the climate control costs in modern buildings simply stem from waste heat, mostly dumped by the lighting system. Conventional incandescent lights exemplify the inefficiencies of the thermal paradigm. They are *considerably* better heaters than illuminators. A filament at 2500 K, assuming blackbody radiation, radiates ~96% of its output in the infrared. Even conventional fluorescent lights dump a great deal of waste heat. Thus, simply converting electricity into visible light with greater efficiency would yield significant energy savings, both directly in minimizing illumination costs, and indirectly in minimizing the costs of dealing with waste heat.

Such considerations, of course, have motivated the recent push toward replacing incandescent lights with low-energy alternatives, including light-emitting diodes (LEDs) that emit white light, and compact fluorescent bulbs. Of course, as the retail consumer now knows, these replacements are also considerably more expensive than traditional light bulbs. The reason for the expense is that they are electronic components, and therefore structured at considerably smaller scales than an incandescent bulb. Hence much the same issues of microfabrication arise in their manufacture as with any electronics, and as nanofabrication matures their costs will decrease greatly.[46]

---

[45]www.energy.gov/energysaver/air-conditioning

[46]Reggiani, A; Farini, A. LEDs and use of white LED for lighting. In *Sustainable Indoor Lighting*. Sansoni, P; Mercatelli, L; Farini, A; eds. GE&T, 2015; Yam, VWW, *WOLEDs and Organic Photovoltaics: Recent Advances and Applications*. GE&T, 2010.

Unwanted solar heating is of course also important, at least in summer. Using *electrochromic* layers to darken windows has attracted attention for decades.[47] Electrochromic materials change color under an applied electric field. Much research has been directed toward lithium ion intercalation into tungsten trioxide (WO_3),[48] already described above (p. 214). On such intercalation, the window becomes reflective, becoming transparent again when the polarity is reversed. Obviously, constructing such a window, particularly of significant area, presents a nanofabrication challenge.

Other electrochromic materials are under investigation, including organic compounds, polymers, and metal coordination compounds like hexacyanometallates (including some related to Prussian blue, a traditional ink).[49] Even titanium dioxide (TiO_2) based materials have been investigated, a spinoff from research on the dye-sensitized solar cells described below—one

---

[47]E.g., Baucke, FGK; Duffy, JA. Darkening glass by electricity. *Chem. Brit. 21,* 643–6, 1985; Oi, T. Electrochromic materials. *Annu. Rev. Mater. Res. 16,* 185–201, 1986; Deb, SK. Opportunities and challenges of electrochromic phenomena in transition metal oxides. *Solar Energy Mater. Solar Cells 25,* 327–38, 1992; Reminiscences on the discovery of electrochromic phenomena in transition metal oxides. *Solar Energy Mater. Solar Cells 39,* 191–201, 1995.

[48]E.g., Passerini, S; Scrosati, B; Gorenstein, A; Andersson, AM; Granqvist, CG. An electrochromic window based on lithium tungsten oxide (Li_xWO_3)/ poly(ethylene oxide)lithium perchlorate/(($PEO)_8LiClO_4$) nickel oxide (NiO). *J. Electrochem. Soc. 136,* 3394–5, 1989; Rauh, RD; Cogan, SF. Design model for electrochromic windows and application to the tungsten trioxide/iridium dioxide system. *J. Electrochem. Soc. 140,* 378–86, 1993; Cogan, Stuart F; Rauh, R. David; Klein, James D; Nguyen, Nguyet M; Jones, Rochelle B; Plante, Timothy D. Variable transmittance coatings using electrochromic lithium chromate and amorphous WO_3 thin films. *J. Electrochem. Soc. 144,* 956–60, 1997; Forslund, Bertil. A simple laboratory demonstration of electrochromism. *J. Chem. Educ. 74,* 962–63, 1997; Zhang, Ji-Guang; Benson, David K; Tracy, C. Edwin; Deb, Satyen K; Czanderna, AW; Bechinger, C. Chromic mechanism in amorphous WO_3 films. *J. Electrochem. Soc. 144,* 2022–26, 1997; also Ref. 49; Zhou, D; Che, B; Kong J; Lu, X. A nanocrystalline tungsten oxide electrochromic coating with excellent cycling stability prepared via a complexation-assisted sol-gel method. *J. Mater. Chem. C 4,* 8041–51, 2016.

[49]Mortimer, RJ. Electrochromic materials. *Annu. Rev. Mater. Res. 41,* 2.1–2.28, 2011.

example of synergies in nanoscale research. As with the "wish list" photovoltaics (p. 253), maybe electrochromic material could be applied in paint, or as polymer sheets dispersed off rolls like carpeting.

In "photochromic" systems, the color change is driven instead by photon absorption, a particularly attractive feature for a self-shading window, or "smart window." In the 1960s, swift self-darkening under the sudden intense illumination of a nuclear explosion was also of military interest for protective sunglasses.[50] One type of photochromic glass relies on reversible photodriven redox reactions of dispersed nanoparticles, typically a silver halide such as silver chloride.[51] An elegant new approach, however, combines electrochromic layers, described above, with a layer of photoactive semiconductor.[52] The current generated by the illuminated semiconductor provides the electric potential to drive intercalation and darkening. To be practical, however, such a system obviously requires low-cost fabrication of large-area nanolayered structures.

High–Temperature Superconductors

At present electricity is commonly sent long distances from its place of generation to its place of use. In part this is unavoidable: sites for hydropower dams, for example, commonly lie far from electrical markets. In fact, when they were built in the 1930s Hoover and Grand Coulee dams lay in remote wilderness. In part, however, this is due to the clumsiness of present technology.

[50]Bertelson, RC. Applications of photochromism. In *Photochromism*. Brown, GH; ed. Wiley-Interscience, 733–840, 1971.

[51]Araujo, R. Inorganic photochromic systems. *Mol. Cryst. Liq. Cryst. Sci. Technol., Sect. A 297*, 1–8, 1997.

[52]Kostecki, R; Mclarnon, FR. Photochromic, electrochromic, photo-electrochromic and photovoltaic devices. Patent (US) 6118572, 2000; Huang, H; Lu, SX; Zhang, WK; Gan, YP; Wang, CT; Yu, L; Tao, XY. Photoelectrochromic properties of NiO film deposited on an N-doped TiO_2 photocatalytical layer. *J. Phys. Chem. Solids.* 70, 745–49, 2009, and references therein.

Because of present "economies of scale," thermal power plants must be large, and due to the pollution they generate it is often politically convenient to site them far from the markets they serve. Many coal-fired generation plants in the US, for example, lie hundreds of miles from their urban California markets (e.g., Valmy and Mojave, Nevada; Page, Arizona). Because of issues of safety and land costs, the ground receivers for solar power satellites (p. 265) are also likely to be located far from their ultimate markets.

In any case, the resulting long-distance transmission of electrical current entails significant losses (as much as 10%, depending on distance) due to electrical resistance. When "high-temperature" superconductors (i.e., substances that become superconducting near liquid nitrogen temperatures (~77 K) rather than near liquid helium temperatures (~4 K)) were discovered in the late 1980s, it was immediately predicted they would revolutionize power transmission.[53] This hasn't happened. In part the mechanisms of high-temperature superconductivity remain poorly understood; furthermore, even if they were fully understood, the routine and economical fabrication of high-temperature superconductors would require a great deal of control at the nanoscale.[54] Here again, the large-scale fabrication of nanostructured materials could make a major contribution to energy efficiency.

---

[53]E.g., Lemonick, M. Superconductors! *Time* pp. 64 ff., 11 May 1987.

[54]E.g., Rao, CNR. Novel materials, materials design and synthetic strategies: recent advances and new directions. *J. Mater. Chem. 9*, 1–14, 1999; Müller, KA; Bussmann-Holder, A; eds. *Superconductivity in Complex Systems. Struct. Bonding 114*, 2005; Lebed, AG; ed. *The Physics of Organic Superconductors and Conductors.* SSMS 110, 2008; Plakida, N. *High-Temperature Cuprate Superconductors: Experiment, Theory, and Applications. Springer Ser. Solid-State Sci. 166*, 2010; Askerzade, I. *Unconventional Superconductors: Anisotropy and Multiband Effects.* SSMS 153, 2012; Saxena, AK. *High-Temperature Superconductors.* SSMS 125, 2012; Uchida, S. *High Temperature Superconductivity. The Road to Higher Critical Temperature.* SSMS 213, 2015.

Ubiquitous Sensing

When extracting from a subsurface natural energy source such as an oil or geothermal reservoir, the details of the subsurface structure are obviously of paramount importance. Subsurface information, however, is also extraordinarily difficult to get, particularly in the detail required. Nanotechnology has obvious applications here.

The extraordinary decrease in the cost of computing over the last few decades has already had a significant effect: the cost of domestic petroleum production from 1984 through 2001 dropped from $14 to $4 per barrel, largely through information technologies.[55] Most of this involves the processing and manipulation of seismic data to picture subsurface structure, and this trend is another that will only accelerate as information manipulation becomes ever cheaper.

Another embryonic trend is cheap distributed sensors based on microtechnology, such as downhole thermometry, from fiber-optic cables, and autonomous micro-flowmeters, which are "throwaways" that can be disseminated in the subsurface.[55] This trend will also accelerate as nanotechnology comes on line.

As in so many cases involving the vagaries of natural systems, geothermal energy is both diffuse and additionally not very uniform. Conventionally a geothermal "field" is exploited somewhat similarly to an oil field. Wells are drilled for steam production, based on geologic and hydrologic models of the subsurface. Hence there is a great deal of practical similarity to oil production: because of incomplete information about subsurface structure, fracture connectivity, flow patterns, and so on, drilling "dry holes" (wells producing insufficient steam to be economic) occurs commonly and is a serious risk. Thus, many of the ideas for information intensive exploitation also apply to geothermal power. Extensive downhole monitoring of temperature, to monitor reservoir conditions, is an obvious first step, as are free-flowing microsensors for (say) determining the fate of reinjected water.

---

[55]See Ref. 20 in Ch. 2.

Energy Storage

Of course, commonly energy is not consumed immediately and typically must be stored—consider hydrocarbons in tank farms, for example, and indeed the storability of conventional fuels is a major part of their convenience. Because so many next-generation technologies are focused on generating electricity, the storage of electricity will be a particularly pressing problem. It consists of two parts, of which the first is the less important:

Portable Power Sources

The inadequacy of current batteries for powering the welter of new portable electronic (and electric) devices is well known, and nanotechnological fabrication can certainly contribute here.

Batteries

A battery is an electrochemical device that generates DC through a coupled set of redox reactions (Box 5.2). The energy storage of the battery is limited both by the redox couple, and by the amount of each available for reaction. In practice this latter constraint means very high internal surface areas are desirable, so that as little as possible of the reacting species end up isolated from further reaction by reaction products. If in addition the battery is rechargable, the reaction must be reversible on application of an opposite potential; furthermore, irreversible changes on each charge-discharge cycle must be minimal for long battery life.

Intercalation-based batteries, generally using the small lithium (Li^+) cation, have attracted a great deal of attention in recent years because of their higher energy densities. Indeed, they have made the welter of current portable electronic devices practical.[56] In such batteries, at least one redox-active electrode has an open crystal structure with voids capable of intercalating

---

[56]Maranas, JK. Polyelectrolytes for batteries: Current state of understanding. In *Polymers for Energy Storage and Delivery: Polyelectrolytes for Batteries and Fuel Cells*, Page, KA; Soles, CL; Runt, J; eds. ACSSS, *1096*, 2012.

lithium cations (Li^+). In so-called "rocking chair" batteries,[57] both electrodes can take up Li^+. During discharge Li^+ is expelled from one electrode and taken up by the other.

In the bulk of commercial batteries, for example, oxidation of cobalt (Co) in $LiCoO_2$ expels lithium (Li^+), which is taken up in a graphite anode. Recharge of the battery involves re-reduction of Co with concomitant take-up of Li^+. In effect an intercalation electrode greatly boosts the surface area at which redox reaction can take place, which improves energy storage while also maintaining recharge capability.

Many fabrication issues arise with such batteries.[58] Obviously a redox-active crystal structure capable of *reversibly* intercalating small ions is required. One mechanism for degradation of such batteries is by irreversible changes in the crystal structure on deintercalation, through mechanisms whose molecular details remain somewhat obscure. The crystal structure must also be fabricated with near-nanoscale precision, and increases in nanoscale control will lead to increases in performance.[59] Moverover, alternative nanoarchitectures leading to higher energy density should become accessible.[60]

Not only is a great deal of research continuing on lithium-based batteries,[61] but other battery designs are the subject

---

[57]E.g., Tarascon, JM; Guyomard, D. The lithium manganese oxide ($Li_{1+x}Mn_2O_4$)/carbon rocking-chair system: a review. *Electrochim. Acta 38*, 1221–31, 1993; Thackeray, MM; Ferg, E; Gummow, RJ; de Kock, A. Transition metal oxides for rocking-chair cells. In *Solid State Ionics IV, Mater. Res. Soc. Symp. Proc. 369*, 17–27, 1995; Exnar, I; Kavan, L; Huang, S.Y; Grätzel, M. Novel 2 V rocking-chair lithium battery based on nano-crystalline titanium dioxide. *J. Power Sources 68*, 720–22, 1997.

[58]Gulbinska, MK. *Lithium-ion Battery Materials and Engineering: Current Topics and Problems from the Manufacturing Perspective.* GE&T, 2014.

[59]E.g., Abu-Lebdeh, Y; Davidson, I; eds. *Nanotechnology for Lithium-Ion Batteries.* NS&T, 2013; Osaka, T; Ogumi, Z; eds. *Nanoscale Technology for Advanced Lithium Batteries.* NS&T, 2014.

[60]E.g., Long, JW; Dunn, B; Rolison, DR; White, HS. Three-dimensional battery architectures. *Chem. Rev. 104*, 4463–92, 2004.

[61]Xie, J; Zhang, Q. Recent progress in rechargeable lithium batteries with organic materials as promising electrodes. *J. Mater. Chem. A 4*, 7091–106, 2016.

of investigation.[62] In particular, sodium-based intercalation is attracting interest.[63] Although the sodium cation (Na^+) is larger, sodium is a much cheaper, more abundant, and non-toxic element; indeed, it is the 5th most abundant in the crust (Table 1.1). Lithium is a rare element, as is cobalt, and the latter, although a micronutrient in trace amounts, is toxic in macroscopic quantities.

Capacitors

Capacitors are, of course, an alternative way to store electricity. The energy is stored electrostatically (Box 5.11). Many designs also have the advantage of being able to discharge very quickly, a so-called pulse power capability. This is important in certain applications (e.g., vehicles) where bursts of power are occasionally needed.

Even though capacitors are less efficient than batteries at energy storage *at the same voltage*, the voltage of batteries is limited by the potential of redox reactions. Capacitors are less limited: by Eq. 5.11-1 (Box 5.11), the energy storage goes up as the square of the voltage. However, the voltage across a capacitor cannot increase arbitrarily because at some voltage the capacitor "breaks down" through charge crossing the gap d. Capacitors thus involve a tradeoff between a larger d to maximize the breakdown voltage, and a smaller d to increase capacitance. In electrolytic capacitors, for example, a thin oxide film is formed atop a metal foil by electro-oxidation, and this film serves as the

---

[62]E.g., McCloskey, BD; Burke, CM; Nichols, JE; Renfrew, SE. Mechanistic insights for the development of Li-O₂ battery materials: addressing Li_2O_2 conductivity limitations and electrolyte and cathode instabilities. *Chem. Commun.* 51, 12701–715, 2015, and refs therein.

[63]Yabuuchi, N; Kubota, K; Dahbi, M; Komaba, S. Research development on sodium-ion batteries. *Chem. Rev.* 114, 11636–82, 2014; Kim, H; Hong, J; Park, K-Y; Kim, H; Kim, S-W; Kang, K. Aqueous rechargeable Li and Na ion batteries. *Chem. Rev.* 114, 11788–827, 2014; Xao, Lifen; Yuliang Cao, Jun Liu; Cathode and Anode Materials for Na-Ion Battery; in Ref. 1, pp. 395–424.

Box 5.11 Capacitors

Capacitors, often called "condensers" in the older literature, store energy electrostatically. Ideally, two two-dimensional conductors, separated by a distance d, contain positive and negative charges, and those charges are held in place by their mutual attraction across that distance. In turn, the conductors must be separated by an insulator so that the charges can't actually cross.

The energy of a capacitor is given by:

$$E = \tfrac{1}{2}\, CV^2 = \tfrac{1}{2}\, QV, \qquad (5.11\text{-}1)$$

where C is the capacitance, V the voltage, and Q the charge. Thus, for a given charge and voltage, a capacitor stores half as much energy as a battery.[1]

The capacitance depends on the area of the conductors, often termed "plates" even though they need not be flat, and the distance d. The value depends on the particular geometry. For a parallel plate capacitor, the capacitance is given by:

$$C = \varepsilon_0 \kappa\, A/d,$$

where ε_0 is the permittivity of free space (a constant), κ is the dielectric constant of the spacing material (= 1 for a vacuum), A is the plate area, and d is the separation distance (all SI units). Obviously, increasing capacitance requires increasing A, and/or increasing κ, and/or decreasing d. All these suggest the potential for nanostructuring in improving capacitor performance; most straightforwardly, values of d on the order of molecular dimensions leads to high capacitance. Nanostructuring also has the potential of increasing dielectric performance, through mininizing defects, which often are the cause of breakdown.

---

[1]Ref. 66 in main text.

dielectric. The thicker the film, the more resistant the capacitor to breakdown, but the smaller its capacitance. In any case, d must be large enough that electron tunneling is unimportant.

Improved dielectrics provide both a way to boost the dielectric constant and minimize d. As defects commonly localize breakdown, too, nanofabrication would also lead to improved performance. Interest in better dielectrics was inspired decades ago by ferroelectric materials such as barium titanium oxide ($BaTiO_3$, "barium titanate"). This compound has a distorted perovskite structure (Box 5.10), which consists of apex-sharing TiO_6 octahedra with the large voids between the octahedra occupied by the barium (Ba^{++}) cations. The TiO_6 octahedra are sufficiently distorted that the Ti^{4+} cations can "rattle around," which leads to very high polarizability (i.e., dipole moment), and hence to a large dielectric constant. Perovskites have inspired further research on better dielectrics, which inevitably will involve nanostructuring.[64]

Double-layer capacitors

More recently, a combination of very high surface area and very small charge separation has been achieved with so-called electric double-layer (EDL) capacitors. A double layer is formed at the interface between a charged surface and an electrolyte, because ions of opposite charge are attracted from the electrolyte to form a compensating layer (Fig. 5.3). This acts like a capacitor in which the distance d (Box 5.11) is of the order of tenths of nanometers. On discharge of such a capacitor, electrical current flows between the electrodes while a compensating ionic current flows in the electrolyte. In conjunction with high-surface-area nanoporous electrodes,[65] this leads to extraordinarily high

---

[64]E.g., Siddabattuni, S; Schuman, TP. Polymer-ceramic nanocomposite dielectrics for advanced energy storage. In *Polymer Composites for Energy Harvesting, Conversion, and Storage*, Li, L; Wong-Ng, W; Sharp, J; eds. ACSSS *1161*, 165–90, 2014.

[65]E.g., Wang, G; Zhang, L; Zhang, J. A review of electrode materials for electrochemical supercapacitors. *Chem. Soc. Rev. 41*, 797–828, 2012.

Electrode Surface

Figure 5.3 Double layer.

A charged surface, such as an electrode, in contact with an electrolyte attracts an "atmosphere" of compensating counter-ions from the solution, forming the so-called double layer. In this illustration the electrode is assumed to be positively charged.

capacitances, on the order of farads per gram. Such capacitors are often termed "supercapacitors"[66] or "ultracapacitors."[67] The term also sometimes is taken to include pseudocapacitors (described below), but usage does not seem to be consistent.[68]

Unfortunately, however, the voltage is limited by the breakdown voltage of the electrolyte (~1.2 V, in case of aqueous solutions). Alternative electrolytes, such as conducting polymers[69] or room-temperature ionic liquids[67] can improve this value considerably.

"Pseudocapacitors" store additional energy through surface redox reactions, a phenomenon termed "pseudocapacitance."[70] Typically they are based on high-surface-area transition metal

[66]Conway, BE. Transition from "supercapacitor" to "battery" behavior in electrochemical energy storage. *J. Electrochem. Soc.* 138, 1539–48, 1991.

[67]Liu, K-C; Anderson, MA. Getting a charge out of microporous oxides: Building a better ultracapacitor. *Mater. Res. Soc. Symp. Proc.* 432 (*Aqueous chemistry and geochemistry of oxides, oxyhydroxides and related materials*, Voigt, JA; Wood, TE; Bunker, BC; Casey, WH; Crossey, LJ, eds.) 221–30, 1997.

[68]E.g., Vatamanu, J; Bedrov, D. Capacitive energy storage: Current and future challenges. *J. Phys. Chem. Lett.* 6, 3594–609, 2015.

[69]E.g., Peled et al., in Ref. 33; Niu, Z; Liu, L; Sherrell, P; Chen, J; Chen, X. Flexible supercapacitors - development of bendable carbon architectures. In *Nanotechnology for Sustainable Energy*, Hu, YH; Burghaus, U; Qiao, S; eds. ACSSS *1140*, 101–41, 2013.

[70]E.g., Huggins, RA. Supercapacitors. *Philos. Trans. Roy. Soc. Lond. A 354*, 1555–66, 1996; Conway, BE; Birss, V; Wojtowicz, J. The role and utilization of pseudocapacitance for energy storage by supercapacitors. *J. Power Sources* 66, 1–14, 1997.

oxides. Specific capacitance 10–100 times higher than simple double-layer capacitors has been observed. They exhibit transitional behavior between capacitors and batteries, and so seem reminiscent of the "batacitor" postulated by science-fiction author Philip José Farmer.[71] Indeed, the electrodes in intercalation-based batteries also have some analogies.[65] Because redox reactions are involved, however, the discharge rate of pseudocapacitors is slower, although still much greater than batteries.

A great deal of research, therefore, is taking place on high-performance capacitors.[72] Many issues arise that could be addressed by better control of fabrication at the nanoscale; e.g., traditional carbon-based electrodes are plagued by high resistance due to the vagaries of the grain-grain contacts on which conduction relies. This has led to research directed at nanofabricating carbon electrodes with a well-defined "hierarchical" nanostructure.[73] Individual capacitor elements can be "stacked" to yield high voltages,[74] but achieving this requires building a nanolayered structure with repeated "sandwiches" of electrode-electrolyte-electrode. Such a nanoarchitecture seems an obvious application of layer-by-layer assembly,[75] and has led to a great deal of research along that line, much focused on

---

[71]Farmer, PJ. *The Fabulous Riverboat*, 1971.

[72]E.g., Kim, M; Ju, H; Kim, J. Oxygen-doped porous silicon carbide spheres as electrode materials for supercapacitors. *Phys. Chem. Chem. Phys. 18*, 3331–38, 2016; Guo, N; Li, M; Wang, Y; Sun, X; Wang, F; Yang, R. N-Doped hierarchical porous carbon prepared by simultaneous-activation of KOH and NH$_3$ for high performance supercapacitors. *RSC Adv. 6*, 101372–379, 2016; Armelin, E; Pérez-Madrigal, MM; Alemán, C; Díaz, DD. Current status and challenges of biohydrogels for applications as supercapacitors and secondary batteries. *J. Mater. Chem. A 4*, 8952–68, 2016.

[73]E.g., Lin, W; Xu, B; Liu, L. Hierarchical porous carbon prepared by NaOH activation of nano-CaCO$_3$ templated carbon for high rate supercapacitors. *New J. Chem. 38*, 5509–14, 2014.

[74]E.g., Bullard, GL; Sierra-Alcazar, HB; Lee, HL; Morris, JL. Operating principles of the ultracapacitor. *IEEE Trans. Magn. 25*, 102–6, 1989.

[75]Xiang, Y; Lu, S; Jiang, SP. Layer-by-layer self-assembly in the development of electrochemical energy conversion and storage devices from fuel cells to supercapacitors. *Chem. Soc. Rev. 41*, 7291–321, 2012.

graphene layers[76] (Box 6.2). And as always, if such capacitors are not to be restricted to niche markets, the nanofabrication must also be carried out *cheaply*.

Fuel cells

Alternatively, perhaps batteries and capacitors are a dead end except in niche markets. Maybe laptop computers and such will be better powered by small fuel cells, which potentially have much higher energy densities. One idea would be a hydrogen storage pack that would exsolve hydrogen in a controlled manner (p. 231). On its exhaustion, it would be swapped out, just like a conventional battery, and could then be recharged with hydrogen. If fuel cells that can use carbon-bearing fuels (e.g., methanol) can be perfected and miniaturized (p. 198), portable electronic devices may instead be simply fueled, rather as one today might fuel a cigarette lighter.[77]

Electrosynthesis

The second, related but perhaps more important electricity storage issue is the lack of a good way of storing *large* amounts

---

[76]Liu, C; Yu, Z; Neff, D; Zhamu, A; Jang, BZ. Graphene-based supercapacitor with an ultrahigh energy density. *Nano Lett.* 10, 4863–68, 2010; Chen, Y; Zhang, X; Yu, P; Ma, Y. Electrophoretic deposition of graphene nanosheets on nickel foams for electrochemical capacitors. *J. Power Sources 195*, 3031–35, 2010; Li, Z; Wang, J; Liu, X; Liu, S; Ou, J; Yang, S. Electrostatic layer-by-layer self-assembly multilayer films based on graphene and manganese dioxide sheets as novel electrode materials for supercapacitors. *J. Mater. Chem.* 21, 3397–403, 2011. Sarker, AK; Hong, J-D. Layer-by-layer self-assembled multilayer films composed of graphene/polyaniline bilayers: High-energy electrode materials for supercapacitors. *Langmuir 28*, 12637–46, 2012; Hsieh, C-T; Hsu, S-M; Lin, J-Y; Teng, H. Electrochemical capacitors based on graphene oxide sheets using different aqueous electrolytes. *J. Phys. Chem. C 115*, 12367–74, 2011; Liu, F; Xue, D. Chemical routes to graphene-based flexible electrodes for electrochemical energy storage. In Ref. 1, pp. 425–55.

[77]E.g., Park, J-E; Shimizu, T; Osaka, T. Nanotechnologies for fuel cells. In *Electrochemical Nanotechnologies*. Osaka, T; Datta, M; Shacham-Diamand, Y; eds. NS&T, 23–33, 2010.

of electricity, the immediate output of a great many energy-conversion systems. It turns out that a related issue is the long-term synthesis of liquid fuels. Because of their ease of handling, owing to their high density coupled with mobility, liquid fuels dominate transportation applications. At present, of course, they are largely obtained from petroleum. In the future alternative fuels will be required, even though they will no longer be "primary" energy sources like petroleum but merely a convenient form for storage and usage.

Electricity can be used to drive chemical reactions "uphill," as in the familiar example of electrolysis of water into hydrogen and oxygen, or the separation of aluminum metal from bauxite ore (p. 244; Box 3.6). More complex transformations are commonly termed *electrosynthesis*, which is a well-established technique in organic electrochemistry.[78] *Electrocatalysis* can be used to distinguish the use of electricity to overcome the activation energy barrier to drive a "downhill" reaction.

Practical electrosynthesis (and electrocatalysis) requires good electrocatalysts. Indeed, as electrosynthesis is formally just the reverse of electricity generation in a fuel cell, it has a broad synergy with fuel-cell research. A so-called regenerative fuel cell can be run in reverse as an electrolyzer.

The direct electrosynthesis of chemical fuels could both provide a way to store electrical energy, and furnish an ongoing source of fuel, as well as providing another route to feedstock chemicals. In particular, a great deal of research effort has focused on the electroreduction of CO_2, which would have the additional advantage of "CO_2 fixation"; i.e., the removal of CO_2 from the atmosphere[79] (cf. Box 5.7). Carbon monoxide (CO) has also

---

[78]E.g., Utley, J. Trends in organic electrosynthesis. *Chem. Soc. Rev. 26*, 157–67, 1997.

[79]E.g., Gattrell, M; Gupta, N; Co, A. A review of the aqueous electrochemical reduction of CO_2 to hydrocarbons at copper. *J. Electroanal. Chem. 594*, 1–19, 2006; Electrochemical reduction of CO_2 to hydrocarbons to store renewable electrical energy and upgrade biogas. *Energy Conv. Mgmt. 48*, 1255–65, 2007; Li, W. Electrocatalytic reduction of CO_2 to small organic molecule fuels on metal catalysts. In Hu, Ref. 80 (Chapter 6), pp. 55–76; Peterson, AA; Abild-Pedersen, F; Studt, F; Rossmeisl J; Nørskov, JK. How copper catalyzes the

received attention,[80] as has the electro-oxidation of methane into

electroreduction of carbon dioxide into hydrocarbon fuels. *Energy Environ. Sci. 3*, 1311–15, 2010; Whipple, DT; Kenis, PJA. Prospects of CO_2 utilization via direct heterogeneous electrochemical reduction. *J. Phys. Chem. Lett. 1*, 3451–58, 2010; Baturina, OA; Lu, Q; Padilla, MA; Xin, L; Li, W; Serov, A; Artyushkova, K; Atanassov, P; Xu, F; Epshteyn, A; Brintlinger, T; Schuette, M; and Collins, GE. CO_2 electroreduction to hydrocarbons on carbon-supported Cu nanoparticles. *ACS Catal. 4*, 3682–95, 2014; Lim, D-H; Jo, JH; Shin, DY; Wilcox, J; Hama, HC; Nama, SW. Carbon dioxide conversion into hydrocarbon fuels on defective graphene-supported Cu nanoparticles from first principles. *Nanoscale 6*, 5087–92, 2014; Lu, X; Leung, DYC; Wang, H; Leung, KH; Xuan, J. Electrochemical reduction of carbon dioxide to formic acid. *ChemElectroChem. 1*, 836–49, 2014; Back, S; Kim, H; Jung, Y. Selective heterogeneous CO_2 electroreduction to methanol. *ACS Catal. 5*, 965–71, 2015; Back, S; Yeom, MS; Jung, Y. Active sites of Au and Ag nanoparticle catalysts for CO_2 electroreduction to CO. *ACS Catal. 5*, 5089–96, 2015; Kortlever, R; Peters, I; Koper, S; Koper, MTM. Electrochemical CO_2 reduction to formic acid at low overpotential and with high Faradaic efficiency on carbon-supported bimetallic Pd-Pt nanoparticles. *ACS Catal. 5*, 3916–23, 2015; Li, Y; Su, H; Chan, SH; and Sun, Q. CO_2 electroreduction performance of transition metal dimers supported on graphene: A theoretical study. *ACS Catal. 5*, 6658–64, 2015; Qiao, J; Liu, Y; Hong, F; Zhang, J. A review of catalysts for the electroreduction of carbon dioxide to produce low-carbon fuels. *Chem. Soc. Rev. 43*, 631–75, 2014; Ren, D; Deng, Y; Handoko, AD; Chen, CS; Malkhandi, S; Yeo, BS. Selective electrochemical reduction of carbon dioxide to ethylene and ethanol on copper(I) oxide catalysts. *ACS Catal. 5*, 2814–21, 2015; Wannakao, S; Artrith, N; Limtrakul, J; Kolpak, AM. Engineering Transition-Metal-Coated Tungsten Carbides for Efficient and Selective Electrochemical Reduction of CO_2 to Methane. *ChemSusChem 8*, 2745–51, 2015; Zhengzheng, C; Zhang, X; Lu, G. Overpotential for CO_2 electroreduction lowered on strained penta-twinned Cu nanowires. *Chem. Sci. 6*, 6829–35, 2015; Choi, SY; Jeong, SK; Kim, HJ; Baek, I-H; Park, KT. Electrochemical reduction of carbon dioxide to formate on tin-lead alloys. *ACS Sustainable Chem. Eng. 4*, 1311–18, 2016; Cui, C; Han, J; Zhu, X; Liu, X; Wang, H; Mei, D; Ge, Q. Promotional effect of surface hydroxyls on electrochemical reduction of CO_2 over SnO_x/Sn electrode. *J. Catal., 343*, 257–65, 2016; Torelli, DA; Francis, SA; Crompton, JC; Javier, A; Thompson, JR; Brunschwig, BS; Soriaga, MP; Lewis, NS. Nickel-gallium-catalyzed electrochemical reduction of CO_2 to highly reduced products at low overpotentials. *ACS Catal. 6*, 2100–4, 2016.

[80]Li, CW; Ciston J; Kanan, MW. Electroreduction of carbon monoxide to liquid fuel on oxide-derived nanocrystalline copper. *Nature 508*, 504–7, 2014; Cheng, T; Xiao, H; Goddard, WA; III. Free-energy barriers and reaction mechanisms for the electrochemical reduction of CO on the Cu(100) surface,

methanol.[81] This last is an example of electrocatalysis because, as discussed elsewhere (Box 5.5), this is a downhill reaction. Indeed, an electrocatalytic approach may have an advantage in providing more control over this reaction, which as noted tends to run away.

Overall, perhaps reactions like those carried out currently thermally by catalysis of syngas (Box 5.4) could instead be carried out at low temperatures by electrosynthesis or electrocatalysis. Indeed, efficient, large-scale, low-temperature electrosynthesis of other commodity chemicals may also prove practical. One example is the ambient-pressure electrosynthesis of ammonia (NH_3) directly from nitrogen (N_2) and hydrogen (H_2).[82] The current Haber process for synthesizing NH_3 is a paragon of the thermal paradigm, involving direct reaction of H_2 and N_2 at high pressure and red heat.

Hydrogen Storage

A critical requirement for the "hydrogen economy" (Box 5.2) is a convenient way of storing hydrogen. One alternative is to store it in convenient hydrogen-bearing compounds, for example by reaction with carbon dioxide (Box 5.7). Such storage also has

---

including multiple layers of explicit solvent at pH 0. *J. Phys. Chem. Lett.* 6, 4767–73, 2015; Hansen, HA; Shi, C; Lausche, AC; Peterson, AA; Nørskov, JK. Bifunctional alloys for the electroreduction of CO_2 and CO. *Phys. Chem. Chem. Phys.* 18, 9194–201, 2016.

[81]E.g., Rocha, RS; Reis, RM; Lanza, MRV; Bertazzoli, R. Electrosynthesis of methanol from methane: The role of V_2O_5 in the reaction selectivity for methanol of a $TiO_2/RuO_2/V_2O_5$ gas diffusion electrode. *Electrochim. Acta* 87, 606–10, 2013.

[82]Garagounis, I; Kyriakou, V; Skodra, A; Vasileiou, E; Stoukides, M. Electrochemical synthesis of ammonia in solid electrolyte cells. *Front. Energy Res.* 2, 1–10, 2014; Lan, R; Alkhazmi, KA; Amar, IA; Tao, S. Synthesis of ammonia directly from wet air at intermediate temperature. *Appl. Catal. B* 152–53, 212–17, 2014; Back, S; Jung, Y. On the mechanism of electrochemical ammonia synthesis on the Ru catalyst. *Phys. Chem. Chem. Phys.* 18, 9161–6, 2016.

the advantage of providing another approach to carbon dioxide fixation.

Another set of approaches envisions storing hydrogen in a solid, from which it could be extracted in a controlled manner. Ideally the solid could be "recharged" easily as well. Unsurprisingly, this is a difficult problem[83] for which an obvious solution has not yet emerged. It seems clear, however, that any solution will involve nanostructuring, fundamentally so that a large surface area for reaction and/or adsorption is available. It may involve reactive nanoparticles,[84] or micro- or nanoporous materials.[85] Even lattices of carbon nanotubes ("buckytubes," Box 6.2) have been suggested.[86]

Nanofabrication and Custom Fabrication

As already noted, since the onset of the Industrial Revolution, the organization of matter has become progressively cheaper. A more recent trend is miniaturization: the dwindling cost of organizing matter at smaller and smaller scales. Nanotechnology will

---

[83]Schuth, F. Challenges in hydrogen storage. *Eur. Phys. J. Spec. Top. 176*, 155, 2009; Lim, KL; Kazemian, H; Yaakob, Z; Daud, WRW. Solid-state materials and methods for hydrogen storage: A critical review. *Chem. Eng. Technol. 33*, 213–26, 2010; Weidenthaler, C; Felderhoff, M. Solid-state hydrogen storage for mobile applications: Quo Vadis? *Energy Environ. Sci. 4*, 2495–502, 2011; Kelly, MT. Perspective on the storage of hydrogen: Past and future. *Struct. Bonding 141*, 169–201, 2011; Broom, DP. *Hydrogen Storage Materials: The Characterisation of Their Storage Properties.* GE&T, 2011.

[84]E.g., Shissler, DJ; Fredrick, SJ; Braun, MB; Prieto, AL. Magnesium and doped magnesium nanostructured materials for hydrogen storage. In Ref. 1, pp. 297–319.

[85]Léon, A. Hydrogen storage. In *Hydrogen Technology: Mobile and Portable Applications*, Léon, A; ed. GE&T, 81–128, 2008. Lee, H; Ihm, J; Cohen, ML; Lourie, SG. Calcium-decorated graphene-based nanostructures for hydrogen storage. *Nano Lett. 10*, 793–8, 2010; Langmi, HW; Ren, J; North, B; Mathe, M; Bessarabov, D. Hydrogen storage in metal-organic frameworks: A review. *Electrochim. Acta 128*, 368–92, 2014.

[86]E.g., Assfour, B; Leoni, S; Seifert, G; Baburin, IA. Packings of carbon nanotubes - new materials for hydrogen storage. *Adv. Mater. 23*, 1237–41, 2011.

continue both these trends, and in those cases it is evolutionary rather than revolutionary. The trends are important to energy applications nonetheless. How cheaply nanostructured materials can be fabricated will be crucial in exploiting diffuse energy resources, as emphasized in the discussions above and below. Conventional microtechnology, for example, is not nearly cheap enough for such bulk applications. A computer is relatively cheap only because the chip at its heart is so small. The present high cost of microfabrication is the major problem that besets photovoltaics, for example (p. 244).

A newer trend that nanotechnology will also accelerate is "custom" rather than "mass" production. Not only is fabrication becoming cheaper, it is also becoming more flexible. This is due, of course, both to increased computer power and to computer-controlled design and manufacture.

An energy-related application of custom fabrication lies in geothermal power. Commonly, the steam from different producing wells in the same geothermal field has significant variations in the temperature. Hence it would be most efficient to use the steam from each individual well to run its own optimized turbine. However, at present turbines are much too expensive to have one atop every well. Conventionally, therefore, the steam from all the wells is combined in a manifold that feeds a single turbine. The additional expense of the plumbing, and the efficiency losses, are overwhelmed by turbine costs. Cheaper fabrication, probably also with superstrength materials, is likely to change this. This, of course, further assumes that next-generation thermoelectric materials (p. 235) don't make turbines obsolete.

Distributed Fabrication: The Demise of Bulk Transportation

In the medium to long term, probably the most important effect of nanotechnology is what might be termed "distributed fabrication": the local construction of artifacts out of local materials. The idea can be summed up by the slogan "Matter as software." Technologically this is a considerably more revolutionary advance, but it merely (again!) replicates what biological systems do already. As noted in Chapter 1, for example, a plant is not assembled by sending off to the leaf factory, and the root factory, and the stem factory, and so on. These items are assembled from ambient sources according to a molecularly encoded digital instruction set: its DNA, and using only the diffuse and intermittent energy of sunlight. Indeed, biology can make extraordinary organization *cheaply*. Consider again the example of firewood, assembled molecule by molecule using sunlight, water, and air, and then casually burned for *fuel!*

As noted in Chapter 3, not only is distributed fabrication already happening with information products, but so-called 3D printers are beginning a trend toward distributed fabrication of physical objects. Perhaps, therefore, as the additional capabilities of nanoscale distributed fabrication come online, their effects will prove to be more evolutionary than revolutionary in this case as well.

Obviously such a capability will also have a huge effect on the global transportation network. The enormous energy demand involved in simply moving around partly organized matter will largely vanish. Of course, undoubtedly there will be a transition period. For example, simple chemical feedstocks are likely to be locally fabricated before complex machinery, much less foodstuffs. But since the transportation of bulk commodities such as raw ores currently accounts for a significant part of global energy costs, even primitive distributed capabilities will have a non-negligible effect.

Diffuse Resources

A great many energy sources (sunlight, geothermal power) are intrinsically diffuse, and so practical collectors must be cheap as well as efficient. This obviously has major implications for their fabrication: it must be as simple and automated as possible. Hence, another result of cheap nanofabrication will be to make collecting energy from diffuse sources much more practical, with enormous potential implications.

Distributed fabrication will most likely accelerate this trend. Biosystems already carry out distributed fabrication of energy-gathering systems; consider photosynthesis by green plants. Indeed, distributed nanofabrication may ultimately be somewhat analogous to agriculture.

Nanotechnology and Solid-State Energy Generation

Technologies applicable to a range of energy conversion applications are described in this section. Technologies specific to a particular application (e.g., photovoltaics) are described later in the appropriate section.

Heat Engines: Thermoelectric Power

Thermoelectric materials

After emphasizing that thermalizing energy is a really silly thing to do, it now seems ironic to talk about using nanotechnology to extract work from heat. Nonetheless, heat engines are not going to vanish entirely. In some cases (e.g., nuclear fission, p. 54) a non-thermal approach is very difficult to envision. In other cases (e.g., OTEC, geothermal power), a thermal gradient exists naturally, and so one might as well exploit such a gradient as efficiently as possible. Waste heat is also not going to vanish entirely, nor is solar heating. Science-fiction writers, for example, have long suggested personal devices powered by body heat.[87]

---

[87]E.g., Heinlein, RA. *Double Star*. Signet, 1956.

The most efficient exploitation technologies will probably involve nanotechnology. Conventional heat engines use a "working fluid," such as steam or combustion products, whose thermal expansion is used to convert heat into mechanical work. Although such technologies are highly mature (consider steam turbines and internal combustion engines), they are high-maintenance and prone to catastrophic failure because of the abundance of macroscopic moving parts. They also require a large temperature difference, and the range of that difference must include a phase transition or other large volume change, such as occurs in the reaction of gasoline vapor with air or the boiling of water into steam. Alternative working fluids—in particular, liquids with lower boiling points than water—are sometimes used to exploit lower temperatures as in geothermal fields, but the increase in complexity makes such approaches marginal at best.

Direct nanoscale conversion of a thermal difference into another form of energy, instead of with a heat engine, is also likely to be more reliable. Such devices already exist in embryonic form: thermoelectric power generators, which generate an electrical potential at a junction between two dissimilar materials, traditionally a p–n junction between two highly doped semiconductors.[88] (If a current is applied in the reverse direction, such junctions act as refrigerators.) Like fuel cells and photovoltaics, these devices have been restricted to niche markets, in this case due to both high cost and low efficiency. The best current conversion factors are only about 10% of the Carnot limit. Even a domestic refrigerator manages ~30% efficiency.

A dimensionless figure of merit ZT, where T is absolute temperature, characterizes thermoelectric materials.[89] For many years the best values have been ~1, but at present they lie in the

---

[88]E.g., DiSalvo, FJ. Thermoelectric cooling and power generation: Review. *Science 285*, 703–6, 1999.

[89]E.g., Mahan, GD. Good thermoelectrics. *Solid State Phys. 51*, 82–157, 1998; DiSalvo, Ref. 88.

range 2–3.[90] A value of $ZT \geq \sim 4$ would lead to a huge number of new applications and might even revolutionize power generation.

Optimizing thermoelectric materials perhaps no better illustrates the promise of nanoscale manufacture, because of the daunting number of parameters that must be optimized, often in inconsistent ways. For example, electrical resistance *and* thermal conductivity must be minimized. For most materials this is an incompatible objective, because the electrons that conduct electricity also conduct heat. Approaches taken include attempting to synthesize structures that minimize phonon (i.e., sound) transmission while leaving the electronic structure relatively unaffected. In any case, it seems likely that higher figures of merit will require nanostructuring, and a great deal of research is taking place along this line.[91]

One of the most interesting applications of thermoelectric materials lies in geothermal power. In few other cases are the difficulties of a working fluid so evident: temperature differences are often too modest, or span the wrong range, or else the fluid is absent completely, as in the case of hot dry rocks. The subsurface "plumbing," both natural and artificial, is also critical but difficult to ascertain, much less modify. Unfortunately, a working fluid probably cannot be eliminated completely. In the case of a geothermal system, the function of the working fluid is not merely mechanical: it also gathers heat from a large volume,

[90]Kong, LB; Li, T; Hng, HH; Boey, F; Zhang, T; Li, S. Waste thermal energy harvesting (I): Thermoelectric effect. In *Waste Energy Harvesting: Mechanical and Thermal Energies*, Kong, LB; Li, T; Hng, HH; Boey, F; Zhang, T; Li, S; eds. LNE, *24*, 263–403, 2014.

[91]Koumoto, K; Wang, Y; Zhang, R; Kosuga, A; Funahashi, R. Oxide thermoelectric materials: A nanostructuring approach. *Annu. Rev. Mater. Res.*, 2010; Shakouri, A. Recent developments in semiconductor thermoelectric physics and materials. *Annu. Rev. Mater. Res.*, 2011; Koumoto, K; Mori, T. *Thermoelectric Nanomaterials: Materials Design and Applications*. SSMS, *182*, 2013; Hewitt, CA; Carroll, DL. Carbon nanotube-based polymer composite thermoelectric generators. In *Polymer Composites for Energy Harvesting, Conversion, and Storage*, Li, L; Wong-Ng, W; Sharp, J; eds. ACSSS *1161*, 191–211, 2014; Wang, X; Wang, ZM; eds. *Nanoscale Thermoelectrics*. LNNS&T 16, 2014.

transferring it to a cooler environment so that a thermal gradient becomes available. Nonetheless, better thermoelectric devices would allow replacing the mechanical turbines used in the current conversion approaches. Not only would such devices minimize the problems with fouling, corrosion, and mechanical mishap that plague conventional geothermal installations, they would allow direct use of lower geothermal gradients, because the phase change necessary to drive a turbine would no longer be required.

Thermoelectric conversion also seems particularly relevant to OTEC, given the small temperature difference involved. In conversion of waste heat, even low efficiencies could be tolerated, if the alternative is no energy recovery at all, as long as the thermoelectric materials themselves are cheap enough.

Piezoelectric Power

Piezoelectric "stacks"

Piezoelectric materials are widely used at present for oscillators and sensors, most notably for nanoscale science in scanning-probe microscopes (SPMs) (Box 4.4). As described in the box, a piezoelectric crystal deforms on the application of an electric potential; or conversely, develops an electrical potential on deformation. Thus they potentially provide a way to transform mechanical work directly into electricity without macroscopic moving parts such as turbines. However, although their potential for power generation has been discussed for decades,[92] it has not been practical as yet. One study has shown that the maximum efficiency of electricity production with conventional packages of PZT (lead zirconate titanate, $Pb(Zr,Ti)O_3$, another perovskite), a commercially available

---

[92]Galasso, FS. *Structure, Properties and Preparation of Perovskite-Type Compounds.* Pergamon Press, 207 p., 1969, pp. 105–6; Goldfarb, M; Jones, LD. On the efficiency of electric power generation with piezoelectric ceramics. Trans. ASME *J. Dyn. Syst. Meas. Contr. 121,* 566–71, 1999.

piezoelectric material, is only ~10%.[93] According to this analysis, a great deal of energy is stored temporarily as strain without causing electrical generation. This energy is simply returned mechanically to the environment.

Quite apart from the potential of developing and fabricating new piezoelectric materials, nanofabrication would help make even the present low-efficiency materials more practical for power generation. Power generation would requires nanostructuring on a large scale, not only for the sheer volume of material required, but because building up significant voltages requires nanolayered structures, in which thin layers of piezoelectric material are interlayered with electrodes of alternating polarity connected in series. Recent research has focused on such nanocomposites, including the incorporation of polymer layers, which makes the structure flexible and more mechanically durable.[94] Some studies have looked at generating alternating current from vibrations, both at small scales[95] and as approach toward recovering waste mechanical energy.[96]

Piezoelectric materials may also prove to be practical for larger-scale power generation; indeed, proposed piezoelectric-based arrays for generating power from surf date back decades.[97]

---

[93]Goldfarb & Jones, Ref. 92.

[94]Setter, N. Trends in ferroelectric/piezoelectric ceramics. In Heywang, W; Lubitz, K; Wersing, W; eds. *Piezoelectricity: Evolution and Future of a Technology.* SSMS *114*, 563–69, 2008.

[95]Baur, C; Apo, DJ; Maurya, D; Priya, S; Voit, W. Advances in piezoelectric polymer composites for vibrational energy harvesting. In *Polymer Composites for Energy Harvesting, Conversion, and Storage*, Li, L; Wong-Ng, W; Sharp, J; eds. ACSSS *1161*, 1–27, 2014; Chen, X; Yao, N; Shi, Y. Energy harvesting based on PZT nanofibers. In *Energy Efficiency and Renewable Energy Through Nanotechnology*, Zang, L; ed. GE&T, 425–38, 2011.

[96]Kong, LB; Li, T; Hng, HH; Boey, F; Zhang, T; Li, S. Waste mechanical energy harvesting (I): Piezoelectric effect. In *Waste Energy Harvesting: Mechanical and Thermal Energies*, Kong, LB; Li, T; Hng, HH; Boey, F; Zhang, T; Li, S; eds. LNE, *24*, 19–133, 2014.

[97]Taylor, GW; Burns, JR. Hydro-piezoelectric power generation from ocean waves. *Ferroelectrics 49*, 101, 1983; Hausler, E; Stein, L. Applications of piezo film for mechanic-electric energy conversion. In *Proceedings of*

For example, the change in tension of a float-tethered cable could be exploited as waves raise the float as they cross it. Similar systems might be employed for low-head hydropower, to avoid such devices as turbines, paddlewheels, and others that use macroscopic rotation, or to exploit the rise and fall of the tides. They also would have considerably less environmental impact than dams. Of course, all such schemes obviously imply that the piezoelectric cables are made extremely cheaply. Once again the importance of cheap fabrication at the nanoscale in exploiting diffuse resources becomes evident.

New Energy Sources

Energy sources, both conventional and unconventional, will now be reviewed briefly, with notes as to the potential applications of nanotechnology. This section generally follows the same order as Chapter 2, but it only highlights aspects that haven't already been mentioned there.

Fossil Fuels and Biomass

These are reduced carbon-containing materials (cf. Box 2.15), and despite their obvious differences thus share a number of characteristics.

Catalysis and syngas

As was noted above, catalysis already underlies most of the petrochemical industry at present. The splitting ("cracking") of long-chain hydrocarbons into the shorter (C_6–C_8) molecules that make up gasoline is one important application, as is the synthesis of many simple organic compounds from syngas (Box 5.4). Higher hydrocarbons can even be built up via Fischer

Tropsch synthesis (Box 5.6). Hence the improvements in catalysis due to nano-designed and nano-fabricated catalysts in these applications will be evolutionary rather than revolutionary.

An obvious improvement is to carry out at the catalyses at lower temperature. It was described above how catalysts capable of cleaving carbon–carbon bonds at low temperatures would make fuel cells much more practical. Such catalysts would also have obvious applications in processing raw or waste materials containing C–C bonds.

To take a more "out-of-the-box" perspective, however, perhaps we should reconsider making syngas in the first place. After all, the various methods of syngas synthesis are themselves high temperature, whether by pyrolysis of biomass, or steam cracking of methane, or CO_2 reforming of methane, or whatever, and hence are thermal technologies. Targeted conversion of the raw materials directly would be more efficient.

Stranded gas and gas-to-liquid technologies

A specific example is stranded gas. As mentioned in Chapter 2, this is natural gas that is remote from markets, usually overseas in fact, such that building a pipeline is impractical. Shipping such gas otherwise is difficult and currently requires liquefying it to cryogenic temperatures. Such "liquified natural gas" (LNG) is obviously extremely hazardous to handle, not least because such a cargo is an obvious terrorist target.

The difficulties of shipping LNG (nearly all liquid methane, CH_4) have sparked interest in so-called gas-to-liquid technologies, the conversion of methane into liquid forms that are more easily and safely shipped and handled. Currently approaches generate syngas from the CH_4, followed by its catalyzed conversion into a liquid form, even into higher hydrocarbons via Fischer–Tropsch reactions. Although reasonably straightforward, this is inefficient for the reasons outlined above.

A much cleaner approach would be to convert the methane directly into liquid compounds, such as methanol (Box 5.5).

Indeed, if the catalytic difficulties can be solved, the direct partial oxidation of methane to methanol would be considerably simpler than cryogenic liquification and most probably considerably safer as well.

Unconventional fossil fuels

Catalysis that could break up long-chain hydrocarbons at modest temperature also seems directly relevant to oil shale and tar sands, as both these deposits are largely composed of such hydrocarbons. Unfortunately, the processing material also has to flow, which is another reason thermal processing has been required. Such deposits may be amenable to bioprocessing,[98] but that lies beyond the scope of this book.

Biomass conversion

As noted in Chapter 2, biological waste is an obvious alternative source of the reduced carbon compounds that now come from petroleum, both for fuels and chemical feedstocks. It has been a subject of study for decades,[99] and it will likely replace petroleum in a few decades more. Most is composed of lignocellulosic material (Box 5.9), and the processing of cellulose into ethanol has already been mentioned (Box 5.8).

---

[98]E.g., Refs. 21 and 24 (Chapter 2).

[99]E.g., Klass, DL; ed. *Biomass as a Nonfossil Fuel Source*. ACSSS, 144, 1981; Sofer, SS; Zaborsky, OR. *Biomass Conversion Processes for Energy and Fuels*. Plenum Press, 1981; Lowenstein, MZ., ed. *Energy Applications of Biomass*. Elsevier, 1985; Rowell, RM; Schultz, TP; Narayan, R; eds. *Emerging Technologies for Materials and Chemicals from Biomass*. ACSSS, 476, 1992; Saha, BC; Woodward, J; eds. *Fuels and Chemicals from Biomass*. ACSSS, 666, 1997; Klass, DL. *Biomass for Renewable Energy, Fuels, and Chemicals*. Academic, 1998; Zhu, J-Y; Zhang, X; Pa, X. *Sustainable Production of Fuels, Chemicals, and Fibers from Forest Biomass*. ACSSS, 1067, 2011; Liebner, F; Rosenau, T; eds. *Functional Materials from Renewable Sources*. ACSSS, 1107, 2012; Deng, J; Li, M; Wang, Y. Biomass-derived carbon: synthesis and applications in energy storage and conversion. *Green Chem*. 18, 4824–54, 2016.

Of course, microbial bioprocessing is a traditional way of dealing with biowaste—consider sewage disposal!—and there is a great deal of research on using microbes to process biowaste into higher-value products. As already noted, though, biotechnology is not a focus here.

Much research has also focused on the conversion of biomass via pyrolysis, to syngas (Box 5.4) and/or other products.[100] As noted above, however, pyrolysis is just another thermal technology, and it's even less efficient here because of the high water content of most biowaste.

For the emphasis here, in recent years there has been much more attention to low-temperature non-biological processing. One such is dissolution in so-called ionic liquids,[101] which have attracted much attention recently as "green" solvents.[102] Large-scale processing of cellulose for such products as ethanol, for example, will require cleaner and less extreme solvents.[103]

Low-temperature processing also lends itself to targeted conversion of biomolecules by specific catalysts, which is also receiving much attention.[104] The catalyzed transformation of

---

[100]E.g., Soltes, Ed J; Milne, Thomas A; eds. *Pyrolysis Oils from Biomass: Producing, Analyzing, and Upgrading.* ACSSS, 376, 1988; Bridgwater, AV. Production of high grade fuels and chemicals from catalytic pyrolysis of biomass. *Catal. Today,* 29, 285–95, 1996; Quaak, P; Knoef, H; Stassen, H. *Energy from Biomass: A Review of Combustion and Gasification Technologies.* World Bank, 1999; Demirbas, A. Thermochemical processes. In *Biorefineries For Biomass Upgrading Facilities,* Demirbas, A. GE&T, 135–92, 2010.

[101]Wang, H; Gurau, G; Rogers, RD. Dissolution of biomass using ionic liquids. *Struct. Bonding 151,* 79–105, 2014.

[102]E.g., Visser, AE; Bridges, NJ; Rog, RD; eds. *Ionic Liquids: Science and Applications.* ACSSS, 1117, 2012; Zhang, S; Wang, J; Lu, X; Zhou, Q. eds. *Structures and Interactions of Ionic Liquids. Struct. Bonding, 151,* 2014.

[103]Liebert, T. *Cellulose Solvents - Remarkable History, Bright Future.* ACSSS 1033, 3–54, 2010; Medronho B; Lindman B. Competing forces during cellulose dissolution: From solvents to mechanisms. *Curr. Opin. Coll. Interf. Sci. 19,* 32–40, 2014.

[104]E.g., Deng, L; Li, J; Lai, DM; Fu, Y; Guo, QX. Catalytic conversion of biomass-derived carbohydrates into γ-valerolactone without using an external H_2 supply. *Angew. Chem. Int. Ed. 48,* 6529–32, 2009; Lin, Y-C;

biological fats and oils to "biodiesel" fuel is a current example (Box 5.12). The production of hydrogen from biomass is another focus.[105] As mentioned below, photocatalyzed hydrogen production from waste biomass seems particularly attractive, and also dovetails with research on the photocatalytic destruction of pollutants, discussed below (p. 322).

Sunlight and Nanotechnology

Next-Generation Photovoltaics

As noted in Chapter 2, conventional photovoltaic (PV) cells convert sunlight directly into electricity by the illumination of a *semiconductor* (Box 2.17) containing a space-charge region, typically a p–n junction. Electron–hole pairs generated in this region travel in opposite directions and are gathered and used to drive an external circuit.

Huber, GW. The critical role of heterogeneous catalysis in lignocellulosic biomass conversion. *Energy Environ. Sci.* 2, 68–80, 2009; Taarning, E; Osmundsen, CM; Yang, X; Voss, B; Andersen, SI; Christensen, CH. Zeolite-catalyzed biomass conversion to fuels and chemicals, *Energy Environ. Sci.* 4, 793–804, 2011; Sievers, C; ed. *Biomass Conversion over Heterogeneous Catalysts: Contributions from the 2011 AIChE Annual Meeting.* TC 55, 2012; Nicholas, KM; ed. *Selective Catalysis for Renewable Feedstocks and Chemicals.* TCC 353, 2014; Bozell, JJ. Approaches to the selective catalytic conversion of lignin: A grand challenge for biorefinery development. In Nicholas above, pp. 229–55; Li, C; Zhao, X; Wang, A; Huber, GW; Zhang, T. Catalytic transformation of lignin for the production of chemicals and fuels. *Chem. Rev. 115*, 11559–624, 2015; Kärkäs, MD; Matsuura, BS; Monos, TM; Magallanes, G; Stephenson, CRJ. Transition-metal catalyzed valorization of lignin: the key to a sustainable carbon-neutral future. *Org. Biomol. Chem. 14*, 1853, 2016.

[105]E.g., Ni, M; Leung, DYC; Leung, MKH; Sumathy, K. An overview of hydrogen production from biomass. *Fuel Process. Technol. 87*, 461–72, 2006; Zhang, Y-HP. Hydrogen production from carbohydrates: A mini-review. In Zhu et al., in Ref. 99; Patel, M; Pant, KK; Mohanty, P. Renewable hydrogen generation by steam reforming of acetic acid over Cu-Zn-Ni supported calcium aluminate catalysts. In *Nanocatalysis for Fuels and Chemicals.* Dalai, AK; ed. ACSSS *1092*, 111–37, 2012.

Box 5.12 Biodiesel

This is a substitute for diesel fuel that can be made from biological fats and oils ranging from deep-fryer waste to pork lard. It has the major advantage that it can be mixed in all proportions with ordinary diesel and used without modification in modern diesel engines. It is also attractive in that it represents yet another transformation of waste material into fuel, and a non-thermal transformation to boot.

Biological fats and oils are generically termed *lipids*, and many (so-called *triglycerides*) consist of three fatty acids bound to a glycerol (glycerine) molecule via three *ester* (see below) links:

$$
\begin{array}{l}
H \\
H-COOR_1 \\
H-COOR_2 \\
H-COOR_3 \\
H
\end{array}
$$

Fatty acids consist of an alphatic hydrocarbon chain (the "R_i" above, numbered 1–3 because they need not be the same) terminated at one end by a carboxylic acid (COOH) group. This give the fatty acid molecule a polar end, because the carboxylic acid group tends to ionize by losing the proton—hence its acidity. In fact, many common surfactants (Box 4.8) are fatty acids. Note also that lipids are "mostly hydrocarbon" already, and in fact it is thought that biological precursors of petroleum are lipids.

Glycerol, $C_3H_5(OH)_3$, is another alcohol; i.e., a hydrocarbon in which one or more hydrogens has been replaced with a hydroxyl (OH) group. Glycerol can be considered as derived from propane (C_3H_8) by replacing one hydrogen on each carbon atom, and in fact the formal name of glycerol is 1,2,3-propanol.

A so-called ester link can be formed by an acid and the hydroxyl group of an alcohol, with the release of a water molecule. In the case of a carboxylic acid, for example, the reaction is:

$$-COOH + HOC- \Rightarrow -COOC- + H_2O.$$

So a triglyeride is a triple ester of glycerol.

Biodiesel consists of *methyl* esters of fatty acids, in which the glycerol is replaced with methanol. The methyl esters have physical properties (boiling point, viscosity, etc.) very similar to the hydrocarbons making up diesel fuel, which is why they can be mixed in all proportions. The process usually used is "transesterfication," in which methanol is "swapped" for glycerol. It involves using strong base to first make highly reactive sodium methoxide from the methanol, which can the displace the glycerol. Of course, as is usual with such corrosive reagents, additional expense is incurred in purifying the products, as well as additional disposal costs. Again the benefits of better catalysis are evident.

Of course, biodiesel has disadvantages as well. Waste fats and oils are unlikely to ever furnish anything but minor part of the demand for diesel fuel, but growing dedicated crops, such as traditional vegetable oil crops such as soybeans, for biodiesel or other fuels competes with food production. Such considerations, of course, would not apply to nonfood crops grown in otherwise unproductive environments, such as algae in brine.[1]

A larger irony, as noted elsewhere, would be the *growing* of fuel crops: using materials assembled molecule by molecule out of the ambient environment, so they can be *burned* in a Carnot-limited engine.

---

[1]Demirbas, A; Demirbas, MF. *Algae Energy: Algae as a New Source of Biodiesel.* GE&T, 2010; D'Addario, EN. Sustainability in carbon capture and utilization. Biodiesel from microalgae. In De Falco et al.; eds., in Ref. 80 (Chapter 6), main text, pp. 95–107.

Present-day solar cells are expensive, however, and that expense results from their fabrication costs. For one thing, a very fine network of wires is need to catch as many charge carriers as possible, and that leads to intricate fabrication issues. Another expense is making the junction in the first place, particularly because a much larger p–n interface is needed than in something like (say) a transistor. The most efficient conventional material is crystalline semiconductor-grade silicon, because electron-hole recombination is minimized, but this material is costly. The repeated cycles of purification through remelting and recrystallization required to make such material represent the thermal paradigm at an extreme. Moreover, the ultrapure silicon must then be doped to yield a large-area p–n junction where the photons are absorbed. And finally, the more perfect the semiconductor crystal, the better the efficiency, because imperfections tend to be "traps" where the holes and electrons can get back together. Since semiconductor-grade silicon is expensive, though, using high-quality crystals makes solar cells even *more* expensive.

Of course, all this is merely another statement about the limitations of current technology. Alternative fabrication approaches could make a large difference in costs,[106] and nanofabrication techniques seem an obvious approach.

Another route to cost minimization is through minimizing the material used. Electron–hole formation is a near-surface process,

---

[106]E.g., Li, J; Yu, HY. Enhancement of Si-based solar cell efficiency via nanostructure integration. In *Energy Efficiency and Renewable Energy Through Nanotechnology*, Zang, L; ed. GE&T, 3–55, 2011; Das, NK; Islam, SM. Conversion efficiency improvement in GaAs solar cells. In *Large Scale Renewable Power Generation: Advances in Technologies for Generation, Transmission and Storage*, Hossain, J; Apel Mahmud, A; eds. GE&T, 53–75, 2014; Hilali, MM; Sreenivasan, SV. Nanostructured silicon-based photovoltaic cells. In Wang & Wang Ref. 111, pp. 131–64; Sagadevan, S. State-of-the-art of nanostructures in solar energy research. In *Advanced Energy Materials*, Tiwari, A; Valyukh, S; eds. Wiley. 69–104, 2014; Sangster, AJ. *Electromagnetic Foundations of Solar Radiation Collection: A Technology for Sustainability*. GE&T, 2014, Ch. 7; Xiao, S; Xu, S. Status and progress of high-efficiency silicon solar cells. In Wang & Wang Ref. 111, pp. 1–58.

so the bulk of the silicon goes unused. This has motivated much research into using thin film materials. Unfortunately, amorphous films, such as are commonly obtained from chemical vapor deposition, have very low efficiencies due to heightened electron–hole recombination. Also, if the semiconductor layer is too thin, photon losses due to transmission become significant, because silicon is an "indirect" semiconductor in which promotion of an electron to the conduction band is formally "forbidden" by the rules of quantum mechanics.[107] Such materials must be 10–100 times thicker than a direct semiconductor to have an equivalent chance of absorbing a photon.

Optimizing the collection of the electrons and holes to driven the external circuit also is a nanofabrication issue. Ideally every electron and every hole would be collected before it has a chance to recombine, and approaching this ideal requires a dense network of fine wires (at present fabricated by conventional microtechnology techniques) on the semiconductor surface.

There also is an intrinsic loss in the very nature of photon-induced electron–hole pair formation in a single semiconductor. The energy of photons with energy less than the bandgap is completely lost because they are not absorbed, while the energy of those with energy greater than the bandgap is partly lost because the additional energy is thermalized, merely going into heating the semiconductor. Thus the bandgap value has a "leveling" effect, and the best bandgap value for a given wavelength distribution of the incoming radiation represents a tradeoff between these loss mechanisms.

In principle, the efficiency of semiconductor collection can be improved by using a stack of semiconductors with bandgaps of different energies, termed multijunction solar cells, and such solar cells are the subject of much research; indeed, recently

---

[107]Tan, MX; Laibinis, PE; Nguyen, ST; Kesselman, JM; Stanton, CE; Lewis, NS. Principles and applications of semiconductor photoelectrochemistry. *Prog. Inorg. Chem. 41*, 21–144, 1993.

an experimental multijunction solar cell achieved over 50% conversion of the incoming sunlight.[108] Obviously, such cells require structuring at the nanoscale, the interface between the layers being of particular importance. The semiconductor itself can also be nanostructured to take advantage of quantum confinement effects to tune the bandgap, as an array of nanowires, for example.[109] Unsurprisingly, quantum dots designs have received much attention along this line.[110]

Photovoltaics based on semiconductors other than silicon are another obvious alternative approach.[111] Gallium arsenide (GaAs) and related compounds are commonly used in multijunction solar cells, as well as in photovoltaics using concentrated sunlight, because of their greater resistance to heating.[112]

Perhaps even more interesting, however, are approaches not based on conventional semiconductors. "Sensitized" semiconductors have attracted much attention in the last couple

[108]Leite, MS; Woo, RL; Munday, JN; Hong, WD; Mesropian, S; Law, DC; Atwater, HA. Towards an optimized all lattice-matched InAlAs/InGaAsP/InGaAs multijunction solar cell with efficiency >50%. *Appl. Phys. Lett. 102*, 033901, 2013.

[109]Garnett, EC; Brongersma, ML; Cui, Y; McGehee, MD. Nanowire solar cells. *Annu. Rev. Mater. Res. 41*, 11.1–11.27, 2011.

[110]Nozik, AJ; Beard, MC; Luther, JM; Law, M; Ellingson, RJ; Johnson, JC. Semiconductor quantum dots and quantum dot arrays and applications of multiple exciton generation to third generation photovoltaic solar cells. *Chem. Rev.* 110, 6873–90, 2010; Nozik, AJ. Nanoscience and nanostructures for photovoltaics and solar fuels. *Nano Lett. 10*, 2735–41, 2010; Wu, J; Wang, ZM. *Quantum Dot Solar Cells*. LNNS&T 15, 2014.

[111]Wang, X; Wang, ZM; eds. *High-Efficiency Solar Cells: Physics, Materials, and Devices*. SSMS 190, 2014; Kirkeminde, A; Gong, M; Ren; S. The renaissance of iron pyrite photovoltaics: Progress, challenges, and perspectives. In Ref. 1, pp. 137–66.

[112]E.g., Luque, AL; Viacheslav. A; eds. Concentrator Photovoltaics. *Spring. Ser. Opt. Sci. 130*, 2007; Alferov, ZI; Andreev, VM; Rumyantsev, VD. III-V Solar cells and concentrator arrays. In *High-Efficient Low-Cost Photovoltaics. Recent Developments*. Petrova-Koch, V; Hezel, R; Goetzberger, A; eds. *Spring. Seri. Opt. Sci. 140*, 101–41, 2009; Sangster, in Ref. 105, Ch. 8; Pérez-Higueras, P; Fernández, EF. *High Concentrator Photovoltaics: Fundamentals, Engineering and Power Plants*. GE&T, 2015.

of decades since the pioneering work of O'Regan & Graetzel.[113] The sensitization is accomplished by surface dyes, which absorb in different wavelengths than the bulk semiconductor. If the energy difference between the ground and excited states of the dye straddles the bottom of the conduction band, an electron from the excited state of the dye can be injected into the conduction band, where it becomes available for photoreduction reactions. Hence, more of the incident radiation becomes usable.

The semiconductor usually used in such cells is titanium dioxide (TiO_2), the same that has been used in the majority of studies on artificial photosynthesis and catalysis (discussed below). The dye is bound to the surface of the TiO_2, and suffused with a thin layer of electrolyte containing a redox couple, often termed a *redox shuttle*. Usually they are triiodide anion (I_3^-) and iodide (I^-). The photogenerated holes generated oxidize the iodide to triiodide, while the returning electrons from the external circuit re-reduce the triiodide to iodide.[114] Such cells can be considered a "light-driven battery," with the redox couple driving the battery continually regenerated by the illumination.

Such devices are intrinsically nanostructured: they require a conducting "window" to hold the electrolyte in, and the dye layer must also be convoluted enough to increase the chances of photon absorption. Indeed, the TiO_2 substrate needs to be nanostructured as well.[115] Noble-metal catalysts have again been used in the early designs, so there is also an ongoing search

---

[113]O'Regan, B; Grätzel, M. A low-cost, high-efficiency solar cell based on dye-sensitized colloidal TiO_2 films. *Nature 353*, 737–40, 1991.

[114]Xu, T. Nanoarchitectured electrodes for enhanced electron transport in dye-sensitized solar cells. In *Energy Efficiency and Renewable Energy Through Nanotechnology*, Zang, L; ed. GE&T, 271–98, 2011; Pastore, M; De Angelis, F. Modeling materials and processes in dye-sensitized solar cells: Understanding the mechanism, improving the efficiency. *Top. Curr. Chem. 352*, 151–236, 2014.

[115]E.g., Bai, Yu; Mora-Seró, I; De Angelis, F; Bisquert, J; Wang, P. Titanium dioxide nanomaterials for photovoltaic applications. *Chem. Rev. 114*, 10095–130, 2014; Liu, Z; Li, L. Passivating the surface of TiO_2 photoelectrodes with

for alternatives here.[116] The dyes and their attachment to the substrate are the subject of much research;[117] indeed, quantum dots are also being investigated as sensitizers.[118]

Although sensitized solar cells definitely require nano-structuring, it is worth noting that their fabrication is likely to be considerably easier than that of conventional semiconductor grade silicon. The latter requires a fabrication facility capable of microelectronics lithography, whereas the dye-sensitized designs seem accessible to the surface functionalization and self-assembly techniques described in the last chapter.

Other alternative photovoltaic designs are based on organic polymers—plastics, in other words.[119] These rely on

---

Nb_2O_5 and Al_2O_3 for high-efficiency dye-sensitized solar cells. In Viswanathan et al., Ref. 137, pp. 201–10.

[116]Wu, M; Ma, T. Low-cost Pt-free counter electrode catalysts in dye-sensitized solar cells. In Ref. 1, pp. 77–87.

[117]Furukawa, S. Dye-sensitized solar cells using natural dyes and nanostructural improvement of TiO_2 film. In *Energy Efficiency and Renewable Energy Through Nanotechnology*, Zang, L; ed. GE&T, 299–316, 2011, and other papers in this volume; Uddin, T; Nicolas, Y; Olivier, C; Toupance, T. Low temperature preparation routes of nanoporous semi-conducting films for flexible dye sensitized solar cells. In *Nanotechnology for Sustainable Energy*, Hu, YH; Burghaus, U; Qiao, S; eds. ACSSS 1140, 143–72, 2013; Guo, W; Ma, T. Nanostructured nitrogen doping TiO_2 nanomaterials for photoanodes of dye-sensitized solar cells. In Wang & Wang, Ref. 111, pp. 55–75; Odobel, F; Pellegrin, Y; Anne, FB; and Jacquemin, D. Molecular engineering of efficient dyes for p-type semiconductor sensitization. In Wang & Wang, Ref. 111, pp. 215–46.

[118]E.g., Fresno, F; Hernández-Alonso, MD. Sensitizers: Dyes and quantum dots. In *Design of Advanced Photocatalytic Materials for Energy and Environmental Applications*, Coronado, JM; Fresno, F; Hernández-Alonso, MD; Portela, R; eds. GE&T, 329–43, 2013; Sudhagar, P; Juárez-Pérez, EJ; Kang, YS; Mora-Seró, I. Quantum dot-sensitized solar cells. In Ref. 1, pp. 89–136; Barceló, I; Guijarro, N; Lana-Villarreal, T; Gómez, R. Recent progress in colloidal quantum dot-sensitized solar cells. In Wu & Wang, Ref. 110, pp. 1–38.

[119]Bundgaard, E; Krebs, F. Development of low bandgap polymers for roll-to-roll coated polymer solar cell modules. In *Energy Efficiency and Renewable Energy Through Nanotechnology*, Zang, L; ed. GE&T, 251–70, 2011, and refs therein; Günes, S. Organic solar cells and their nanostructural improvement. In *Energy Efficiency and Renewable Energy Through Nanotechnology*,

more subtle charge-separation mechanisms that are still incompletely understood,[120] which underscores the importance of nanostructure at the molecular level. They offer the promise of a considerably more convenient form for photovoltaics, e.g. as flexible sheets than the present rigid modules.[121]

Dispersed collection

Sunlight is the *sine qua non* of a dispersed energy resource, so not only do we need nanoscale fabrication, we need *cheap* nanoscale fabrication. This is particularly pressing because power generation requires that substantial area be exposed to the sun. Land is a commodity in short supply in most of the industrialized world, particularly in urban areas where electricity demand is highest. In such areas, an obvious source of underutilized area is

---

Zang, L; ed. GE&T, 171–225, 2011; Angmo, D; Espinosa, N; Krebs, F. Indium tin oxide-free polymer solar cells: Toward commercial reality. In Ref. 1, pp. 189–225; Beljonne, D; Cornil, J; eds. *Multiscale Modelling of Organic and Hybrid Photovoltaics.* TCC 352, 2014; Chen, F-C; Chou, C-H; Chuang, M-K. High-performance bulk-heterojunction polymer solar cells. In Ref. 1, pp. 167–87; Jemison, RC; McCullough, RD. Techniques for the molecular design of push-pull polymers towards enhanced organic photovoltaic performance. In *Polymer Composites for Energy Harvesting, Conversion, and Storage*, Li, L; Wong-Ng, W; Sharp, J; eds. ACSSS, 1161, 71–109, 2014; Lin, Y-H; Verduzco, R. Synthesis and process-dependent film structure of all-conjugated copolymers for organic photovoltaics. In *Polymer Composites for Energy Harvesting, Conversion, and Storage*, Li, L; Wong-Ng, W; Sharp, J; eds. ACSSS, 1161, 49–70, 2014; Sun, S-S; Harding, A. Dye-sensitized polymer composites for sunlight harvesting. In *Polymer Composites for Energy Harvesting, Conversion, and Storage*, Li, L; Wong-Ng, W; Sharp, J; eds. ACSSS, 1161, 29–47, 2014; Youn, H; Kim, H; Guo, LJ. Low-cost fabrication of organic photovoltaics and polymer LEDs. In Ref. 1, pp. 227–65.

[120]E.g., Gao, F; Inganäs, O. Charge generation in polymer-fullerene bulk-heterojunction solar cells. *Phys. Chem. Chem. Phys. 16*, 20291–304, 2014.

[121]van Bavel, SS; Loos, J. On the importance of morphology control for printable solar cells. In *Energy Efficiency and Renewable Energy Through Nanotechnology*, Zang, L; ed. GE&T, 227–49, 2011.

rooftops. Covering every roof with a solar array would obviously go a long way toward resolving issues of both land and power. For this reason several countries have targeted rooftop installations, including Japan, Germany, and the US. In the state of Nevada, for example, a political battle erupted in 2016 when subsidies for residential rooftop solar installations were to be phased out.[122] Roads are another potential source of area that is underutilized at present, although they are a considerably more demanding environment and would probably require self-cleaning materials, possibly through photocatalytic coatings.

The form of the PV cells could also be considerably more convenient than the present rigid modules. One "wish list" item would be a PV material, probably a polymer, that could be dispersed off rolls like carpeting or wrapping paper. Alternatively, perhaps it could be applied in a colloidal suspension like paint. Distributed power generation, often considered desirable for security reasons as much as for its more benign ecological footprint, would be a direct and beneficial consequence of such technologies.

Photovoltaic electrical generation shares the disadvantages of other forms of electrical generation, compounded by the intermittency of sunlight. As noted above, the storage of electrical energy is already an issue even with power sources less fickle than sunlight. Transmission of electricity can remain an issue because many of the places where sunlight is most attractive as an energy source, such as deserts, are remote from population centers. Of course, as already described solution of these problems leads to synergies with other technologies such as high-temperature superconductors and electrical storage. An alternative approach, though, is not to generate electricity in the first place, as addressed below.

---

[122]Hidalgo, J. The solar battle on Nevada rooftops. Reno, Nevada (USA) *Gazette-Journal*, 24 May 2016.

Artificial Photosynthesis

It's curious that "solar power" is so often thought of exclusively in terms of the direct conversion to electricity. Nature doesn't use solar energy that way, and there is no reason humanity has to. Via photosynthesis, green plants store energy of sunlight directly into chemical bonds. Indeed, as noted this is the ultimate source of most of the fuels we depend on today. Moreover, as already discussed, conversion to electrical energy has the serious disadvantages of lack of storability and lack of transportability, and both problems are exacerbated by the intermittency of sunlight.

Converting sunlight directly into chemical bonds, as green plants do, solves all three problems at a stroke. Indeed, early speculations about using solar energy envisioned a photochemical approach in which dyes or pigments were used to trap sunlight, in emulation of natural photosynthesis,[123] and similar dye-based approaches continue to attract attention even now.[124]

Much more research has been carried out, however, on semiconductor-based photosynthesis, which has a literature stretching back to the early 1970s, when it was discovered that ultraviolet (UV) illumination of titanium dioxide in water yielded free hydrogen.[125] In the last couple of decades, there has been an explosion of interest on this topic, and only some highlights of this literature can be given.

Semiconductor-based photosynthesis is fundamentally different from natural photosynthesis, which relies on dye-based absorption in conjunction with a nanostructured molecular chain

---

[123]E.g., Ref. 27 (Chapter 2).

[124]E.g., Karkas, MD; Verho, O; Johnston, EV; Åkermark, B. Artificial photosynthesis: Molecular systems for catalytic water oxidation. *Chem. Rev.* *114*, 11863–2001, 2014.

[125]Fujishima, A; Honda, K. Electrochemical photolysis of water at a semiconductor electrode. *Nature 238*, 37–8, 1972.

to carry out charge separation. In semiconductors, by contrast, the electron–hole pair generated by photon absorption is used to drive coupled redox reactions rather than power an electric circuit. The photogenerated electron is a powerful reducing agent, while the hole is a powerful oxidizing agent[126] (Box 5.13).

Familiar semiconductors such as silicon are not well suited for such applications, for a couple of reasons. The bandgap is too narrow and the band edges are not positioned optimally. Silicon is also too chemically reactive to tolerate aqueous solutions in the presence of air, much less in the case of direct water oxidation.

Although early efforts were made to passivate silicon and other semiconductors so that they could be used in aqueous solution,[127] much subsequent work has been carried out with wide-bandgap oxide semiconductors such as zinc oxide (ZnO), tin dioxide (SnO_2), and strontium titanium oxide ("strontium titanate," $SrTiO_3$, another perovskite). Such semiconductors were already noted above in the context of sensitized "redox couple" photovoltaic cells.

---

[126]E.g., Nozik, AJ. Photoelectrochemistry: Applications to solar energy conversion. *Annu. Rev. Phys. Chem. 29*, 189–222, 1978; Gerischer, H. Solar photoelectrolysis with semiconductor electrodes. In *Solar Energy Conversion: Solid-State Physics Aspects*, Seraphin, BO; ed. TAP *31*, 115–72, 1979; Bard, AJ; Fox, MA. Artificial photosynthesis: Solar splitting of water to hydrogen and oxygen. *Acc. Chem. Res. 28*, 141–5, 1995; Lewis, N. Artificial photosynthesis - "wet" solar cells can produce both electrical energy and chemical fuels. *Am. Sci. 83*, 534–41, 1995.

[127]Bocarsly, AB; Walton, EG; Wrighton, MS. Use of chemically derivatized n-type silicon photoelectrodes in aqueous media. Photooxidation of iodide, hexacyanoiron(II), and hexaammineruthenium(II) at ferrocene-derivatized photoanodes. *J. Am. Chem. Soc. 102*, 3390–8, 1980; Bolts, JM; Bocarsly, AB; Palazzotto, MC; Walton, EG; Lewis, NS; Wrighton, MS. Chemically derivatized n-type silicon photoelectrodes. Stabilization to surface corrosion in aqueous electrolyte solutions and mediation of oxidation reactions by surface-attached electroactive ferrocene reagents. *J. Am. Chem. Soc. 101*, 1378–85, 1979; Wrighton, MS; Austin, RG; Bocarsly, AB; Bolts, JM; Haas, O; Legg, KD; Nadjo, L; Palazzoto, MC. Design and study of a photosensitive interface: a derivatized n-type silicon photoelectrode. *J. Am. Chem. Soc. 100*, 1602–3, 1978.

Box 5.13 Semiconductor Photochemistry

As described (Box 2.17), absorption of a photon by a semiconductor in the vicinity of a space-charge region yields an electron-hole pair that then separates. In conventional photovoltaic devices they are then used to drive an external circuit.

However, they can instead be used to drive chemical reactions. The hole is a strong oxidizing agent, whereas the electron is a strong reducing agent, and they can bring about reactions that are energetically "uphill." Of course, the electron-hole pair must have enough energy to drive the desired reaction, but the bandgap must also be positioned properly. For reduction to occur, the energy at the bottom edge of the valence band, where the electron is promoted, must lie above that of the species to be reduced. In other words, it must be energetically favorable to transfer the electron to carry out the reduction. Similarly, a species to be oxidized must have an energy below the top of the conduction band, so that it's energetically favorable for the species to be oxidized to lose an electron into the CB (Fig. 2.2). The exact values depend not just on the details of the semiconductor, but also on the composition of the contacting solution.[1]

For example, the important case of breaking down water ("water splitting") ideally is as follows. A hole (h^+) must remove an electron from a water molecule, to leave a hydrogen ion and a hydroxyl radical (OH·):

$$H_2O + h^+ \rightarrow H^+ + OH\cdot.$$

The hydroxyl radical is written with a superscripted dot to show that it has an unpaired electron. It is an extremely reactive species; two can react to yield hydrogen peroxide, which in turn breaks down to water and oxygen:

[1]See, e.g., Ref. 107 in main text.

$$OH^{\cdot} + OH^{\cdot} \rightarrow H_2O_2 \rightarrow H_2O + \tfrac{1}{2}\,O_2,$$

Conversely, the hydrogen ion can react with a photogenerated electron (e⁻) to yield hydrogen gas:

$$H^+ + e^- \rightarrow \tfrac{1}{2}\,H_2.$$

These are the desired reactions, but the species are so reactive that back-reaction to reform H_2O is a major source of inefficiency. These are the sorts of issues that will have to be addressed with better catalysts and nanostructuring. Indeed, if the hydrogen peroxide intermediate product could be isolated before breaking down, it would be a high-value byproduct in its own right.

By far the majority of this work, however, has employed titanium dioxide (TiO_2), which is cheap, nontoxic, and abundant, as well as stable indefinitely in contact with H_2O in the presence of oxygen (O_2). In fact, in some sense TiO_2 is considered the "benchmark" photocatalyst.[128] The bandgap of TiO_2 is 0.512 aJ (~3.2 eV), which corresponds to a wavelength of ~390 nm. This lies in the near ultraviolet (UV), which is why TiO_2 powder appears bright white to the naked eye.

Most of this research has focused on "water splitting"— the decomposition of water with the release of hydrogen gas[129]

---

[128]Walter, MG; Warren, EL; McKone, JR; Boettcher, SW; Mi, Q; Santori, EA; Lewis, NS. Solar water splitting cells. *Chem. Rev.* 110, 6446–73, 2010; Zhu, J; Chakarov, D; Zäch, M. Nanostructured materials for photolytic hydrogen production. In *Energy Efficiency and Renewable Energy Through Nanotechnology*, Zang, L; ed. GE&T, 441–86, 2011; Coronado, JM; Hernández-Alonso, MD. The keys of success: TiO_2 as a benchmark photocatalyst. In *Design of Advanced Photocatalytic Materials for Energy and Environmental Applications*, Coronado, JM; Fresno, F; Hernández-Alonso, MD; Portela, R; eds. GE&T, 85–101, 2013.

[129]E.g., Schneider, J; Kandiel, TA; Bahnemann, DW. Solar photocatalytic hydrogen production: Current status and future challenges. In Viswanathan

(Box 5.13). Unsurprisingly, in the light of all the discussions above, catalysts are critical for practical photohydrogen production. Early workers had found that dispersed surface deposits of a catalyst such as platinum greatly improve photoredox efficiency, probably through facilitating hole–electron separation as well as through direct electrocatalytic activity,[130] and the action of platinum catalysts has recently been modeled in detail.[131] Once again, then, there is a strong incentive to seek alternatives to noble-metal catalysts.[132] There is also an extensive literature on nanostructuring the TiO_2, which can increase its reactivity and efficiency.[133]

et al. (Ref. 137); Schneider, J; Matsuoka, M; Takeuchi, M; Zhang, J; Horiuchi, Y; Anpo, A; Bahnemann, DW. Understanding TiO_2 photocatalysis: Mechanisms and materials. *Chem. Rev. 114*, 9919–86, 2014.

[130]E.g., Sakata, T; Kawai, T; Hashimoto, K. Photochemical diode model of platinum/titanium dioxide particle and its photocatalytic activity. *Chem. Phys. Lett. 88*, 50–4, 1982; Disdier, J; Herrmann, JM; Pichat, P. Platinum/titanium dioxide catalysts. A photoconductivity study of electron transfer from the ultraviolet-illuminated support to the metal and of the influence of hydrogen. *Faraday 1 79*, 651–60, 1983; Hope, GA; Bard, AJ. Platinum/titanium dioxide (rutile) interface. Formation of ohmic and rectifying junctions. *J. Phys. Chem. 87*, 1979–84, 1983; Kobayashi, T; Yoneyama, H; Tamura, H. Role of platinum overlayers on titanium dioxide electrodes in enhancement of the rate of cathodic processes. *J. Electrochem. Soc. 130*, 1706–11, 1983; Ohtani, B; Iwai, K; Nishimoto, S-I; Sato, S. Role of platinum deposits on titanium(IV) oxide particles: Structural and kinetic analyses of photocatalytic reaction in aqueous alcohol and amino acid solutions. *J. Phys. Chem. B 101*, 3349–59, 1997.

[131]Meng, Q; Yao, Y; Kilin, D. Anions vs. cations of $Pt_{13}H_{24}$ cluster models: Ab initio molecular dynamics investigation of electronic properties and photocatalytic activity. In *Nanotechnology for Sustainable Energy*, Hu, YH; Burghaus, U; Qiao, S; eds. ACSSS *1140*, 173–85, 2013.

[132]E.g., Wang, G; Lu, X; Li, Y. Low-cost nanomaterials for photoelectrochemical water splitting, in Ref. 1, pp. 267–95; Kärkäs, MD; Björn Åkermark, B. Water oxidation using earth-abundant transition metal catalysts: opportunities and challenges. *Dalton 45*, 14421–61, 2016.

[133]Chen, X; Selloni, A. Introduction: Titanium dioxide (TiO_2) nanomaterials. *Chem. Rev. 114*, 9281–82, 2014, and other papers in this issue; Ye, M; Lv, M; Chen, C; Iocozzia, J; Lin, C; Lin, Z. Design, fabrication, and modification of cost-effective nanostructured TiO_2 for solar energy applications. In Ref. 1, pp. 9–54; cf. also Ref. 114.

Nanostructuring is also another approach to better utilization of the incident sunlight. A wide-bandgap semiconductor such as TiO_2 absorbs <12% of the incident sunlight, because relatively few solar photons lie in the UV. Much as in the case of photovoltaic designs discussed above, a number of approaches have been taken to boost the absorption.[134] These include layered semiconductors with different bandgaps, doping of the semiconductor to change its absorption,[135] and sensitization with surface deposits, including quantum dots.[136]

Alternative semiconductors to TiO_2, indeed to oxide semiconductors generally, also remain a topic of current research. They offer potential advantages in terms of light absorption and the location of the bandgap.[137]

---

[134]Chen, X; Shen, S; Guo, L; Mao, SS. Semiconductor-based photocatalytic hydrogen generation. *Chem. Rev. 110*, 6503–70, 2010.

[135]E.g., Jensen, S; Kilin, D. Electronic properties of silver doped TiO_2 anatase (100) surface. In *Nanotechnology for Sustainable Energy*, Hu, YH; Burghaus, U; Qiao, S; eds. ACSSS *1140*, 187–218, 2013; Asahi, R; Morikawa, T; Irie, H; Ohwaki, T. Nitrogen-doped titanium dioxide as visible-light-sensitive photocatalyst: Designs, developments, and prospects. *Chem. Rev. 114*, 9824–52, 2014.

[136]Fernando, KAS; Sahu, S; Liu, Y; Lewis, WK; Guliants, EA; Jafariyan, A; Wang, P; Bunker, CE; Sun, Y-P. Carbon quantum dots and applications in photocatalytic energy conversion. *ACS Appl. Mater. Interfaces 7*, 8363–76, 2015.

[137]E.g., Viswanathan, B; Subramanian, V; Lee, JS; eds. *Materials and Processes for Solar Fuel Production*. NS&T *174*, 2014; Huda, MN. Theoretical modeling of oxide-photocatalysts for PEC water splitting. In Viswanathan et al. (eds.) above; Maeda, K; Takata, T; Domen, K. (Oxy)nitrides and oxysulfides as visible-light-driven photocatalysts for overall water splitting. In *Energy Efficiency and Renewable Energy Through Nanotechnology*, Zang, L; ed. GE&T, 487–529, 2011; Osterloh, FE; Parkinson, BA; Recent developments in solar water-splitting photocatalysis, *MRS Bull. 36*(01), 17–22, 2011; Huang, S; Lin, Y; Yang, J-H; Yu, Y. CdS-based semiconductor photocatalysts for hydrogen production from water splitting under solar light. In *Nanotechnology for Sustainable Energy*, Hu, YH; Burghaus, U; Qiao, S; eds. ACSSS *1140*, 219–41, 2013; Schneider et al., Ref. 129; Kwolek, P; Pilarczyk, K; Tokarski, T; Lapczynska, M; Michal Pacia, M; Szacilowski, K. Lead molybdate - a promising material for optoelectronics and photocatalysis. *J. Mater. Chem. C 3*, 2614–23, 2015.

Of course, although the direct photoproduction of hydrogen would dovetail nicely with the proposed hydrogen economy, the disadvantages of H_2 storage and delivery remain, particularly in the near term. Such considerations have motivated investigations into the semiconductor-mediated photoreduction of carbon dioxide or inorganic carbonate to simple organic molecules,[138] which could be used as fuels or feedstocks. This would provide yet another approach to carbon dioxide fixation, and in addition inorganic carbonate is abundant in rocks such as limestone (p. 332). Moreover, even though water-miscible fuels such as alcohols

---

[138]E.g., Halmann, M; Katzir, V; Borgarello, E; Kiwi, J. Photoassisted carbon dioxide reduction on aqueous suspensions of titanium dioxide. *Solar Energy Mater. 10*, 85–91, 1984; Thampi, KR; Kiwi, J; Grätzel, M. Methanation and photo-methanation of carbon dioxide at room temperature and atmospheric pressure. *Nature 327*, 506–8, 1987; Raphael, MW; Malati, MA. The photocatalyzed reduction of aqueous sodium carbonate using platinized titania. *J. Photochem. Photobiol. A 46*, 367–77, 1989; Goren, Z; Willner, I; Nelson, AJ; Frank, AJ. Selective photoreduction of carbon dioxide/bicarbonate to formate by aqueous suspensions and colloids of palladium-titania *J. Phys. Chem. 94*, 3784–90, 1990; Heleg, V; Willner, I. Photocatalyzed CO_2-fixation to formate and H2-evolution by eosin-modified Pd-TiO_2 powders. *Chem. Commun.* 2113–14, 1994; Yamashita, H; Nishiguchi, H; Kamada, N; Anpo, M; Teraoka, Y; Hatano, H; Ehara, S; Kikui, K; Palmisano, L; Sclafani, A; Schiavello, M; Fox, MA. Photocatalytic reduction of CO_2 with H_2O on TiO_2 and Cu/TiO_2 catalysts. *Res. Chem. Intermed. 20*, 815–23, 1994; Kuwabata, S; Uchida, H; Ogawa, A; Hirao, S; Yoneyama, H. Selective photoreduction of carbon dioxide to methanol on titanium dioxide photocatalysts in propylene carbonate solution. *Chem. Commun.* 829–30, 1995; Teramura, K; Tanaka, T; Photocatalytic reduction of CO_2 using H_2 as reductant over solid base photocatalysts. In Hu, Ref. 80 (Chapter 6), pp. 15–24; Centi, G; Perathoner, S. Nanostructured electrodes and devices for converting carbon dioxide back to fuels: Advances and perspectives. In *Energy Efficiency and Renewable Energy Through Nanotechnology*, Zang, L; ed. GE&T, 561–83, 2011; Yoshida, H. Heterogeneous photocatalytic conversion of carbon dioxide. In *Energy Efficiency and Renewable Energy Through Nanotechnology*, Zang, L; ed. GE&T, 531–59, 2011; Ganesh, I. Conversion of carbon dioxide to methanol using solar energy: A brief review. *Materials Sciences and Applications, 2*, 1407–15, 2011; Kumar, B; Llorente, M; Froehlich, F; Dang, T; Sathrum, A; Kubiak, CP. Photochemical and photoelectrochemical reduction of CO_2. *Annu. Rev. Phys. Chem. 63*, 541–69, 2012; Viswanathan, B. Reduction of carbon dioxide: Photo-catalytic route to solar fuels. In Viswanathan et al. Ref. 137.

are difficult to separate from the aqueous synthesis environment, as discussed above water-alcohol mixtures should be directly usable in next-generation fuel cells. Ammonia production via the semiconductor-mediated photoreaction of nitrogen (N_2) with hydrogen-bearing molecules has also been the subject of some studies,[139] as well as nitrogen fixation in general.[140]

Hydrogen production from alternative compounds is also being investigated. In some sense this is an "easier" problem, because hydrogen is less tightly bound in most organic molecules than in water.[141] This also dovetails with the applications of photocatalysis in environmental remediation, discussed in Chapter 6 (p. 322). It also is another approach to processing waste biomass (p. 242), particularly in aqueous solution or suspension, and indeed could provide another example of "waste into resource."

Photocatalytic conversion of other compounds is also under study. Methane conversion to methanol via photocatalytic

---

[139]Schrauzer, GN; Strampach, N; Hui, LN; Palmer, MR; Saleshi, J. Nitrogen photoreduction on desert sands under sterile conditions. *Proc. Natl. Acad. Sci. USA 80*, 3873–76, 1983; Ileperuma, OA; Weerasinghe, FNS; Lewke Bandara, TS. Photoinduced oxidative nitrogen fixation reactions on semiconductor suspensions. *Solar Energy Mater. 19*, 409–14, 1989.

[140]E.g., Kisch, H. Nitrogen photofixation at nanostructured iron titanate films. In *Energy Efficiency and Renewable Energy Through Nanotechnology*, Zang, L; ed. GE&T, 585–99, 2011; Schrauzer, GN, Photoreduction of nitrogen on TiO_2 and TiO_2-containing minerals. In *Energy Efficiency and Renewable Energy Through Nanotechnology*, Zang, L; ed. GE&T, 601–23, 2011.

[141]E.g., Sakata, T; Kawai, T. Hydrogen production from biomass and water by photocatalytic processes. *Nouv. J. Chim. 5*(5–6), 279 81, 1981; Fu, X; Long, J; Wang, X; Leung, DYC; Ding, Z; Wu, L; Zhang, Z; Li, Z; Fu, X. Photocatalytic reforming of biomass: A systematic study of hydrogen evolution from glucose solution. *Int. J. Hydrog. Energy 33*, 6484–91, 2008; Fukumoto, S; Kitano, M; Takeuchi, M; Matsuoka, M; Anpo, M. Photocatalytic hydrogen production from aqueous solutions of alcohol as model compounds of biomass using visible light-responsive TiO_2 thin films. *Catal. Lett. 127*, 39–43, 2009; Kaneco, S; Miwa, T; Hachisuka, K; Katsumata, H; Suzuki, T; Verma, SC; Sugihara, K. Photocatalytic hydrogen production from aqueous alcohol solution with titanium dioxide nanocomposites. In *Nanocatalysis for Fuels and Chemicals*. Dalai, AK; ed. ACSSS 1092, 25–36, 2012.

oxidation[142] would be of particular interest. As discussed above, the direct conversion of methane into methanol with atmospheric oxygen would be a breakthrough technology for the utilization of stranded gas. Moreover, many of the stranded gas deposits are located in areas with abundant sunlight.

Nanofabrication and semiconductor photosynthesis

Although cheap nanofabrication also is likely to be necessary to make artificial photosynthesis practical, a number of issues are considerably simpler or even irrelevant. Junction fabrication is not required because a Schottky barrier is set up automatically at the wetted semiconductor interface with the aqueous solution, due to the difference in Fermi levels between water and the semiconductor. Micro- or nanoscale wiring to collect the generated holes and electrons is also not needed. To be sure, as reviewed above the photocatalysts themselves will require nanostructuring, and those issues are now less well-defined than they are for photovoltaics, because of the lesser maturity of the field. Indeed, many questions about the most advantageous technological approaches still remain.

Nonetheless, in the author's opinion artificial photosynthesis seems a much more flexible and powerful way of harnessing solar energy, because of its automatic solution of three pressing problems: the fitfulness of solar irradiation, the storability of sunlight-derived energy, and the looming depletion of convenient chemical fuels. Of course, breakthroughs in the storage of electrical energy, as discussed above, might change this assessment, particularly if large new sources of electricity become available (as with solar power satellites, p. 265). Nonetheless, semiconductor-based artificial photosynthesis seems extremely promising for otherwise attractive solar-power sites, such as deserts, that are remote from demand centers. Moreover, not only is semiconductor photosynthesis

---

[142]Taylor, CE. Methane conversion via photocatalytic reactions. *Catal. Today* 84, 9–15, 30 August 2003.

considerably more efficient than the natural version, it also is not nearly so water-intensive, nor does the water feedstock have to be pure—brines will work fine. In any case, deserts are likely

---

Box 5.14 On Deserts

Why are deserts traditionally valueless? After all, they get lots of solar energy! Of course, it's because they're waterless. But why is that such an issue? It's not so much because humans and animals require some water; it's because plants require *lots* of water. A smidgen of water is consumed in photosynthesis, but the amount is utterly trivial. The water a plant takes in is also its "blood," needed to transport nutrients and waste products. A rough calculation, based on high-productivity biomass crops and the rainfall they require, is that less than one H_2O molecule in 400 is split by photosynthesis. The other 400-odd just pass through the plant. If plants, the primary producers. can't grow, the whole foundation of traditional livelihoods—indeed, of life itself—doesn't exist. And if water isn't present, plants can't exist.

Even now deserts aren't worth much. If they're not *too* deserty, there may be enough plant life to support some grazing—although it's easy to overdo, as overgrazed lands from the ancient Levant to the 20th century western US show all too clearly. If they have minerals worth the energy and infrastructure investment, they may sprout mines. In wealthy societies deserts are handy for proving grounds and (in *really* wealthy societies) for recreation. But overall, they're little more valuable than they were several thousand years ago.

Making fuels with sunlight could be the first really new economic activity for deserts since their use for military testing ranges, and would have the additional advantage of not competing with more conventionally valuable land. Of course, if hydrogen is being made by splitting water, *some* water is still needed as raw material. But the amount is trivial by comparison with that required by plants. Besides, the water need not be pure; brines (seawater, saline lake water, even the waste brine from oil fields) will work just fine. Plants, on the other hand, require much purer water.

Because hydrogen is almost as difficult to store and ship as is electricity itself (Box 5.3), it might make sense to make other fuels instead, such as methanol (Box 5.5). The carbon in the methanol could come from CO_2 in the air, or from limestone (made up of calcium carbonate, $CaCO_3$), which is a common rock in many deserts. Of course, drawing CO_2 from the atmosphere would also help stabilize atmospheric CO_2 content just as does natural photosynthesis. Finally, what would make a solar-fuel installation even *more* attractive economically would be to make a high-value non-gaseous oxidizer like H_2O_2, instead of just oxygen, as a by-product.

There's more than a little irony here, too: with such technologies the desert-rich nations of the Mideast could continue to export fuel indefinitely. For example, if a net production of 100 watts per square meter for 12 hours a day is assumed, which makes reasonable allowances for inefficiencies even in a technologically mature system, the ~10 million barrels of oil that Saudi Arabia exports daily is equivalent to a piece of desert about 120 km on a side.

There's *lots* more desert than that around the Persian Gulf. If the hydrogen required all comes from water, about a billion gallons a day is required—or only about 3000 acre-feet ($3,700,000$ m^3). Even with evaporation losses, that will still be just a few percent of the water required for a comparably sized agricultural operation. Furthermore, the water need not be fresh: seawater, or the brines the "oil" wells in those once-great fields are now producing copiously, will work fine. Of course, *other* desert areas could also become energy competitors: Australia, Chile, the vast parts of the Sahara *not* underlain by any oil fields, the southwest US, and so on. Deserts are going to go from being wastelands to highly valuable real estate.

to become considerably more valuable real estate than has been historically the case (Box 5.14).

Sunlight as a Distributed Power Source

If "solar power" shouldn't be automatically considered in terms of conversion into electricity, perhaps a further consideration is why convert it at all. Of course, direct use of solar heat for low-grade process heating dates back to antiquity (e.g., drying food), but it can be also used for such applications as space heating. Indeed, this is merely another aspect of passive energy management, already discussed above.

Solar heat could be used to drive thermal engines directly, as in current experimental projects. As mentioned above, should such direct use of solar heat prove convenient, thermoelectric materials are likely to be more attractive than heat engines using working fluids and macroscopic moving parts.

Other applications of direct solar energy use light as a "reagent" for powering chemical processes. Photocatalysis for environmental remediation through destruction of pollutants is becoming an established technique, as described in Chapter 6 (p. 322). Also described in Chapter 6 are so-far nascent approaches to using light as a "switch," to trigger the adsorption/desorption of solutes from a surface, for example (p. 305).

Solar Power Satellites (SPSs)

A trivial calculation shows that the entire Earth intercepts only $\sim 5 \times 10^{-10}$th of the $\sim 4 \times 10^{26}$ watts radiated by the Sun. This "missed" solar energy represents an enormous potential energy resource. Indeed, many years ago Dyson[143] speculated that an extremely advanced civilization might trap all its star's outgoing energy by surrounding it completely with a shell.

[143]Dyson, FJ. Search for artificial stellar sources of infrared radiation. *Science* 131, 1667–8, 1960.

This is obviously not of near-term relevance for humanity. Nonetheless, sunlight passing by Earth seems accessible to near-term technology. So-called solar power satellites (SPSs, sunsats) were first discussed seriously almost 50 years ago[144] and have been (and continue to be[145]) the subject of studies worldwide. Such satellites would intercept sunlight in Earth orbit and beam it down to receivers on Earth, probably via collimated microwaves.

Space holds many advantages as a location for solar energy collection. First, the incident sunlight is at least 30% stronger, not having been partly reflected or absorbed by the atmosphere. Obviously there is no weather, which decreases maintenance costs as well. Depending on the orbit details, the satellites are also not subjected to a day-night cycle, but only short and intermittent eclipses. Against these must be set the high costs of access and construction, as well as potential public fears of beamed power.

The economics of such satellites are analogous to those of large hydropower dams: an enormous up-front capital expenditure is required before any payback is received. Once the expenditure is made, however, the ongoing costs are minimal. Hence the costs are dominated by the amortization of the investment.

Nanotechnology and SPS

The same economies of fabrication that will make ground-based PVs considerably more attractive also make SPS more feasible. The economies of mass from superstrength materials would be particularly valuable because of the necessity to decrease costs to orbit. Thus the economics of space- vs. ground-based solar power seem poised particularly finely between the economics of production, the

---

[144]Glaser, PE. Power from the Sun: Its future. *Science 162*, 857–61, 1968.
[145]Glaser, PE; Davidson, FP; Csigi, KI; eds. *Solar Power Satellites: A Space Energy System for Earth,* Praxis Publishing, Chichester, 654 p. 1998; Sangster, in Ref. 106, Ch. 9.

tradeoffs of competing technologies (e.g., photovoltaics vs. photosynthesis), and space access costs.

Applications of Nanotechnology to Fission Energy

Nuclear fission clearly has a number of serious disadvantages as an energy source. Not only is there the problem of nuclear waste (Box 2.13), but accidents such as those at Chernobyl and Fukushima underscore the potential hazards of the reactors themselves. However, the real hazard of nuclear energy is probably malice; the consequences of deliberate misuse are so extraordinarily severe that the motivation hardly matters, whether for purported political goals or just out of sheer malevolence.

Conventional nuclear energy also seems archaic in other respects: it's highly centralized, probably unavoidably, and is a thermal technology with a strong 19th-century flavor. Nonetheless, fission plants could come online quickly for electricity production, and have the advantage of producing no greenhouse gases.

In an ironic working of the "law of unintended consequences," however, nuclear reactor designs have attracted little attention over the last few decades, due to the lack of interest in the expansion of nuclear energy. There is little doubt that safer and better reactors could be designed, but there has hardly been an incentive to do so.[146]

In any case, though, fission is obviously a thermal technology, and it's hard to see how it can be otherwise. The energies of nuclear reactions lie orders of magnitude above those of chemical bonds, so even molecularly structured assemblages seem unlikely to be able to control nuclear reactions at the atomic level. However, this does not mean nanotechnology is irrelevant to fission. Most obviously, thermoelectric power generation may prove practical (p. 235).

---

[146]Dyson, FJ. *Disturbing the Universe*, 283 pp., Colophon Books/Harper & Row, 1979, pp. 94–106.

Mining and Extraction

The big application of nanotechnology toward fission fuels, however, probably lies in their extraction and separation. Fission fuels consist of the two rare metals uranium and thorium (Box 5.15) and they are "fossil," too, in a sense: they also are dug out of the Earth and their supply is ultimately finite. Thus the considerations of concentration and grade (p. 85) apply to these deposits just as they do to non-energy-related metals. Of course, the absolute quantities in the crust are enormous by everyday standards.

As described in Chapter 6, a major resource-related application of nanotechnology lies in the low-energy molecular separation of elements, most directly from aqueous solution (p. 277). As for other elemental commodities, therefore, nanotechnology could vastly increase the supply of uranium and thorium. Leaching of low-grade uranium- or thorium-bearing rocks, for hydrometallurgical extraction, has already been carried out,[147] and selective "uranophiles" (highly selective agents for complexing the uranyl ion, UO_2^{2+}) have been the subject of research.[148] Uranium also exists in natural

---

[147]E.g., Sayed, SA; Sami, TM; Abaid, AR. Extraction of uranium from "Abu-Tartur" phosphate aqueous leachate solution. *Sep. Sci. Technol. 32*, 2069, 1997.

[148]Araki, K; Hashimoto, N; Otsuka, H; Nagasaki, T; Shinkai, S. Molecular design of a calix[6]arene-based super-uranophile with C3 symmetry. High UO_2^{2+} selectivity in solvent extraction. *Chem. Lett.* 829–32, 1993; Jacques, V; Desreux, JF. Complexation of thorium(IV) and uranium(IV) by a hexaacetic hexaaza macrocycle: Kinetic and thermodynamic topomers of actinide chelates with a large cavity ligand. *Inorg. Chem. 35*, 7205–10, 1996; Nagasaki, T; Shinkai, S. Synthesis and solvent extraction studies of novel calixarene-based uranophiles bearing hydroxamic groups. *Perkin 2*, 1063–6, 1991; Shinkai, S; Shiramama, Y; Satoh, H; Manabe, O; Arimura, T; Fujimoto, K; Matsuda, T. Selective extraction and transport of uranyl ion with calixarene-based uranophiles. *Perkin 2*, 1167–71, 1989; Xu, J; Raymond, KN. Uranyl sequestering agents: Correlation of properties and efficacy with structure for UO22+ complexes of linear tetradentate 1-methyl-3-hydroxy-2(1H)-pyridinone ligands. *Inorg. Chem. 38*, 308–15, 1999.

Box 5.15 Fission Fuels

The practical fissionable metals for conventional nuclear power are uranium-235 (^{235}U), plutonium-239 (^{239}Pu), and uranium-233 (^{233}U). Of these, the only naturally occurring isotope is ^{235}U, and it makes up only 0.7% of natural uranium. It is extremely difficult to separate out; isotopes are difficult to separate in any case, as described in Chapter 6, and the small proportional mass difference exacerbates the difficulty. Traditionally a major industrial plant has been required for ^{235}U enrichment.

Uranium-238 and Plutonium

The abundant uranium isotope ^{238}U is not fissionable by slow neutrons. On absorbing a neutron, however, it will transform into fissionable plutonium-239 (^{239}Pu) by a chain of short-lived *beta decays* (Box 2.14) over the course of a few weeks. Journalistic hyperbole notwithstanding, ^{239}Pu is *not* "the most toxic substance known," but it certainly is extremely hazardous. It is effectively "eternal" on human timescales, with a half-life of 24,110 years. Because it is a distinct element, too, it is relatively easy, although exceedingly hazardous without a fully equipped industrial plant, to separate from the unchanged uranium. Ordinary wet-chemical techniques can be used, particularly if the safety of the personnel involved is not an issue. Plutonium has been extensively used in nuclear weapons, but it is not routinely used in conventional power plants, although an oxide mixture containing 5% plutonium recycled from nuclear warheads is being used in some power reactors.[1]

Quite apart from its toxicity, a major concern with a "plutonium economy" is nuclear proliferation. If plutonium is being extracted and processed on a large scale, clandestine diversion of some for a nuclear explosive has seemed altogether too easy. Much the same considerations, of course, apply to nuclear waste in general, especially given the next-

---

[1]USDOE-EIA, 2000.

generation molecular separation technologies discussed in Chapter 6. A so-called "breeder" reactor avoids a specific plutonium-generation step by making ^{239}Pu from ^{238}U directly inside the reactor, and then fissioning the ^{239}Pu in situ. Although a subject of interest for decades, it is still at best an immature technology. Research was abandoned by the US in the 1970s, but continues elsewhere.

Thorium and Uranium-233

Another approach to nuclear fission uses thorium-232 (^{232}Th), the only long-lived isotope ($t_{1/2} \sim 10$ billion years) of thorium, as a nuclear fuel precursor. Like ^{238}U, it is not fissionable by slow neutrons, but it will absorb a neutron to form ^{233}Th, which beta-decays over several days to fissionable ^{233}U.[2] (The reaction is analogous to the formation of ^{239}Pu from ^{238}U.) Although there has been low level research on thorium as an energy source for decades, ^{233}U has not been employed on other than an experimental basis, although in recent years interest has dramatically increased.[3] India, in particular, with large thorium resources is mounting a large effort.[4] In any case thorium remains a "backstop" energy resource, particularly since the crustal abundance of thorium (7.2 ppm) is about four times that of uranium.[5]

---

[2]E.g., Glasstone, S. *Sourcebook on Atomic Energy.* D. Van Nostrand, 1958, p. 452.

[3]Hargraves, R; Moir, R. Liquid fluoride thorium reactors. *Am. Sci. 98,* 304-13, July-Aug, 2010.

[4]Das, D; Bharadwaj, SR. *Thoria-based Nuclear Fuels: Thermophysical and Thermodynamic Properties, Fabrication, Reprocessing, and Waste Management.* GE&T, 2013, and refs therein.

[5]Ref. 1 in Table 1.1.

brines; indeed, extraction of UO_2^{2+} from seawater has been investigated.[149]

---

[149]Aihara, T; Goto, A; Kago, T; Kusakabe, K; Morooka, S. Rate of adsorption of uranium from seawater with a calix[6]arene adsorbent. *Sep. Sci. Technol.* 27, 1655, 1992; Kanno, M. Design and cost studies on the extraction of uranium

A potential "down side" relevant to nuclear proliferation should be mentioned. Protactinium-231 (half-life ~3 × 10⁴ years), which occurs naturally in small amounts in uranium-bearing materials from the decay of ^{235}U, could also be extracted. This isotope is the only naturally occurring one besides ^{235}U with possible utility in a nuclear device,[150] and unlike ^{235}U it could be purified without the necessity of separating isotopes.

Reprocessing of Nuclear Fuel and Separation of Nuclear Waste

The result of nuclear fission is a mishmash of fission products, the lighter nuclei formed by the fission of the uranium atoms, mixed together with a set of heavier uranium and actinide isotopes formed by absorption of stray neutrons (Box 2.13). Indeed, it's just the debris from another thermal technology. It's also ironic that radwaste is dangerous—i.e., highly radioactive— because it still contains a lot of energy! A fair amount (~1% ^{235}U, as well as ~5% ^{239}Pu generated in situ[151]) of unreacted fuel also remains, but it is unusable because of competitive absorption of neutrons by the other nuclides.

Obviously, it would be attractive to separate the components of nuclear waste, both for fuel recovery and also to recover potentially valuable radionuclides. However, separating— "reprocessing"—nuclear waste is both difficult and hazardous. Indeed, such complex mixtures are difficult to deal with even when not strongly radioactive. The current procedures involve dissolution in strong acid and then separation via a long and complicated sequence of steps using precipitation, ion exchange, solvent extraction, and so on, with the additional difficulty that

---

from seawater. *Sep. Sci. Technol. 16*, 999–1017, 1981; Kelmers, AD. Status of technology for the recovery of uranium from seawater. *Sep. Sci. Technol. 16*, 1019–35, 1981; Yamashita, H; Fujita, K; Nakajima, F; Ozawa, Y; Murata, T. Extraction of uranium from seawater using magnetic adsorbents. *Sep. Sci. Technol. 16*, 987–98, 1981.

[150]Franta, J. Letter. *Phys. Today 47*, 84, Dec 1994.

[151]E.g., USDOE-EIA, 2000.

all reagents and materials used themselves become contaminated with radioactive material.

Separation from such solutions could alternatively be carried out molecularly, by systems like those described in the next chapter (p. 277). Indeed, besides the "uranophiles" described above, effort has been directed toward finding actinide-specific binding agents, although the focus of these studies has been therapeutic applications.[152] This is a more difficult problem, however, because of radiation damage to the nanomechanisms. They will have to be robust and ultimately probably self-repairing. If such systems can be developed, though, they will make the reprocessing of nuclear waste considerably more practical. Indeed, possibly each reactor installation could reprocess its own waste. Such systems would also render irrelevant the idea of "isolating" nuclear waste for ca. 10,000 years, as has been the stated US policy for decades. Indeed, it *will* be reprocessed one of these days; and probably sooner rather than later.[153] (That 10,000 year target number is the age at which the radioactivity has decayed to roughly the level of natural uranium ores.)

Such technologies, however, will also have unfortunate political ramifications. They are likely to make it much easier for small groups to extract nuclear materials from radwaste such as that from conventional power reactors. Separation of elemental plutonium (or ^{233}U from irradiated thorium) should be relatively straightforward, even if the lifetimes of the nanomechanisms are limited due to radiation damage. These concerns become particularly cogent with the recent global concerns about terrorism.

---

[152]E.g., Raymond, KN; Freeman, GE; Kappel, MJ. Actinide-specific complexing agents: their structural and solution chemistry. *Inorg. Chim. Acta 94*, 193–204, 1984; Kappel, MJ; Nitsche, H; Raymond, KN. Specific sequestering agents for the actinides. 11. Complexation of plutonium and americium by catecholate ligands. *Inorg. Chem. 24*(4), 605–11, 1985.

[153]A perceptive student of mine once observed that stashing nuclear waste in long-term "permanent" storage sites merely amounts to building mines for the future.

Isotope Separation

Isotope separation is fundamental to nuclear materials. Separating isotopes, however, is much more difficult than separating elements, and at least in the near term there seem to be few ways that nanotechnology can yield major improvements, as will be described in Chapter 6 (p. 314). If nanotechnology approaches to isotope separation prove feasible, however, this is an obvious application. Unfortunately, if for example ^{235}U could be readily separated from ^{238}U it would lead to similar political ramifications as the cheap separation of plutonium.

Applications of Nanotechnology to Fusion Energy

As noted in Chapter 2, conventional approaches to fusion energy embody the thermal paradigm: get the material hot enough so that a few nuclei slam together and fuse. The approach is seen most extremely, of course, in the hydrogen bomb, triggered by a fission explosion!

Whether nanotechnological fabrication would significantly help the advent of practical fusion reactors, in particular via non-thermal approaches, is unclear. Obviously, the same considerations of relative energy differences between the nuclear reactions vs. those of chemical bonds apply just as strongly here as they do in the case of fission energy. Possibly, however, alternatives to the "brute-force" fusion by thermal plasmas, such as laser ignition or electrostatic confinement, will benefit from fabrication at nanoscale precision. One such different conceptual approach is to accelerate particles into target nuclei. For example, the reaction

$$^1H + {}^{11}B \rightarrow 3\,{}^4He$$

has been the subject of patents.[154] (^{11}B, boron-11, is one of the stable isotopes of boron.) To add to its conceptual elegance, such

---

[154]Farnsworth, PT. US Patent 3,258,402; Bussard, RW. Method and apparatus for controlling charged particles. US Patent 4826646, 1989.

Box 5.16 Fusion Fuels

As noted in Chapter 2 (Box 2.12), very light nuclei—in principle, anyway—can be fused into heavier nuclei to yield energy, and the Sun derives its energy from fusing 4 ordinary hydrogen nuclei (protons) into a helium-4 nucleus. The obstacle is getting the fusing nuclei close enough, against the mutual repulsion of their protons, so that the nuclear strong force takes over.

Certain other light nuclei are much easier to fuse than ordinary hydrogen. They include "heavy" hydrogen (^2H, deuterium, whose nucleus consists of a proton and a neutron), lithium, and boron. These nuclides are rare in the Earth's crust, and indeed in the Universe as a whole; On a per-atom basis, boron makes up 0.0019% of the crust; lithium 0.0060%.[1] Nonetheless, Earth's complement is still enormous by human standards. The Earth's oceans, for example, contain something like 10^{17} kilograms of deuterium, even though it makes up less than 1 atom in 5000 in terrestrial hydrogen.

Both lithium and boron are particularly amenable to separation from aqueous solutions using one or another of the technologies described in Chapter 6. However, both will also require isotopic separation as well, as each consists of two stable isotopes (^6Li and ^7Li, and ^{10}B, ^{11}B). As emphasized, although different isotopes have very similar chemical properties, they have very different nuclear properties. In any case, of course, deuterium will require isotopic separation from ordinary hydrogen. As discussed in Chapter 6, isotopic separation is a considerably more difficult problem despite its intrinsic low thermodynamic cost, but if controlled nuclear fusion can be achieved, the payoff will be worth it.

---

[1]Ref. 1 in Table 1.1.

fusion need not merely yield heat because the alpha-particles (^{4}He nuclei) could be braked electrostatically, and thus their energy be converted to electricity directly.[155] Evidently, however, this approach has proved more difficult than expected. Certainly no commercial applications exist.[156]

"Nuclear catalysis" has also been intermittently studied since the 1950s. A *muon* is a particle that's for all intents and purposes a "heavy electron." When a muon orbits a nucleus, then, the effective size of the resulting "muonic" atom is much smaller, and this can let a nucleus get close enough to another for fusion to occur. In effect the muon cancels the charge on the nucleus. Unfortunately, the muon is an unstable particle, and so, although "muon-catalyzed" fusion has been observed, it's been far below the level needed for practical applications.

Finally, the claims for "cold fusion," advanced some two decades ago, should be mentioned. In this case fusion of deuterium (^{2}H) was claimed to occur at room temperature when the deuterium was absorbed into palladium metal, which has an extremely high affinity for hydrogen. Its existence is doubted by most scientists. If, however, the irreproducibility of the claimed nuclear reactions result from the vagaries of the nanoscale structure of the host material,[157] fabricating reproducible nanostructures will be necessary for its documentation, much less its practical application.

---

[155]Bussard, RW; Jameson, LW. From SSTO to Saturn's moons: Superperformance fusion propulsion for practical spaceflight. In *Fusion Energy in Space Propulsion*, T. Kammash, ed., Washington, DC: American Institute of Aeronautics and Astronautics, 1995.

[156]Such "Bussard reactors" are even the subject of science-fair projects these days, based on the author's experience as a judge in 2009. In fact, it's not always appreciated that such reactors are a way to generate hazardous nuclear radiation using only parts gotten from the local hobby shop. Any backyard experimenters should exercise extreme caution.

[157]Marwan, J; Krivit, SB; eds. *Low-Energy Nuclear Reactions Sourcebook*. ACSS 998, 2008; *Low-Energy Nuclear Reactions and New Energy Technologies Sourcebook Volume 2* ACSSS 1029, 2009.

Separation of Fusion Fuels

The obvious application of nanotechnology to fusion, as for fission, again lies in the cheap separation of nuclear materials. Also as with fission, nuclear reactions depend on the particular nuclide, not on the element, so that isotope separation is again a critical issue (p. 314). The prospects of separating deuterium from hydrogen more cheaply seem best, because they have the largest relative mass difference of any two stable isotopes. Against this, of course, is the fact that deuterium makes up such a small percentage of natural hydrogen. Furthermore, other nuclides that are most promising for fusion, such as lithium-7 (^{7}Li) and boron-11 (^{11}B), not only will also require isotope separation, but are in addition all isotopes of rare elements. However, both elements have been enriched in certain natural brines (p. 116), from which they are already extracted with current technology, so their separation by the more efficient molecular extraction technologies described in Chapter 6 (p. 277) would be straightforward.

This concludes the chapter on energy applications. In the next chapter, applications of nanotechnology to material resources and environmental remediation will be discussed.

Chapter 6

Mineral Resources, Pollution Control, and Nanotechnology

In this chapter we turn to applications of nanotechnology to materials resources. Besides the applications in extraction and pollution control, already touched on in Chapter 3, maturing nanotechnology is likely to cause sweeping changes in the mix of desired materials. In particular, structural metals are likely to become obsolete. This bold statement will be buttressed later in the chapter, after discussion of molecular separation technologies.

Nanotechnology and Molecular Separation

As was shown in Chapter 3, a single, fundamental technological problem pervades much of engineering: the separation of one kind of atom (or molecule) from an arbitrary background of other kinds. If we *want* what we extract, it's resource extraction; if we don't, it's the problem of pollution control or purification. It was also shown that, despite widespread beliefs, element extraction is *not* a fundamentally expensive procedure, and the astonishing separation capabilities of organisms are an obvious illustration of this fact. In turn, biosystems are capable of their

Nanotechnology and the Resource Fallacy
Stephen L. Gillett
Copyright © 2018 Pan Stanford Publishing Pte. Ltd.
ISBN 978-981-4303-87-3 (Hardcover), 978-0-203-73307-3 (eBook)
www.panstanford.com

remarkable feats of separation because they use molecular mechanisms that literally carry out separations molecule by molecule. The diatoms mentioned in Chapter 3 (Fig. 3.5), for example, bind individual molecules of dissolved silicic acid (H_4SiO_4) in the ambient water with specialized complexing agents, whence they are brought into the cell. Such molecular separation is both far more efficient and capable of extraction from considerably lower concentrations than techniques based on phase changes, particularly when the latter are driven by heating and cooling (i.e. melting, crystallization).

Some technical approaches to non-thermal separation were briefly reviewed in Chapter 3. All would be much more effective, and in particular more selective, if they could be better structured at molecular level, and we now turn to a discussion of the improvements that could be possible, as well as some new approaches.

Semipermeable Membranes

As described in Chapter 3, semipermeable membranes are "molecular filters": some molecules can pass, while others cannot. Although they already have a host of applications in separating gas mixtures and aqueous solutions, better membranes promise a vast new array of applications. For example, better membranes for the pervaporation of ethanol from ethanol–water mixtures are the subject of intense research. As noted, such membrane technology would obviate the perceived necessity to purify ethanol derived from biomass by distillation. As also emphasized in Chapter 3, among the most important separation problems is the extraction of solutes from aqueous solutions. This is basic to pollution control and purification, which are clearly not amenable to separation by thermally driven phase changes, but for which a number of membrane processes (electrodialysis, reverse osmosis) also exist (cf. Box 6.1).

Despite their obvious differences in detail, all semipermeable membranes share similar issues for their improvement, which

Box 6.1 Desalination and Water Purification

The availability of fresh water is critical to human activities and indeed to life itself. The dwindling supplies of fresh water throughout the world, therefore, are a source of serious concern and have inspired a number of suggestions, ranging from better water conservation to the transport of huge icebergs from the Antarctic. These proposals have gained greater urgency because water shortages are likely to be exacerbated by climate change.

A long-standing suggestion is to desalinate seawater, by far the largest reservoir of water on the planet. Although desalination is carried out now, it is restricted to niche markets, and has not been practical in general. Indeed, it, and water purification generally, are widely believed to be intrinsically energy intensive processes. However, this again proves to reflect the clumsy technologies employed.

To quantify the energy savings possible, the fundamental thermodynamic costs of seawater desalination can be calculated. Seawater is assumed for simplicity to consist of an aqueous solution containing only aqueous sodium cations ($Na^+(aq)$) and chloride anions ($Cl^-(aq)$). The free energy is concentration dependent:

$$\Delta G = \Delta G_0 + RT \ln \{Na^+\} \{Cl^-\}$$

where ΔG_0 ($= -8.5$ kJ/mol[1]) is the free energy of formation of the aqueous ions in their standard state from solid NaCl, R the gas constant, T the absolute temperature, and the braces indicate the activities of the aqueous ions. For seawater, $\{Na^+\} \sim 0.332$ and $\{Cl^-\} \sim 0.346$,[2] and hence $\Delta G \sim -13.9$ kJ/mol NaCl. Since there is roughly 30 g (~ 0.5 mole) NaCl in 1 kg seawater, the fundamental limiting cost of purifying 1 kg of seawater is ~ 6.9 kJ. By contrast, purification by boiling and

[1]From Robie & Hemingway, 1995.
[2]Berner, RA. *Principles of Chemical Sedimentology.* McGraw-Hill, 240 pp., 1971, p. 47.

recondensation, at atmospheric pressure, is roughly 350 times as great.

Even though minor constituents have been neglected, therefore, desalination is clearly not an inherently expensive process; it is merely so if carried out with phase changes such as with distillation. Approaching this thermodynamic limit, however, requires carrying out the separation atomistically under isothermal conditions, and that implies processes using nanostructured materials. Semipermeable membranes provide one approach, as does capacitive deionization, as are both discussed in the main text.

come down to better control at the nanoscale. Ideally, such membranes are strong, with large nanodefect-free surface areas, and with the pores all uniform to maximize permeate selectivity. Additionally, in the case of membranes that must be permeable to charged species, as in electrodialysis, the charged groups in the membrane that cause the selectivity should be consistently arranged in the pores. It further follows that the nanofabrication must also be carried cheaply enough for the applications to be economic.

To address such issues, research is being carried out using in many cases using one or another of the techniques briefly reviewed in Chapter 4.[1] In the last few years, graphene (Box 6.2) has received particular attention as the "ultimate" membrane material due to its great potential strength.[2]

---

[1]E.g., Jackson, EA; Hillmyer, MA. Nanoporous membranes derived from block copolymers: From drug delivery to water filtration. *ACS Nano* 4, 3548–53, 2010; Zhao, J; Pan, F; Li, P; Zhao, C; Jiang, Z; Zhang, P; Cao, X. Fabrication of ultrathin membrane via layer-by-layer self-assembly driven by hydrophobic interaction towards high separation performance. *ACS Appl. Mater. Interfaces* 5, 13275–83, 2013; Zhang, G; Ruan, Z; Ji, S; Liu, Z. Construction of metal-ligand-coordinated multilayers and their selective separation behavior. *Langmuir* 26, 4782–9, 2010.

[2]Jiang, D; Cooper, VR; Dai, S. Porous graphene as the ultimate membrane for gas separation. *Nano Lett.* 9, 4019–24, 2009.

Box 6.2 Carbon: Diamond, Graphite, and More

Schoolchildren learn about the drastically different forms ("polymorphs") of pure elemental carbon, diamond and graphite. Diamond is one of the hardest substances known. It is transparent but with a high index of refraction which gives the crystals their well-known fire. Graphite, on the other hand, is black, flaky, and weak, making up the bulk of pencil lead. It's even used as a lubricant! It's also a conductor, while diamond is an insulator.

These striking macroscopic contrasts, of course, are due to the different ways the carbon atoms are linked to each other in each substance. In diamond the carbon atoms are covalently linked into a three-dimensional framework, with each carbon being bonded to four others that lie at the corners of a regular tetrahedron. This is an example of "sp^3 hydridization"; the outer four bonding orbitals blend into a symmetric pattern, a regular tetrahedron where the bonding electrons reside (Box 4.5). This hydridization also accounts for the tetrahedral shape of the methane molecule, and indeed it is widespread in carbon compounds. It is also strong: the carbon-carbon bond in diamond, for example, has a tensile strength of 10.6 nanonewtons (nN)/bond. Moreover, because the bonds are covalent, they also offer strong resistance to a shear stress between the atoms. The diamond C—C bond has a shear strength of 6.7 nN/bond,[1] nearly as great as its tensile strength.

In graphite, by contrast, the carbon atoms are bonded in hexagonal sheets. Each carbon atom has three coplanar nearest neighbors, where the angle between the central carbon and any two of the neighbors is 120°, leading to the hexagonal pattern. This bonding is termed "sp^2;" only three of the low energy orbitals are hybridized to give the symmetrical 120° pattern. The remaining orbital forms a so-called "π- (pi)-bond", where electrons are delocalized between the carbon atoms. It is exactly analogous to the "aromatic" bonding in

---

[1] Drexler, in Ref. 3 (Chapter 4) (main text), pp. 142–3.

molecules such as benzene. The delocalization strengthens the bonding, because the extra electrons give the bonds some double-bond character. It also accounts for the electrical conductivity and the opacity, because the electrons are free to move, somewhat in the matter of electrons in metallic bonding. Conversely, the sheets are only weakly bonded to each other, being nearly free to slide around. This accounts for graphite's flakiness and lubricity.

Diamond and graphite have been known for centuries and have been well understood for decades. It hardly would have been expected that new forms of pure carbon would be discovered. Yet in the last 25 years two separate Nobel prizes have been awarded for the discovery and characterization of new forms of carbon. The first, awarded in 1996 to Richard Smalley, Harry Kroto, and Robert Curl, was for buckminsterfullerene and related compounds.[2] Buckminsterfullene, C_{60}, has perfect icosahedral symmetry, the carbons being arranged in a pattern like the alternating hexagons and pentagons on a soccer ball. The name comes from the resemblance of the pattern to Buckminster Fuller's geodesic domes. Related nearly spherical molecules also exist, being generically termed "fullerenes."

For the purposes of nanotechnology, so-called nanotubes (or "buckytubes") are probably more important, and are in any case the subject of much research.[3] These conceptually can be derived from a fullerene by cutting it in half and "stitching in" new rings of carbons between the halves. The molecule then becomes elongated into a tube—a buckytube. There further is no theoretical limit to the elongation.

Most immediately, if they can be made long enough, buckytubes have obvious applications at superstrength fibers. They seem to have ultimate tensile strengths comparable

---

[2]www.nobelprize.org/nobel_prizes/chemisty/laureates/1996/index.html.
[3]E.g., Jorio, A; Dresselhaus, G; Dresselhaus, MS; eds. *Carbon Nanotubes. Top. Appl. Phys. 111*, 2008.

to diamond (~100 GPa[4]). However, as their bulk density will presumably be similar to graphite's (~2.2 g/cm[3]), over a third less than diamond's (3.5 g/cm[3]), their specific strength will be substantially (>35%) larger.

Another conceptual way to make a buckytube is to consider a single sheet of graphite: "graphene." Snip out a long strip of the sheet, roll the strip up into a long tube, and fasten the sides together. It was long thought that an isolated graphene sheet would be unstable, but isolated graphene was the basis for the 2010 Nobel Prize in Physics to Andre Geim and Konstantin Novoselov.[5] Indeed, not only is free graphene stable, it is attracting a host of attention for various applications,[6] some of which (e.g., semipermeable membranes) have obvious resource and environmental relevance.

---

[4]E.g., Yakobson, BI; Smalley, R. Fullerene nanotubes: C1,000,000 and beyond. *Am. Sci. 85*, 324–37, 1997.

[5]Geim, AK; Novoselov, KS. The rise of graphene. *Nat. Mater. 6*, 183–91, 2007.

[6]Geim, AK. Graphene: Status and prospects. *Science 324*, 1530–34, 2009. Zhu, Y.; Murali, S; Cai, W; Li, X; Suk, JW; Potts, JR; Ruoff, RS. Graphene and graphene oxide: Synthesis, properties, and applications. *Adv. Mater. 22*, 3906–24, 2010; Ferrari, AC; et al. Science and technology roadmap for graphene, related two-dimensional crystals, and hybrid systems. *Nanoscale 7*, 4598–810, 2015.

Membranes based on molecular sieves (Box 6.3) are also receiving increased attention.[3] They are more chemically resistant, especially at high temperature, and furthermore have pores whose uniformity is enforced by the crystal structure. Although conventional molecular-sieve syntheses yield tiny crystals, membranes fabricated on supports, commonly nanoporous

---

[3]E.g., Bowen, TC; Noble, RD; Falconer, JL. Fundamentals and applications of pervaporation through zeolite membranes. *J. Membr. Sci. 245*, 1–33, 2004, and references therein; Nenoff, TM; Dong, J. Highly selective zeolite membranes. *Ordered Porous Solids* 365–86, Elsevier, 2009; Algieri, C; Barbieri, G; Drioli, E. Chapter 17. Zeolite membranes for gas separations. In *Membrane Engineering for the Treatment of Gases 2*, 223–52, RSC, 2011.

Box 6.3 Molecular Sieves and Zeolites

Molecular sieves are crystalline compounds that have open framework structures such that molecular-scale voids and tunnels, whose size and shape depend on the particular sieve, extend throughout the crystal. The first known molecular sieves were *zeolites*, framework aluminosilicates similar to feldspars (Box 3.5) except for their open structures. Indeed, some zeolites occur naturally as minerals, and the name "zeolite," which comes from the Greek for "boiling stone," was given by the pioneering mineralogist Cronstedt in 1750. When a natural zeolite is heated, water in the voids fizzes out spectacularly. Of course, synthetic molecular sieves are not restricted to the silicon and aluminum that occur as framework cations in natural zeolites, and oxide frameworks containing phosphorus, boron, and even metal ions such as iron and titanium have been synthesized. Other metal phosphates, titanium and zirconium in particular, form 3D framework structures, and have attracted some attention for carrying out separations.[1]

Molecular sieves are an excellent example of present proto-nanotechnology[2] and they have been mentioned repeatedly throughout this book. As discussed in the main text, an important current application is as catalysts, especially in the petrochemical industry, and indeed much of this research on synthesizing new molecular sieves has been motivated by the search for better catalysts. The molecular-scale voids in the sieve structure lead to molecular recognition, as different sieves can be very specific indeed for absorbing certain molecules and rejecting others, depending on whether the species fits into the holes. Such specificity is also a major factor in their applications in separation, as adsorbers and semipermeable membranes, as noted in the main text. They were also among the first ion-exchangers.

[1]See, e.g., Ref. 9 in main text.
[2]E.g., Niwa, M; Katada, N; Okumura, K. *Introduction to Zeolite Science and Catalysis.* Springer Ser. Mater. Sci., 2010.

With better control on their nanoscale fabrication, they are also likely to find more application as solid electrolytes (also called ionic conductors), substances that can conduct a flow of cations. Solid electrolytes are critical for practical fuel cells (Chapter 5) and may find application in separation as well.

Finally, another embryonic application of molecular sieves is as a crystallographically uniform molecular framework support for novel nanomaterials. Using *chimie douce* techniques (p. 154), for example, not just new ions but clusters of atoms can be introduced into the framework. For one example, zeolites can provide a nanoscale framework for artificial photosynthesis, in one case by imposing molecular scale organization on dyes, and in the other by providing a vastly enlarged surface area.[3] In another example, a group has precipitated cadmium sulfide clusters—groups consisting of 4 cadmium and 4 sulfur atoms into the voids of zeolite Y.[4] The "crystal" spacing of these clusters is obviously very different from that in a bulk CdS crystal, so the electronic interactions between the clusters are also very different. Such materials are essentially three-dimentional quantum dot arrays.

[3]Calzaferri, G. Artificial photosynthesis *Top Catal. 53*, 130–40, 2010.
[4]E.g., Ozin, GA; Ozkar, S. Intrazeolite topotaxy. *Adv. Mater. 4*, 11–22, 1992.

alumina, have yielded usable materials.[4] Nonetheless, the junctions between the crystals act as defects, and the ability to nanofabricate large defect-free surfaces remains a key goal.

[4]E.g., Dong, JH; Lin, YS. In situ synthesis of P-type zeolite membranes on porous alpha-alumina supports. *Ind. Eng. Chem. Res. 37*, 2404, 1998; Gu, X; Dong, JH; Nenoff, TM. Synthesis of defect-free FAU-type zeolite membranes and separation for dry and moist CO_2/N_2 mixtures. *Ind. Eng. Chem. Res. 44*, 937–44, 2005; Julbe, A. Zeolite membranes – synthesis, characterization and application. *Stud. Surf. Sci. Catal. 168*, 181–219, 2007; Alomair, AA; Al-Jubouri, SM; Holmes, SM. A novel approach to fabricate zeolite membranes for pervaporation processes. *J. Mater. Chem. A 3*, 9799–806, 2015.

Indeed, the nanofabrication challenges for all semipermeable membranes are much as with superstrength materials (p. 324): that is, the problem of building structures over macroscopic distances that remain defect-free at the molecular level. It can be expected, therefore, that synergies will again exist in the research relevant to all these applications.

Electrolysis with solid electrolytes

Solid electrolytes were mentioned in Chapter 5 in the context of fuel cells and electrochromic materials. A quite different use of a solid electrolyte that has attracted recent attention is in electrometallurgy,[5] where a solid electrolyte layer separates the electrode surface from the molten electrolyte during electrolysis. It thereby minimizes electrode back-reactions, a serious source of efficiency losses. As noted in Chapter 3, electrolytic processes are capable of much greater theoretical efficiencies than thermally driven phase changes, and this may be one path to realizing such efficiencies.

"Binding" Approaches

Binding approaches such as ion exchange, adsorption, or complexation, in which the species to be extracted becomes attached to a substrate of some sort, were touched on in Chapter 2. The paramount importance of selectivity in most applications was also noted. Selectivity obviously involve a large degree of molecular recognition, and so the relevance of not just nanofabrication, but molecular design is highlighted.

The factors leading to recognition are often more subtle than merely the fit of the molecule to the substrate, as mentioned

---

[5]Krishnan, A; Lu, XG; Pal, UB. Solid oxide membrane process for magnesium production directly from magnesium oxide. *Metall. Mater. Trans. B* 36B, 463–73, Aug 2005; Pal, UB; Powell, AC., IV. The use of solid-oxide-membrane technology for electrometallurgy. *JOM-J. Miner. Met. Mater. Soc.* 59, 44–9, 2006.

previously. For example, the use of a molecular sieve (one of several zeolites) to enrich air in oxygen was mentioned in Chapter 3 (p. 130). The sieve acts as a specific absorber for nitrogen, not because nitrogen molecules fit better into the voids of the crystal, but because they interact more strongly with the charge-balancing cations in those voids.[6]

Similar applications using molecular sieves and related substances for various kinds of gas separation are already commercially available, as mentioned in Chapter 3, but a great deal of new research is occurring as well. One example is the use of metal-organic frameworks, which are open structures similar to molecular sieves but with potentially even greater flexibility in their design.[7] Another study investigated tailoring the pore size of zeolites.[8]

Ion exchangers

As mentioned in Chapter 3, ion exchange materials have electrically charged molecular groups to which ions can bind, and dissolved ions having greater affinities with the binding sites replace the species already there. Ion exchange is thus a mechanism for the selective adsorption of charged solutes, and much the same issues of selectivity and solute recognition are as with adsorption in general. Indeed, molecular sieves such as zeolites have long been used as ion exchangers. Other issues are much as with membranes containing charged species, as in electrodialysis, and indeed the same polymers (often termed *resins*) are commonly used. In particular, the ions in solution

[6]Papai, I; Goursot, A; Fajula, F; Plee, D; Weber, J. Modeling of N_2 and O_2 adsorption in zeolites. *J. Phys. Chem. 99*, 12925–32, 1995.

[7]E.g., Czaja, AU; Trukhan, N; Muller, U. Industrial applications of metal-organic frameworks. *Chem. Soc. Rev. 38*, 1284–93, 2009; Li, J-R; Sculley, J; Zhou, H-C., Metal-organic frameworks for separations. *Chem. Rev. 112*, 869–932, 2012.

[8]Chudasama, CD; Sebastian, J; Jasra, RV. Pore-size engineering of zeolite Z for the size/shape selective molecular separation. *Ind. Eng. Chem. Res. 44*, 1780–6, 2005.

obviously must have access to the charged sites, and this is an aspect that could be addressed with better nanofabrication.

The importance of selectivity continues to motivate a great deal of empirical research,[9] and although better selectivity also could be addressed with better nanofabrication, this of course requires that the mechanisms of selectivity be understood. Overall, many of the issues of selective recognition of the target species in adsorption-based approaches are similar to the issues with designer catalysts—cf. molecular sieves—and so there should be synergies with these applications.

In any case, the major disadvantage of exchange, as mentioned in Chapter 3, is that it doesn't change the total number of solutes.

Complexing agents

The use of agents for binding specific solutes was also touched on in Chapter 3, with the example of the use of hydoximes in the solvent extraction of copper (p. 126). This, of course, is also an application of molecular recognition (Box 4.7), and it proves to be a very fruitful approach indeed for binding specific solutes. A vast number of such agents have been synthesized by many research groups, and only a bare overview can be given here.

"Crown ethers," among the simplest of a group of molecules generically termed "macrocycles"[10] for their ringlike structures,

---

[9] E.g., Clearfield, A; Poojary, DM; Behrens, EA; Cahill, RA; Bortun, AI; Bortun, LN. Structural basis of selectivity in tunnel type inorganic ion exchangers. In *Metal Ion Separation & Preconcentration; Progress & Opportunities.* Bond, AH; ed. ACSSS 716, 168–82, 1999; Pan, BC; Zhang, QR; Zhang, WM; Pan, BJ; Du, W; Lv, L; Zhang, QJ; Xu, ZW; Zhang, QX. Highly effective removal of heavy metals by polymer-based zirconium phosphate: A case study of lead ion. *J. Colloid Interface Sci. 310,* 99–105, 2007. Zhang, Q; Pan, B; Pan, B; Zhang, W; Jia, K; Zhang, Q. Selective sorption of lead, cadmium and zinc ions by a polymeric cation exchanger. *Environ. Sci. Technol. 42,* 4140–45, 2008; Thakkar, R; Chudasama, U. Synthesis and characterization of zirconium titanium phosphate and its application in separation of metal ions. *J. Hazard. Mater. 172,* 129–37, 2009.

[10] Davis, F; Higson, S. *Macrocycles: Construction, Chemistry and Nanotechnology Applications.* John Wiley & Sons, 608 pages, 2011.

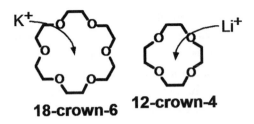

Figure 6.1 Crown ethers.

Selectivity depends on the fit of the ion inside the ring. Li^+ is the lithium cation; K^+ is the potassium cation.

furnish an illustrative example.[11] Chemically, an "ether" contains hydrocarbon groups linked by an oxygen atom. The anesthetic ether of years ago, for example, is diethyl ether, consisting of two ethyl (C_2H_5) groups linked by an oxygen: $(C_2H_5)_2O$. A "crown" ether is a ring-shaped ether, and such compounds are highly effective complexing agents for many metal ions.[12] They can also be strongly selective, in part due to the differing fits of ions into the ring ("crown"), and the selectivity can be strongly varied by changing the crown size, by adding side groups, and/or by substituting other atoms into the crown, which brings more subtle factors of the electronic structure into play (Fig. 6.1).[13] For example, the compound formed by replacing four of the oxygens in 18-crown-6 with sulfur (Fig. 6.2) forms complexes with silver ion in preference to sodium ion by a factor of over 10 billion. In fact, this substituted crown ether is the one used to recover palladium in the system described below.

---

[11]Gokel, GW; Leevy, WM; Weber, ME. Crown ethers: sensors for ions and molecular scaffolds for materials and biological models. *Chem. Rev. 104*, 2723–50, 2004.

[12]E.g., Bradshaw, JS; Izatt, RM. Crown ethers: The search for selective ion ligating agents. *Acc. Chem. Res. 30*, 338–45, 1997, and refs. therein.

[13]E.g., Fenton, DE. Metal ion recognition: The story of an oxa-aza macrocycle. *Pure Appl. Chem. 65*, 1493–8, 1993; Ref. 12.

Although similar complexing agents have attracted attention for therapeutic treatment of toxic metal poisoning,[14] merely complexing an ion is generally of limited value for *separation* because the complexed ion remains in solution. Now the whole complex must now be extracted, which may or may not be an easier problem. The example of the solvent extraction of copper (p. 126) shows one approach to such extraction, in which the complexing agent is more soluble in a coexisting liquid than in the aqueous solution, and research has been directed toward finding novel agents for use in solvent extraction, such as base metals[15] and radwaste.[16]

A perhaps more general approach to separation, however, is to bind the macrocycle to a substrate, in effect making a highly specific adsorbing surface. For example, modified calixarenes, a different and more complex class of macrocycles that has attracted enormous attention,[17] have been attached to a polymer "backbone" to make a highly specific adsorber for certain toxic anions such as dichromate and nitrite.[18]

---

[14]Ref. 151 (Chapter 5); Abu-Dari, K; Hahn, FE; Raymond, KN. Lead sequestering agents. 1. Synthesis, physical properties, and structures of lead thiohydroxamato complexes. *J. Am. Chem. Soc. 112*, 1519–24, 1990; Abu-Dari, K; Karpishin, TB; Raymond, KN. Lead sequestering agents. 2. Synthesis of mono- and bis(hydroxypyridinethione) ligands and their lead complexes. Structure of bis(6-(diethylcarbamoyl)-1-hydroxy-2(1H)-pyridine-2-thionato-O,S)lead(II). *Inorg. Chem. 32*, 3052–5, 1993.

[15]Tasker, P; Gasperov, V. Ligand design for base metal recovery. In *Macrocyclic Chemistry: Current Trends and Future Perspectives*. Springer, 365–82, 2005.

[16]Moyer, B; Birdwell, J; Bonnesen, P; Delmau, L. Use of macrocycles in nuclear-waste cleanup: A realworld application of a calixcrown in cesium separation technology. In *Macrocyclic Chemistry: Current Trends and Future Perspectives*. Springer. 383–405, 2005.

[17]E.g., Lhoták, P. Anion receptors based on calixarenes. *Top. Curr. Chem. 255*, 65–95, 2005; Davis & Higson, Ref. 10.

[18]Memon, S; Akceylan, E; Sap, B; Tabakci, M; Roundhill, DM; Yilmaz, M. Polymer supported calix[4]arene derivatives for the extraction of metals and dichromate anions. *J. Polym. Environ. 11*, April, 67–74, April 2003; Akceylan, E; Yilmaz, A; Yilmaz, M. Synthesis and properties of calix[4] arene polymers containing amide groups: Exploration of their extraction properties towards dichromate and nitrite anions. *Macromol. Res. 21*, 1091–96, 2013.

1,4,7,10-tetrathia
-18-crown-6

Figure 6.2 Substituted crown ether.

Because of the replacement of oxygen (O) by sulfur (S), this macrocycle has a selectivity for silver ion (Ag^+) over sodium ion (Na^+) of $>10^{10}$. It also has high affinity for palladium (Pd), as described in the text.

The substrate to which the macrocycle is attached need not be an organic polymer. Another group has "tethered" (covalently bound, via a hydrocarbon linkage) the substituted crown ether above (Fig. 6.2) to a silica surface.[19] Such functionalized surfaces are now being used for recovering platinum and palladium selectively from catalytic converter scrap.[20] The catalytic converters are dissolved in acid, and the tethered macrocycles specifically bind the precious metals, while the other, much more abundant metals (iron, aluminum, etc.) remain in solution. In the second decade of the 21st century, palladium prices have varied between roughly $US600 and $800 per troy ounce. Other such applications include nuclear waste separation,[21] and quantitative analysis.[22] This last application also illustrates how specific

---

[19]Izatt, RM; Bradshaw, JS; Bruening, RL; Tarbet, BJ; Bruening, ML. Solid-phase extraction of ions using molecular recognition technology. *Pure Appl. Chem. 67*, 1069–74, 1995; Izatt, RM; Bradshaw, JS; Bruening, RL. Selective separation using supported devices. *Compreh. Supramol. Chem. X*, Reinhoudt, DN; ed. Pergamon, pp. 1–11, 1996.

[20]Anonymous. New separation technique. *Industry Week 246*, 52, 1 Dec 1997.

[21]E.g., Izatt et al., 1996, Ref. 19.

[22]Goken, GL; Bruening, RL; Krakowiak, KE; Izatt, RM. Metal-ion separations using SuperLig or AnaLig materials encased in Empore cartridges and disks. In *Metal Ion Separation & Preconcentration; Progress & Opportunities*. Bond, AH; ed. ACSSS 716, 251–9, 1999.

complexation of solutes by macrocycles has found use in solute sensing. Although not directly relevant to separation, such an application certainly underscores the utility of nanotechnological approaches in environmental remediation.

Although already practical in such applications, where the expense is justified by the high value or extreme toxicity of the extractants, or simply because very little material is required, as in solute sensing, broader application will require cheaper syntheses of the complexing compounds. At present they are made by the elaborate, inefficient, and expensive conventional synthetic techniques mentioned in Chapter 4; indeed, typically the compounds themselves are considerably more valuable than the metals they are to extract, and the process is cost-effective only because they can be reused multiple times. More cost-effective syntheses are likely to result from the improvements in synthesis and nanofabrication outlined in Chapter 4, and provide yet another incentive for their development.

The "Elution Problem"

Even with much cheaper syntheses, however, all binding-based approaches have a fundamental disadvantage: they all require a *desorption* step. Once the exchanger, or adsorber, or complex has been fully loaded with the species to be extracted, it must be "regenerated": the extracted solute must be removed from the substrate so that it can be used again. (Alternatively, in some cases the fully loaded exchanger can just be discarded, as is sometimes practical for waste disposal when dealing with small amounts of an extremely toxic solute such as mercury. But at best it's an inelegant solution to the problem, and it is completely impractical in general. After all, in most applications the goal is to recover what's been extracted.)

Typically regenerating the complexing agent takes extreme chemical measures. As mentioned in Chapter 3 (p. 124), this is already familiar with ion exchangers: once completely charged with extracted calcium, for example, water softeners are regenerated by flushing a concentrated brine through them. The result is an ion exchanger whose active sites are again occupied by sodium ions, so that it is again ready to exchange calcium from dilute solution. This has been achieved, however, at the cost of generating a huge volume of secondary waste brine. Thus, not only do ion exchangers not change the total solute content, but in practice they worsen overall pollution problems.

Moreover, in complexation-based systems such as that for palladium described above (p. 291), the elution problem is even worse. In general, the more selective and quantitative the binding of a solute to a substrate, the more tightly it is bound and so the more difficult it is to *un*bind. Typically, therefore, regenerating the exchanger requires extreme chemical conditions such as elution with strong acids or bases. In the palladium recovery system, for example, highly concentrated acid must be used to flush out the complexed palladium.

Obviously, such regeneration steps generate a much larger volume of waste water (brine or acid waste) that now becomes a serious disposal problem. Indeed, further separation step(s) are now typically required; hence overall, the problem has been made worse, as there is now *more* solution than what was started with! Separation requiring elution can still be practical when recovering relatively small quantities of something very valuable like palladium, but its applications are obviously limited.

Switchable Binding

"Switchable" binding is a way to solve the elution problem: under one set of conditions binding occurs, but changing some variable (illumination, electrical potential, etc.) causes the solute to unbind. Once again, biology has anticipated technology: hemoglobin, for example, binds strongly to oxygen (O_2) in the lungs, but under the different chemical conditions elsewhere in the body it gives up the oxygen to the tissues. The hemoglobin molecule actually changes its configuration in doing so.

Indeed, switchable binding is characteristic of biological systems. Nutrients or raw materials brought into the cell, or waste products excreted from the cell, must be bound, transported, and then unbound again. Consider again the silica structures built by the diatoms (Fig. 3.5), in which individual dissolved silicate species are first gathered from the environment and then assembled into a silicate framework.

Electrosorption

Perhaps the simplest example of switchable binding is electrosorption, which is based on straightforward electrostatic attraction and repulsion. Charging an electrode attracts out ions having the opposite charge; reversing the charge of the electrode desorbs the ions again. Electrosorption has seemed a promising technique for water purification since the 1960s,[23] but only recently have improvements in electrode materials made it practical,[24] and commercial units are now available under the name capacitative deionization (CDI).[25] The technique is elegantly simple in conception: charged electrodes attract dissolved ions out of water to establish a double layer. When the electrodes are filled to capacity, they can then be regenerated merely by switching their polarity.

---

[23] Johnson, AM; Newman, J. Desalting by means of porous carbon electrodes. *J. Electrochem. Soc. 118*, 510–7, 1971.

[24] Farmer, JC; Fix, DV; Mack, GV; Pekala, RW; Poco, JF. Capacitive deionization of NaCl and $NaNO_3$ aq. solutions with carbon aerogel electrodes. *J. Electrochem. Soc. 143*, 159–69, 1996.

[25] Weinstein, L; Dash, R. Capacitive deionization: Challenges and opportunities. *Desalination & Water Reuse* 34–7, November-December, 2013; Suss, ME; Porada, S; Sun, X; Biesheuvel, PM; Yoon, J; Presser, V. Water desalination via capacitive deionization: what is it and what can we expect from it? *Energy Environ. Sci. 8*, 2296, 2015.

To extract a significant amount of material a high-surface-area electrode is obviously required. A kinship to double-layer capacitors (p. 225) is also obvious. Indeed, some of the same high-surface-area materials, such as carbon aerogels, are used for electrosorption electrodes. Also, because the completely filled electrode is essentially a charged capacitor, in theory a great deal of the electrical energy can be recovered when the electrode polarity is switched to desorb the ions, although this hasn't yet proved practical in the existing commercially available systems.

Redox-switchable binding

Another obvious approach is "redox" switching, in which the affinity of the substrate for the solute depends on whether it is reduced or oxidized (Box 2.15). As described above, complexation relies on molecular recognition, so it is not surprising that changes in a molecule's electronic structure might drastically affect its affinity for solutes—not least because of simple electrostatic effects!

A great many such systems have been synthesized over the last few decades.[26] Many are directed toward sensing

---

[26]Some recent references include: Higuchi, H; Matsufuji, T; Oshima, T; Ohto, K; Inoue, K; Tsend-Ayush, T; Gloe, K. Selective extraction of silver ion with proton-switchable calix[4]arene. *Chem. Lett.* 34, 80–1, 2005; Yanilkin, VV; Nastapova, NV; Mamedov, VA; Kalinin, AA; Gubskaya, VP. Redox-switchable binding of the Mg^{2+} ions by 21,31-diphenyl-12,42-dioxo-7,10,13-trioxa-1,4(3,1)-diquinoxaline-2(2,3),3(3,2)-diindolysine-cyclopentadecaphane. *Russ. J. Electrochem.* 43, 770–5, 2007; Yanilkin, VV, Stepanov, AS, Nastapova, NV, Mustafina, AR, Burilov, VA, Solovieva, SE, Antipin, IS, Konovalov, AI. Electroswitchable binding of $[Co(dipy)_3]^{3+}$ and $[Fe(dipy)_3]^{2+}$ sulfonato(thia) calix[4]arenes. *Russ. J. Electrochem.* 46, 1263–79, 2010; Yanilkin, VV; Mustafina, AR: Stepanov, AS; Nastapova, NV; Nasybullina, GR; Ziganshina, AYu; Solovieva, SE; Konovalov, AI. Electricoswitchable bonding of metal ions and complexes by calixarenes. *Russ. J. Electrochem.* 47, 1082, 2011; Yang, Y-W; Sun, Y-L; Song, N. Switchable host-guest systems on surfaces. *Acc. Chem. Res.* 47, 1950–60, 2014; Wang, Q; Cheng, M; Zhao, Y; Yang, Z; Jiang, J; Wang, L; Pan, Y. Redox-switchable host-guest systems based on a bisthiotetrathiafulvalene-bridged cryptand. *Chem. Commun.* 50, 15585–8, 2014; Pochorovski, I; Diederich, F. Development of redox-switchable

applications,[27] which, as noted above, are both important and involve relatively little material.

Quinones, for example, a well-known redox system in organic chemistry, have been functionalized with cation-binding groups.[28] Quinones readily and reversibly undergo reduction to

---

resorcin[4]arene cavitands. *Acc. Chem. Res. 47*, 2096–105, 2014; Kannappan, R; Bucher, C; Saint-Aman, E; Moutet, J-C; Milet, A; Oltean, M; Métay, E; Pellet-Rostaing, S; Lemaire, M; Chaix, C. Viologen-based redox-switchable anion-binding receptors. *New J. Chem. 34*, 1373–86, 2010. Dsouza, RN; Pischeland, U; Nau, WM. Fluorescent dyes and their supramolecular host/guest complexes with macrocycles in aqueous solution. *Chem. Rev. 111*, 7941–80, 2011. Also see Davis & Higson (Ref. 10) under the various classes of macrocycles, and Ref. 11. Older but still useful reviews include: Kaifer, A; Mendoza, S. Redox-switchable receptors. *Compreh. Supramol. Chem. I*, G. Gokel, ed. Pergamon, pp. 701–32, 1996; Shinkai, S. Switchable guest-binding receptor molecules. *Compreh. Supramol. Chem. I*, Gokel, G; ed. Pergamon. pp. 671–700, 1996.

[27]E.g., Wei, P; Li, D; Shi, B; Wang, Q; Huang, F. An anthracene-appended 2:3 copillar[5]arene: synthesis, computational studies, and application in highly selective fluorescence sensing for Fe(III) ions. *Chem. Commun. 51*, 15169–72, 2015.

[28]E.g., Delgado, M; Gustowski, DA; Yoo, HK; Gatto, VJ; Gokel, GW; Echegoyen, L. Contrasting one- and two-cation binding behavior in syn- and anti-anthraquinone bibracchial podand (BiP) mono- and dianions assessed by cyclic voltammetry and electron paramagnetic resonance spectroscopy. *J. Am. Chem. Soc. 110*, 119–24, 1988; Delgado, M; Wolf, RE, Jr; Hartman, JAR; McCafferty, G; Yagbasan, R; Rawle, SC; Watkin, DJ; Cooper, SR. Redox-active crown ethers. Electrochemical and electron paramagnetic resonance studies on alkali metal complexes of quinone crown ethers. *J. Am. Chem. Soc. 114*, 8983–91, 1992; Echegoyen, LE; Yoo, HK; Gatto, VJ; Gokel, GW; Echegoyen, L. Cation transport using anthraquinone-derived lariat ethers and podands: the first example of electrochemically switched on/off activation/deactivation. *J. Am. Chem. Soc. 111*, 2440–3, 1989; Echegoyen, L; Lawson, RC; Lopez, C; de Mendoza, J; Hafez, Y; Torres, T. Synthesis and electrochemical complexation studies of 1,8-bis(azacrown ether) anthraquinones. *J. Org. Chem. 59*, 3814–20, 1994; Gomez-Kaifer, M; Reddy, PA; Gutsche, CD; Echegoyen, L. Electroactive calixarenes. 1. Redox and cation binding properties of calixquinones. *J. Am. Chem. Soc. 116*, 3580–7, 1994; D'Souza, F. Molecular recognition via hydroquinone-quinone pairing: Electrochemical and singlet emission behavior of (5,10,15-Triphenyl-20-(2,5-dihydroxyphenyl) porphyrinato)zinc(II)-quinone complexes. *J. Am. Chem.*

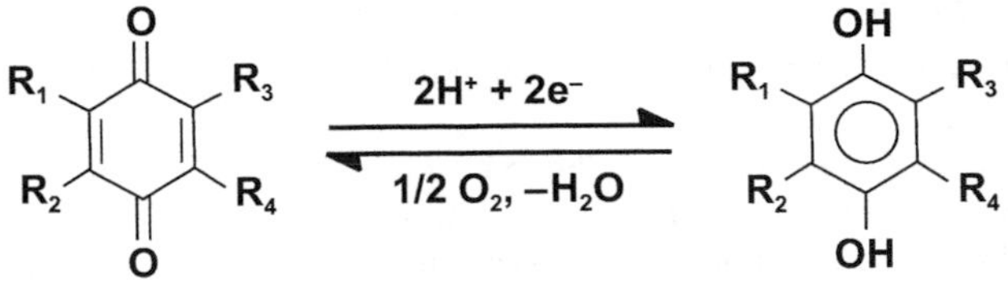

Figure 6.3 Quinone ↔ hydroquinone.

A simple example showing the basics of the quinone-hydroquinone redox pair. Quinones (left) are readily reduced to the hydroquinone form (right), and this redox switching is already extensively employed by biological electron-transfer mechanisms. Note that the quinone is defined by the presence of double-bonded oxygens on opposite sides of the 6-membered carbon ring. The R_i are various sidegroups. Note also that the reduced form, with the oxygens converted into diametrically opposite —OH groups, is aromatic; i.e., the 6-membered ring is now a benzene ring. Incorporation of a (hydro)quinone group into a macrocycle can greatly alter its affinity for incorporated ions, depending on its oxidation state.

the "hydroquinone" form (Fig. 6.3), and in fact are the basis of many biological electron-transfer reactions. The stoichiometry is

$$Q + 2\,e^- + 2\,H^+ \qquad \leftrightarrow \qquad QH_2$$

quinone (oxidized) hydroquinone (reduced)

Incorporation of a quinone group into a macrocycle causes the macrocycle's affinity for a cation to depend on the oxidation state of the quinone. Typically the reduced form shows higher affinity, as in effect the hydrogen of the hydroquinone is displaced by the cation.

Soc. 118, 923–4, 1996; Gokel, GW; Meisel, JW. Synthetic receptors for alkali metal cations. In *Synthetic Receptors for Biomolecules: Design Principles and Applications*, Smith, BD, ed. *Monographs in Supramolecular Chemistry* 14, RSC, 2015.

A great deal of research has also been focused on functionalized ferrocenes,[29] especially for analytical applications.[30] The

[29]E.g., Hall, CD; Tucker, JHR; Chu, SYF. Electroactive cryptands containing metallocene units. *Pure Appl. Chem.* 65, 591–4, 1993; Beer, PD; Wild, KY. New bis-ferrocenyl dibenzo-18-crown-6 ligands that can electrochemically sense group 1 and 2 metal cations. *Polyhedron* 775–80, 1996; Tendero, MJL; Benio, A; Martinez-Manez, R; Soto, J; Paya, J; Edwards, AJ; Raithby, PR. Tuning of the electrochemical recognition of substrates as a function of the proton concentration in solution using pH-responsive redox-active receptor molecules. *Dalton* 343–51, 1996; Plenio, H; Aberle, C. Oxaferrocene cryptands as efficient molecular switches for alkaline and alkaline earth metal ions. *Organometallics* 16, 5950–7, 1997; Hall, CD; Truong, T-K-U; Nyburg, SC. The synthesis and structure of a redox-active cryptand containing both aromatic and phenanthroline units within the macrocyclic structure. *J. Organomet. Chem.* 547, 281–6, 1997; Allgeier, AM; Slone, CS; Mirkin, CA; Liable-Sands, LM; Yap, GPA; Rheingold, A. Electrochemically controlling ligand binding affinity for transition metals via RHLs: The importance of electrostatic effects. *J. Am. Chem. Soc.* 119, 550–9, 1997; Plenio, H; Burth, D; Vogler, R. 1,2-FDTA: A ferrocene-based redox-active EDTA analogue with a high Ca^{2+}/Mg^{2+} and Ca^{2+}/Sr^{2+} selectivity in aqueous solution. *Chem. Ber.* 130, 1405–9, 1997; Beer, PD; Smith, DK. Tunable bis(ferrocenyl) receptors for the solution-phase electrochemical sensing of transition-metal cations. *Dalton* 417–24, 1998; Lloris, JM; Martinez-Manez, R; Padilla-Tosta, M; Pardo, T; Soto, J; Tendero, MJL. Open-chain polyazaalkane ferrocene-functionalised receptors for the electrochemical recognition of anionic guests and metal ions in aqueous solution. *Dalton* 3657–62, 1998; Su, N; Bradshaw, JS; Zhang, XX; Savage, PB; Krakowiak, KE; Izatt, RM. Syntheses of diaza-18-crown-6 ligands containing two units each of 4-hydroxyazobenzene, benzimidazole, uracil, anthraquinone, or ferrocene groups. *J. Heterocyclic Chem.* 36, 771–5, 1999; See also Davis & Higson (Ref. 10), esp. pp. 29 ff.

[30]E.g., Audebert, P; Cerveau, G; Corriu, RJP; Costa, N. Modified electrodes from organic-inorganic hybrid gels formed by hydrolysis-polycondensation of some trimethoxysilylferrocenes. *J. Electroanal. Chem.* 413, 89–96, 1996; Moutet, J-C; Saint-Aman, E; Ungureanu, M; Visan, T. Electropolymerization of ferrocene bis-amide derivatives: a possible route to an electrochemical sensory device. *J. Electroanal. Chem.* 410, 79–85, 1996; Blonder, R; Katz, E; Cohen, Y; Itzhak, N; Riklin, A; Willner, I. Application of redox enzymes for probing the antigen-antibody association at monolayer interfaces: Development of amperometric immunosensor electrodes. *Anal. Chem.* 68, 3151–57, 1996; Bruening, ML; Zhou, Y; Aguilar, G; Agee, R; Bergbreiter, DE; Crooks, RM. Synthesis and characterization of surface-grafted, hyperbranched polymer films containing fluorescent, hydrophobic, ion-binding, biocompatible, and electroactive groups. *Langmuir* 13, 770–8, 1997; Anicet, N; Anne, A; Moiroux, J; Saveant, J-M. Electron transfer in organized

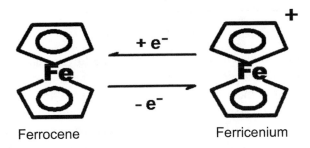

Ferrocene   Ferricenium

Figure 6.4 Ferrocene.

The "iron sandwich" compound ferrocene (dicyclopentadienyliron(II), abbreviated Fc) undergoes ready and reversible oxidation to the ferricenium cation, Fc^+. The two 5-membered rings are the cyclopentadienyl anion ($C_5H_5^-$), which is aromatically bonded as indicated by the enclosed circle (cf. Fig. 3.7).

"iron sandwich" compound ferrocene (Fc; Fig. 6.4, Box 6.4) is a very stable group that has been incorporated into a vast number of compounds.

A few groups have demonstrated prototype systems using functionalized ferrocenes for separation. One group, for example, demonstrated that a functionalized ferrocene could be a carrier in switchable liquid membrane transport.[31]

A potentially important example is the redox-switched binding and release of pertechnate ion (TcO_4^-) by modified ferrocenes.[32] Technetium (Tc, at. no. $Z = 43$) has no stable

[31] Saji, T; Kinoshita, I. Electrochemical ion transport with ferrocene functionalized crown ether. *Chem. Commun.* 716–7, 1986.

[32] Clark, JF; Clark, DL; Whitener, GD; Schroeder, NC; Strauss, SH. Isolation of soluble ^{99}Tc as a compact solid using a recyclable, redox-active, metal-

---

assemblies of biomolecules. Construction and dynamics of avidin/biotin co-immobilized glucose oxidase/ferrocene monolayer carbon electrodes. *J. Am. Chem. Soc. 120,* 7115 6, 1998; Collinson, SR; Gelbrich, T; Hursthouse, MB; Tucker, JHR. Novel ferrocene receptors for barbiturates and ureas. *Chem. Commun.* 555–6, 2001; Pratt, MD; Beer, PD. Anion recognition and sensing by mono- and bis-urea substituted ferrocene receptors. *Polyhedron 22,* 649–53, 2003; Scavetta, E; Mazzoni, R; Mariani, F; Margutta, RG; Bonfiglio, A; Demelas, M; Fiorilli, S; Marzocchi, M; Fraboni, B. Dopamine amperometric detection at a ferrocene clicked PEDOT:PSS coated electrode. *J. Mater. Chem. B 2,* 2861–7, 2014, and references therein.

Box 6.4 Ferrocene

The "iron sandwich" compound ferrocene (often abbreviated Fc; Fig. 6.4) is a very stable group that has been incorporated into a vast number of compounds. It consists of an iron atom between two cyclopentadienyl rings, a 5-membered aromatic hydrocarbon ring. Ferrocene itself is electrically neutral, with the iron atom in the Fe^{2+} oxidation state, and in this form is hydrophobic, as would be expected both from its hydrocarbon exterior and its neutrality. Ferrocene, however, can undergo a highly reversible oxidation to the "ferricenium" cation, losing one electron in the process:

$$Fc \Rightarrow Fc^+ + e^-$$

and in this form can make water-soluble salts with a characteristic baby-blue color. This reversible redox switching has led to the incorporation of the ferrocene group into many compounds, including ones directed toward sensing and separation, as discussed in the main text.

isotopes. The isotope ^{99}Tc, however, is a long-lived ($t_{1/2} \sim 2 \times 10^5$ years) fission product that is a source of serious environmental

complex extractant. *Environ. Sci. Technol.* 30, 3124–7, 1996; Clark, JF; Chamberlin, RM; Abney, KD; Strauss, SH. Design and use of redox-recyclable organometallic extractants for the cationic radionuclides ^{137}Cs$^+$ and ^{90}Sr^{2+} from waste solutions. *Environ. Sci. Technol.* 33, 2489–91, 1999; Strauss, SH. Redox-recyclable extraction and recovery of heavy metal ions and radionuclides from aqueous media. In *Metal Ion Separation & Preconcentration; Progress & Opportunities.* Bond, AH; ed. ACSSS 716, 156–65, 1999; Chambliss, CK; Odom, MA; Morales, CML; Martin, CR; Strauss, SH. A strategy for separating and recovering aqueous ions: Redox-recyclable ion-exchange materials containing a physisorbed, redox-active, organometallic complex. *Anal. Chem.* 70, 757–65, 1998; Dorhout, PK; Strauss, SH. The design, synthesis, and characterization of redox-recyclable materials for efficient extraction of heavy element ions from aqueous waste streams. In *Inorganic Materials Synthesis.* Winter, CH; Hoffman, DM; eds. ACSSS 727, 53–68, 1999.

concern. In aerated water it readily oxidizes to the soluble and mobile pertechnate ion, so methods of extracting low concentrations of dissolved pertechnate are obviously of great interest.

Ferrocene is already a hydrophobic molecule, but these researchers rendered it even more hydrophobic by attaching branched hydrocarbon chains at two points on each cyclopentadienyl ring. This molecule could also be adsorbed to a silica gel surface.[33] In the ferricenium form, this modified ferrocene forms ion pairs with large, low-charge, relatively hydrophobic anions such as pertechnate, so such anions are preferentially adsorbed onto this functionalized surface. On reduction of the ferricenium, however, the ion pairs break up so that the anion can be extracted and concentrated. Re-oxidation of the ferrocene to ferricenium then regenerates the adsorbing surface.

As for ordinary "unswitchable" complexing agents, current syntheses are cumbersome and expensive, and this will hinder large-scale applications in bulk separation. Furthermore, to make a switchable adsorber or exchanger the redox-active molecules will need to be tethered to some sort of substrate, as described in the previous section for unswitchable macrocycles. Probably the most important issue is that a practical system will need to undergo many—thousands if not millions—of redox cycles before needing replacement, and developing systems with this level of robustness is a challenge. Catalysts, of course, have similar issues, as described in the previous chapter, but they usually are not "switchable" themselves.

"Switchable membranes"

Much of the research on redox switched systems has been directed toward modeling of biological membrane transport. Transport across cell membranes, which are hydrophobic, typically involves complexation of the solute to be transported

---

[33]Chambliss et al., 1998, Ref. 32.

with a hydrophobic, switchable "carrier," which releases the complexed solute on crossing the membrane.[34] If such a system can be emulated technologically sufficiently cheaply, it would have obvious applications in separation involving solvent extraction or in so-called liquid membranes, in which the membrane is an immiscible liquid phase separating two aqueous phases.

Of course, conceptually solid semipermeable membranes also could be made "switchable," and indeed one group has reported electroswitchable transport of ions across a nanoporous membrane functionalized with redox-active groups.[35]

Intercalation-based systems (ESIX)

A different approach to redox-switched binding is through electrochemically switched intercalation and release of ions into an open crystal structure. Under the name "electrically switched ion exchange" (ESIX), this has attracted much recent attention, especially for cesium ion (Cs^+) extraction. That system is based on an electrode coated with cesium nickel hexacyanoferrate ($CsNiFe(CN)_6$),[36] which has the perovskite structure with cyanide ions replacing oxygen ions (Box 5.10), and with only half of the large cavities occupied. On reduction of some of the

---

[34]Fyles, TM. Synthetic ion channels in bilayer membranes. *Chem. Soc. Rev.* 36, 335–47, 2007; Sakai, N; Matile, S. Synthetic ion channels. *Langmuir 29,* 9031–40, 2013; Gokel, GW; Negin, S. Synthetic ion channels: From pores to biological applications. *Acc. Chem. Res. 46,* 2824–33, 2013.

[35]Small, LJ; Wheeler R; Spoerke, ED. Nanoporous membranes with electrochemically switchable, chemically stabilized ionic selectivity. *Nanoscale 7,* 16909–20, 2015.

[36]Lilga, MA; Orth, RJ; Sukamto, JPH; Rassat, SD; Genders, JD; Gopal, R. Cesium separation using electrically switched ion exchange. *Sep. Purif. Technol. 24,* 451–66, 2001; Sun, B; Hao, XG; Wang, ZD; Guan, GQ; Zhang, ZL; Li, YB; Liu, SB. Separation of low concentration of cesium ion from wastewater by electrochemically switched ion exchange method: Experimental adsorption kinetics analysis. *J. Hazard. Mater.* 233–4, 177–83, 2012; Chen, R; Tanaka, H; Kawamoto, T; Asai, M; Fukushima, C; Na, H; Kurihara, M; Watanabe, M; Arisaka, M; Nankawa, T. Selective removal of cesium ions from wastewater using copper hexacyanoferrate nanofilms in an electrochemical system. *Electrochim. Acta 87,* 119–25, 2013.

iron atoms from Fe^{3+} to Fe^{2+}, the cesium cation is preferentially drawn into the remaining large cavities to maintain charge balance. This intercalation is surprisingly selective even in the presence of other singly charged cations like sodium (Na^+), typically an abundant background ion. Reoxidizing the Fe^{2+} on the electrode by reversing the potential expels the intercalated cesium. As the isotope ^{137}Cs, which is both highly radioactive and has a half-life just long enough (about 30 years) to remain a hazard for decades, is of great concern in fission waste, the motivation for selective cesium removal at low concentrations is obvious. Another system simultaneously extracts iodide (I^-) by including in addition an electrode functionalized with polypyrrole.[37] Long-lived iodine-129 (^{129}I, $t_{1/2} \approx 15.7$ million years) is another fission product of great concern as an environmental contaminant.

ESIX systems have also been the object of study for other solutes. A similar reversible intercalation system has been proposed for alkali metal cations using Prussian blue (potassium iron hexacyanoferrate, $KFe_2(CN)_6$), which obviously could be of great interest for desalination.[38] To illustrate the difficulties with a practical system, though, irreversible behavior was found to occur on rubidium (Rb^+) and thallium (Tl^+) intercalation, which was attributed to changes in channel sizes in the hexacyanoferrate structure on intercalation.[39]

Another group has demonstrated switchable lithium intercalation into an electrode functionalized with lambda (λ)-

---

[37]Liao, S; Xue, C; Wang, Y; Zheng, J; Hao, X; Guan, G; Abuliti, A; Zhang, H; Ma, G. Simultaneous separation of iodide and cesium ions from dilute wastewater based on PPy/PTCF and NiHCF/PTCF electrodes using electrochemically switched ion exchange method. *Sep. Purif. Technol. 139*, 63–9, 2015.

[38]Ikeshoji, T. Separation of alkali metal ions by intercalation into a Prussian blue electrode. *J. Electrochem. Soc. 133*, 2108–9, 1986.

[39]Doštal, A; Kauschka, G; Reddy, SJ; Scholz, F. Lattice contractions and expansions accompanying the electrochemical conversions of Prussian blue and the reversible and irreversible insertion of rubidium and thallium ions. *J. Electroanal. Chem. 406*, 155–63, 1996.

manganese dioxide (MnO_2).[40] λ-MnO_2 has a "defect" spinel structure.[41] In this structure, one-third of the possible tetrahedral sites are not occupied. In the load cycle the electrode is made negative, which reduces one-third of the manganese cations from Mn^{4+} to Mn^{3+}. This causes the small lithium ion (Li^+) to be drawn into the empty sites to maintain charge balance. Lithium ions are small enough to fit into these sites, but other, larger cations cannot. In particular, sodium ion (Na^+) is too large to be intercalated in this way. As sodium is much more common than lithium, the intercalation also provides an elegant way of separating out lithium. Reoxidation of the manganese cations to Mn^{4+} by reversing the electrode potential expels the lithium again. Although this system is the subject of a patent for selective removal of lithium from dilute brines,[42] it evidently as yet has not found practical applications. Other groups have develop ESIX systems for heavy metal removal,[43] fluoride,[44] and perchlorate (ClO_4^-).[45] A similar system, using an electrode coated with redox-active polymers electropolymerized onto the surface, was shown to extract lead reversibly depending on electrode polarity.[46] These latter studies also underscore what

[40]Kanoh, H; Ooi, K; Miyai, Y; Katoh, S. Electrochemical recovery of lithium ions in the aqueous phase. *Sep. Sci. Technol. 28*, 643–51, 1993.

[41]Hunter, JC. Preparation of a new crystal form of manganese dioxide: λ-MnO_2. *J. Solid State Chem. 39*, 142–47, 1981.

[42]Kanoh, H; Ooi, K; Miyai, Y; Katoh, S. Method and electrode for electrochemical recovery of lithium values from aqueous solution. US Patent 5198081, 7 pp, 1993.

[43]Wang, Z; Feng, Y; Hao, X; Huang, W; Feng, X. A novel potential-responsive ion exchange film system for heavy metal removal. *J. Mater. Chem. A 2*, 10263–72, 2014.

[44]E.g., Cui, H; Qian, Y; An, H; Sun, C; Zhai, J; Li, Q. Electrochemical removal of fluoride from water by PAOA modified carbon felt electrodes in a continuous flow reactor. *Water Res. 46*, 3943–50, 2012.

[45]Zhang, S; Shao, Y; Liu, J; Aksay, IA; Lin, Y. Graphene-polypyrrole nanocomposite as a highly efficient and low cost electrically switched ion exchanger for removing ClO_4 from wastewater. *ACS Appl. Mater. Interfaces 3*, 3633–7, 2011.

[46]Admassie, S; Elfwing, A; Skallberg A; Inganäs, O. Extracting metal ions from water with redox active biopolymer electrodes. *Environ. Sci.: Water Res. Technol. 1*, 326–31, 2015.

has already been mentioned in passing: that environmental applications are likely to be the initial economic driver for such "cutting-edge" technologies.

Ion intercalation was already mentioned in the context of batteries (p. 221). Indeed, λ-MnO_2 has also been proposed as an electrode for Li batteries. Crystal structures capable of intercalating ions are also obviously closely akin to solid electrolytes (p. 209), and the intercalation of hydrogen ions into tungsten bronzes as an approach to fuel-cell electrolytes was also noted (p. 214). Indeed, ESIX provides another example of electrode functionalization, mentioned above with respect to electrocatalysts (p. 189), so developmental synergies with these other applications should exist.

Light-driven switching

An attractive "reagent" for switching the binding state of a solute is light. Merely shining sunlight on a surface to desorb its solutes would obviously be a lot cleaner and "greener" than flushing it with (say) strong acid solutions. It would also provide an example of using sunlight directly as a distributed power source (p. 265).

A number of prototype systems employing photoswitchable binding of a solute species have been described.[47] In one set of approaches, a complexing group such as a crown ether is linked to the active molecule so that the degree of binding depends on the state of the photoactive molecule.[48] Another approach

---

[47]See also Davis & Higson (Ref. 10) under the various classes of macrocycles. Shinkai (Ref. 26) gives a still useful review.

[48]E.g., Barrett, G; Corry, D; Creaven, BS; Johnston, B; McKervey, MA; Rooney, A. Anion- and solvent-dependent photochemical decomplexation of sodium salt complexes of a calix[4]arene tetraester. *Chem. Commun. 363*, 1995; Kimura, K; Mizutani, R; Yokoyama, M; Arakawa, R; Matsubayashi, G; Okamoto, M; Doe, H. All-or-none type photochemical switching of cation binding with malachite green carrying a bis(monoazacrown ether) moiety. *J. Am. Chem. Soc. 119*, 2062–63, 1997; Martin, MM; Plaza, P; Meyer, YH; Badaoui, F; Bourson, J; Lefevre, J-P; Valeur, B. Steady-state and picosecond spectroscopy of Li^+ or Ca^{2+} complexes with a crowned merocyanine.

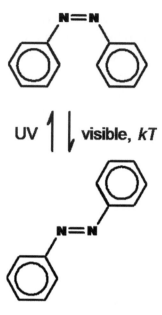

Figure 6.5 Azobenzene.

UV illumination causes rotation around the double bond to the sterically hindered (cis-) form. Longer-wavelength illumination, as in the visible, or exposure to ambient temperatures (indicated by kT, Boltzmann's constant times the ambient absolute temperature, which is a measure of average energy), cause reversion to the open (trans-) form.

is to employ molecules that change shape ("photoisomerize") on absorbing a photon. Azobenzene (Fig. 6.5), for example, in its ground state strongly resists rotation around the nitrogen–nitrogen double bond. On absorbing a photon, though, the molecule goes into an excited state in which the bond is free to rotate. When the excited state drops back to the ground state again, the double bond can "latch" into a position 180°

Reversible photorelease of cations. *J. Phys. Chem.* **100**, 6879–88, 1996; Stauffer, MT; Knowles, DB; Brennan, C; Funderburk, L; Lin, F-T; Weber, SG. Optical control over Pb^{2+} binding to a crown ether-containing chromene. *Chem. Commun.* 287–8, 1997; Tucker, JHR; Bouas-Laurent, H; Marsau, P; Riley, SW; Desvergne, J-P. A novel crown ether-cryptand photoswitch. *Chem. Commun.* 1165–6, 1997.

from its original orientation, and this is the basis for many photoswitchable systems.[49] Obviously, if the azobenzene group is incorporated into a macrocycle,[50] this completely changes the shape of the ring. Since, as described above, the binding of macrocycle to a solute strongly depends on the fit in the ring, changing the ring size and shape greatly changes the effectiveness of the binding. Similarly, photoisomerization around a C=C double bond has also been exploited,[51] as have other systems.[52]

---

[49]E.g., Li, Z; Liang, J; Xue, W; Liu, G; Liu, SH; Yin, J. Switchable azo-macrocycles: from molecules to functionalisation. *Supramol. Chem. 26*, 54–65, 2014. McCoy, TM; Liu, ACY; Tabor, RF. Light-controllable dispersion and recovery of graphenes and carbon nanotubes using a photo-switchable surfactant. *Nanoscale 8*, 6969–74. 2016.

[50]E.g., Blank, M; Soo, LM; Wassermann, NH; Erlanger, BF. Photoregulated ion binding. *Science 214*, 70–2, 1981; Shinkai, S; Minami, T; Kusano, Y; Manabe, O. A new "switched-on" crown ether which exhibits a reversible all-or-none ion-binding ability. *Tetrahedron Lett. 23*, 2581–4, 1982; Photoresponsive crown ethers. 8. Azobenzenophane-type switched-on crown ethers which exhibit an all-or-nothing change in ion-binding ability. *J. Am. Chem. Soc. 105*, 1851–6, 1983; Shinkai, S; Miyazaki, K; Manabe, O. Photoresponsive crown ethers. Part 18. Photochemically switched-on crown ethers containing an intra-annular azo substituent and their application to membrane transport. *Perkin 1* 449–56, 1987; Akabori, S; Miura, Y; Yotsumoto, N; Uchida, K; Kitano, M; Habata, Y. Synthesis of photoresponsive crown ethers having a phosphoric acid functional group as anionic cap and their selective complexing abilities toward alkali metal cations. *Perkin 1* 2589–94, 1995; Fuerstner, A; Sedidel, G; Kopiscke, C; Krueger, C; Mynott, R. Syntheses, structures, and complexation properties of photoresponsive crownophanes. *Liebigs Ann.* 655–62, 1996.

[51]E.g., Irie, M; Kato, M. Photoresponsive molecular tweezers. Photoregulated ion capture and release using thioindigo derivatives having ethylenedioxy side groups. *J. Am. Chem. Soc. 107*, 1024–8, 1985; Marquis, D; Henze, B; Bouas-Laurent, H; Desvergne, JP. Synthesis and cation complexing properties of a new type of photoactive coronands. Towards photocontrol of Na$^+$ complexation. *Tetrahedron Lett. 39*, 35–8, 1998.

[52]E.g., Kim, Y; Ko, YH; Jung, M; Selvapalam, N; Kim, K. A new photo-switchable "on-off" host-guest system. *Photochem. Photobiol. Sci. 10*, 1415–9, 2011; Nilsson, JR; O'Sullivan, MC; Li, S; Anderson, HL; Andreasson, J. A photoswitchable supramolecular complex with release-and-report capabilities. *Chem. Commun. 51*, 847–50, 2015; Wiktorowicz, S; Aseyev, V; Tenhu; H. Novel photo-switchable polymers based on calix[4]arenes. *Polym. Chem. 3*, 1126–29, 2012.

Several research groups[53] have investigated light-driven cation binding based on spiropyrans and spiroxazines, a class of photoactive organic compounds that has attracted particular attention.[54] The molecular "backbone" of these molecules rearranges so dramatically on ultraviolet illumination that a solution containing the molecule changes color when illuminated, a behavior termed "photochromism" (Fig. 6.6). Cleavage of the C–X bond at the "spiro" (tetrahedrally coordinated) carbon between carbons 1' and 3 causes free rotation around the 1'-3 axis to yield a structure with a double-bonded linkage like a merocyanine dye. This utterly rearranges the electronic structure of the molecule so that its absorption spectrum changes drastically, and in turn this causes the color change. Typically these molecules return to the *spiro-* form in the dark, but also

---

[53]E.g., Atabekyan, LS; Chibisov, AK. Complex formation of spiropyrans with metal cations in solution: A study by laser flash photolysis. *J. Photochem. 34*, 323–31, 1986; Atabekyan, L; Chibisov, A. Spiropyrans complexes with metal ions. Kinetics of complexation, photophysical properties and photochemical behavior. *Mol. Cryst. Liq. Cryst. Sci. Technol., Sect. A 246*, 263–64, 1994; Gorner, H; Chibisov, AK. Complexes of spiropyran-derived merocyanines with metal ions. Thermally activated and light-induced processes. *Faraday 94*, 2557–64, 1998; Inouye, M; Akamatsu, K; Nakazumi, H. New crown spirobenzopyrans as light-and ion-responsive dual-mode signal transducers. *J. Am. Chem. Soc. 119*, 9160–5, 1997; Kimura, K; Kaneshige, M; Yamashita, T; Yokoyama, M. Cation complexation, photochromism, and reversible ion-conducting control of crowned spironaphthoxazine. *J. Org. Chem. 59*, 1251–6, 1994; Sasaki, H; Ueno, A; Anzai, J; Osa, T. Benzo-15-crown-5 linked spirobenzopyran. I. Photocontrol of cation-binding ability and photoinduced membrane potential changes. *Bull. Chem. Soc. Jpn. 59*, 1953–6, 1986.

[54]E.g., Bertelson, RC. Photochromic processes involving heterolytic cleavage. In *Photochromism*, Brown, GH; ed. Wiley-Interscience, pp. 49–431, 1971; Guglielmetti, R. 4n+2 systems: spiropyrans. In *Photochromism: Molecules and Systems*, Duerr, H; Bouas-Laurent, H; eds. Elsevier, pp. 314–466, 1986; Chu, NYC. 4n+2 systems: spirooxazines. In *Photochromism: Molecules and Systems*, Duerr, H; Bouas-Laurent, H; eds. Elsevier, pp. 493–509, 1986; Berkovic, G; Krongauz, V; Weiss, V. Spiropyrans and spirooxazines for memories and switches. *Chem. Rev. 100*, 1741, 2000; Lukyanov, B; Vasilyuk, G; Mukhanov, E; Ageev, L; Lukyanova, M; Alexeenko, Y; Besugliy, S; Tkachev, V. Multifunctional spirocyclic systems. *Int. J. Photoenergy* Article ID 689450, 2009.

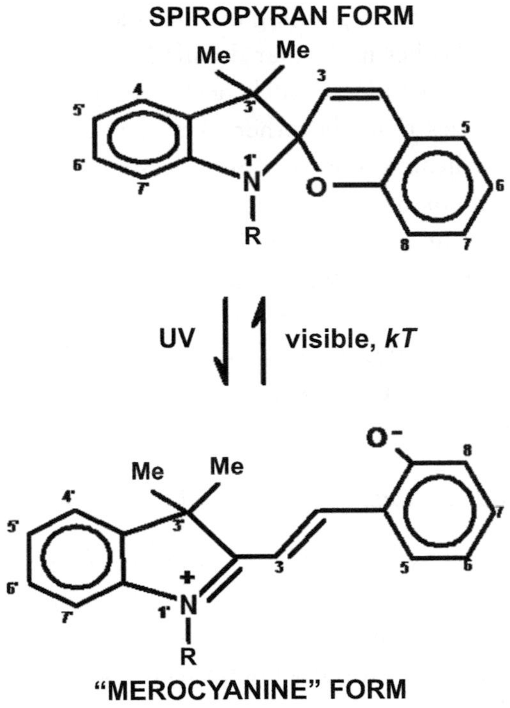

SPIROPYRAN FORM

Figure 6.6 Spiropyran photochromism.

Absorption of an ultraviolet (UV) photon flips the molecule into the open, "merocyanine" form. Visible light or room temperature (kT) flip it back into the spiropyan form. With appropriate side groups at R and the numbered positions, reversible light-driven ion binding can occur, as described in the text. "Me" is the methyl (CH_3) group.

commonly under longer-wavelength illumination. The reversible photochromism of these compounds has attracted much interest for possible applications in information storage,[55] in analysis, and as biological signaling models,[56] but light-mediated switching of binding specifically for solute extraction has been a particular

---

[55]E.g., Irie, M. Photochromism: Memories and switches—introduction. *Chem. Rev. 100*, 1683, 2000.

[56]Inouye, M. Artificial-signaling receptors for biologically important chemical species. *Coord. Chem. Rev. 148*, 265–83, 1996.

focus of one study.[57] Depending on the sidegroups (R', R," etc. in Fig. 6.6) and other molecular details (e.g., whether X = O, S, etc.), one isomer can be a much better complexing agent for a particular species than the other. Another study[58] investigated a more complicated photochromic system, with separation suggested as a possible application.

The issues in all these molecular switching systems are much as with the redox-switchable molecular systems described above: syntheses must be more economic, tethering of the photoswitchable molecules to a substrate will be required, and a practical system must be capable of reversibly switching over thousands of cycles.

Oxide semiconductors

A different approach uses the absorption of light by a semi-conductor surface. As in semiconductor photosynthesis (p. 254), the electron–hole pair generated by an absorbed photon can drive redox reactions at the semiconductor surface. In essence a redox switch is being driven with light.

For a simple example, many researchers have demonstrated the photoreduction of metal ions out of solution by illuminated semiconductors, usually oxides,[59] and this has even been

---

[57]Alward, MR. The synthesis, analysis and applications of photoreversible metal-ion chelators. Dissertation, University of Pittsburgh, Pittsburgh, PA, USA, 1998.

[58]Han, M; Michel, R; He, B; Chen, Y-S; Stalke, D; John, M; Clever, GH. Light-triggered guest uptake and release by a photochromic coordination cage. *Angew. Chem. Int. Ed. 52*, 1319–23, 2013.

[59]E.g., Brown, JD; Williamson, DL; Nozik, AJ. Moessbauer study of the kinetics of iron^{3+} photoreduction on titanium dioxide semiconductor powders. *J. Phys. Chem. 89*, 3076–80, 1985; Curran, JS; Domenech, J; Jaffrezic-Renault, N; Philippe, R. Kinetics and mechanism of platinum deposition by photoelectrolysis in illuminated suspensions of semiconducting titanium dioxide. *J. Phys. Chem. 89*, 957–63, 1985; Tanaka, K; Harada, K; Murata, S. Photocatalytic deposition of metal ions onto TiO_2 powder. *Solar Energy 36*, 159–61, 1986; Serpone, N; Borgarello, E; Barbeni, M; Pelizzetti, E; Pichat, P; Hermann, J-M; Fox, MA. Photochemical reduction of gold(III) on semiconductor dispersions of TiO_2 in the presence of CN⁻ ions: disposal

suggested as a recovery method for precious metals.[60] These processes merely underscore the value of switchability, however, because eluting the photoreduced metals to regenerate the semiconductor surface has so far proven to be uneconomic.

Switchable systems have been reported by several groups, in which the reduction induced by the photogenerated electron spontaneously reverses when illumination is discontinued due to re-oxidation by atmospheric oxygen (O_2). One group[61] demonstrated that illumination of titanium dioxide (TiO_2) in deaerated solutions, in the presence of certain organic compounds, causes precipitation of an ill-defined purple cuprous (Cu^+) complex from photoreduction of cupric ion (Cu^{++}) in solution. On exposure to air oxidation by O_2 re-oxidizes the cuprous (Cu^+) ion to cupric (Cu^{++}), which goes back into

of CN⁻ by treatment with hydrogen peroxide. *J. Photochem. 36*, 373–88, 1987; Jacobs, JWM; Kampers, FWN; Rikken, JMG; Blue-Lieuwma, CWT; Koningsberger, DC. Copper photodeposition on TiO_2 studied with HREM and EXAFS. *J. Electrochem. Soc. 136*, 2914, 1989; Eliet, V; Bidoglio, G. Kinetics of the laser-induced photoreduction of U(VI) in aqueous suspensions of TiO_2 particles. *Environ. Sci. Technol. 32*, 3155–61, 1998; Chenthamarakshan, CR; Yang, H; Savage, CR; Rajeshwar, K. Photocatalytic reactions of divalent lead ions in UV-irradiated titania suspensions. *Res. Chem. Intermed. 25*, 861–76, 1999.

[60]E.g., Borgarello, E; Harris, R; Serpone, N. Photochemical deposition and photorecovery of gold using semiconductor dispersions. A practical application of photocatalysis. *Nouv. J. Chim. 9*, 743–7, 1985; Borgarello, E; Serpone, N; Emo, G; Harris, R; Pelizzetti, E; Minero, C. Light-induced reduction of rhodium(III) and palladium(II) on titanium dioxide dispersions and the selective photochemical separation and recovery of gold(III), platinum(IV), and rhodium(III) in chloride media. *Inorg. Chem. 25*, 4499–503, 1986; Herrmann, JM; Disdier, J; Pichat, P. Photocatalytic deposition of silver on powder titania: consequences for the recovery of silver. *J. Catal. 113*, 72–81, 1988.

[61]Foster, NS; Noble, RD; Koval, CA. Reversible photoreductive deposition and oxidative dissolution of copper ions in titanium dioxide aqueous suspensions. *Environ. Sci. Technol. 27*, 350–6, 1993; Foster, NS; Koval, CA; Noble, RD. Reversible photodeposition and dissolution of metal ions. US Patent 5332508, 14 pp., 1994; Foster, NS; Lancaster, AN; Noble, RD; Koval, CA. Effect of organics on the photodeposition of copper in titanium dioxide aqueous suspensions. *Ind. Eng. Chem. Res. 34*, 3865–71, 1995.

solution. They have proposed this system for copper extraction and recovery.

Similarly, another group[62] demonstrated reversible photo-reduction of ferricenium to ferrocene from acidic solution onto TiO_2. On standing in the dark for several hours, the ferrocene would spontaneously reoxidize to ferricenium through reaction with atmospheric O_2. The TiO_2 had been functionalized by a covalently bound surface hydrocarbon layer, and so on photoreduction the hydrophobic ferrocene molecules would dissolve into this layer (Fig. 6.7). This hindered the re-oxidation sufficiently that the ferricenium could be removed from the solution essentially quantitatively so long as illumination was maintained. However, the system cycled only a few times before breaking down, and in any case, was applicable only to dissolved ferricenium, although with further investigation it may prove of wider applicability.

In a somewhat similar system, ultraviolet irradiation triggered the adsorption of azo dyes onto titania xerogels, the system spontaneously reversing in the dark.[63] These authors noted that such a system could conceivably be "tuned" for different solutes by changing the parameters of the sol–gel synthesis of the xerogels.

Finally, other environmental switches have been investigated. Nanoporous silica particles, for example, have been functionalized with "nanovalves" that act as gates on the nanopores, such that release of molecules out of the pores can occur when the "valves" are opened.[64] Depending on the molecular construction of the nanovalves, the valve opening can be triggered by light, redox

---

[62]Muraoka, M; Gillett, SL; Bell, TW. Reversible photoinsertion of ferrocene into a hydrophobic semiconductor surface: A chemionic switch. *Angew. Chem. Int. Ed. 41*, 3653–6, 2002.

[63]Zhang, S; Peng, Y; Jiang, W; Liu, X; Song, X; Pan, B; Yu, H-Q. Light-triggered reversible sorption of azo dyes on titanium [titania] xerogels with photo-switchable acetylacetonato anchors. *Chem. Commun. 50*, 1086–8, 2014.

[64]Sun, Y-L; Yang, Y-W; Chen, D-X; Wang, G; Zhou, Y; Wang, C-Y; Stoddart, JF. Mechanized silica nanoparticles based on pillar[5]arenes for on-command cargo release. *Small 9*, 3224–9, 2013; Fahrenbach, AC; Warren, SC; Incorvati,

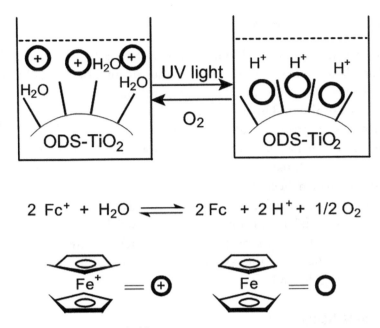

$$2\ Fc^+ + H_2O \rightleftharpoons 2\ Fc + 2\ H^+ + 1/2\ O_2$$

Figure 6.7 Ferrocene switching system.

This is the system described by Muraoka et al. (Ref. 62 in the main text). "Fc" is ferrocene, the "iron sandwich" $Fe(C_5H_5)_2$ (Box 6.4, Fig. 6.4, and the sketch at the lower right below the chemical reactions). It's represented in the diagram above by heavy open circles, as indicated by the "−" sign. Similarly, Fc^+ is ferricenium, represented by heavy circles containing a plus sign. "ODS-TiO$_2$" is titanium dioxide (TiO$_2$) whose surface has been covered with a covalently bound hydrocarbon "fuzz," represented by the "hairs" sticking out of the surface. Electrons generated by shining light on the TiO$_2$ react with ferricenium in solution to reduce it to ferrocene, which then dissolves in the hydrocarbon fuzz atop the TiO$_2$. In the dark the ferrocene re-oxidizes with atmospheric oxygen (O_2) and returns to solution.

potential, or even other environmental variables such as pH. The immediate application of these particles is for highly directed drug delivery, but as the technology matures, and becomes amenable to scale-up, it should be broadly applicable to general extraction problems as well.

JT; Avestro, A-J; Barnes, JC; Stoddart, JF; Grzybowski, BA. Organic switches for surfaces and devices. *Adv. Mater. 25*, 331–48, 2013.

Switchable Solvents

An elegant recent development is that of switchable *solvents*.[65] The details vary, but in essence the affinity of the solvent molecules for the solute is changed by making them more or less hydrophilic. This causes the solute to precipitate, and so provides an attractive alternative to thermal-based separation processes such as distillation or evaporation.[66] In an alternative system, the "unswitched" form consists of two immiscible liquids, which become miscible on switching. Such a system lends itself to an elegant variation on solvent extraction. The usual "switch" is carbon dioxide gas, which reacts with the solvent to change its properties. The reactions are reversible, however, such that the solvent reverts to its original state on outgassing the CO_2.[67]

Isotope Separation

Isotope separation is crucial for nuclear materials because the nuclear properties are what is relevant. This has already been commented on in the case of nuclear fuels, waste reprocessing, and the potential separation of radionuclides for research or medical use (Boxes 2.13, 5.15, 5.16).

The utility of isotopes ranges well beyond nuclear materials, however. Stable isotopes are widely used as tracers in biological,

---

[65] Jessop, PG. Switchable solvents. In 10th Green Chemistry Conference. Barcelona – Spain. 2013; Mercer, SM; Jessop, PG. "Switchable water": Aqueous solutions of switchable ionic strength. *ChemSusChem 3*, 467–70, 2010.

[66] Boyd, AR; Champagne, P; McGinn, PJ; MacDougall, KM; Melanson, JE; Jessop, PG. Switchable hydrophilicity solvents for lipid extraction from microalgae for biofuel production. *Bioresour. Technol. 118*, 628–32, 2012; Holland, A; Wechsler, D; Patel, A; Molloy, BM; Boyd, AR; Jessop, PG. Separation of bitumen from oil sands using a switchable hydrophilicity solvent. *Can. J. Chem. 90*, 805–10, 2012.

[67] Jessop, PG; Mercer, SM; Heldebrant, DJ. CO_2-triggered switchable solvents, surfactants, and other materials. *Energy Environ. Sci. 5*, 7240–53, 2012.

geochemical, and environmental systems. They are preferred for biological tracing at present because of their ease of handling due to the absence of radiological safety issues. Indeed, geochemical and environmental studies routinely exploit the fractionation of isotopes by natural processes. If isotopically enhanced material could be obtained cheaply enough, however, tracking processes such as pollutant pathways by stable isotopic tracing would also become of great interest. Isotopically enhanced sulfur, for example, has been proposed as a method of definitively tracing the source of the sulfate in acid rain.[68] This in turn would imply having cheap isotopically enhanced material available in ton-scale lots.

In no place, however, does a dichotomy between the trivial thermodynamic costs and the enormous practical costs of separation loom so large as in the separation of isotopes. Because isotopes have essentially identical chemical properties, their mixtures are well-described by the mixing entropy alone (Box 3.9). For deuterium in hydrogen (mole fraction $x = 0.00015$), for example, the normalized mixing entropy, found by Eq. 3.8-1, is

$$S_m' = 81.5 \text{ J/K mol}.$$

At 25°C (298 K), this corresponds to a limiting energy cost TS of 24.3 kJ/mol.

Vastly more energy than this is spent in purifying deuterium. As is well known, practical techniques for isotope separation mostly rely on the repetition of physical processes such as diffusion or phase changes, so that small effects due to the mass difference slowly accumulate. Such separation also relies on having a molecular species with the appropriate physical properties (e.g., a vapor phase or phase change at a convenient temperature). Electromagnetic separators ("calutrons"), essentially industrial-

---

[68]Mills, TR. Practical sulfur isotope separation by distillation. *Sep. Sci. Technol.* 25, 1919, 1990.

scale mass spectrometers, are more generally applicable but even more energy-intensive.

A less "brute-force" approach would be to exploit the vibrational spectrum of a molecule, that is, the specific frequencies (wavelengths) at which it absorbs or emits light. Because of the laws of quantum mechanics, the exact energy levels of the spectrum depend on the isotopes involved, such that species containing different isotopes absorb or emit at slightly different frequencies. Thus, illumination of a mixture of isotopic species by radiation of sufficiently precise wavelength could excite only one isotopic species and cause its separation. Lasers are capable of radiating light with sufficient precision, and investigations of "spectroscopic separation" of isotopes date back decades.[69] The molecular species proves to be very important, since most molecules have too many vibrational modes for clean separation, the number of modes depending on both the molecular symmetry and the number of different isotopic species of each element.

For example, sulfur hexafluoride (SF_6) has been the subject of many studies. The molecule has octahedral symmetry, so all six fluorine atoms are symmetrically equivalent. Since ^{19}F is the only stable fluorine isotope, the vibrational modes thus depend

---

[69]E.g., Balling, LC; Wright, JJ. Use of angular-momentum selection rules for laser isotope separation. *Appl. Phys. Lett.* 29, 411–3, 1976; Arisawa, T; Maruyama, Y; Suzuki, Y; Kato, M; Wakaida, I; Akaoka, K; Miyabe, M; Ohzu, A; Sugiyama, A. Atomic vapor laser isotope separation. *Optoelectronics: Devices and Technologies 8*, 203, 1993; Kojima, H; Fukumi, T; Nakajima, S; Maruyama, Y; Kosasa, K. Laser isotope separation of ^{13}C by an elimination method. *Chem. Phys. Lett.* 95, 614–7, 1983; Energy efficiency in the multistage laser separation of ^{13}C by an elimination method. *Appl. Phys. B 30*, 143–8, 1983; Okada, Y; Kato, S; Satooka, S. Laser isotope separation of heavy elements by infrared multiphoton dissociation of metal alkoxides. *Spectrochim. Acta 46*, 643, 1990; van der Veer, WE; Uylings, PHM. Laser isotope separation via coherent excitation of Zeeman or hyperfine levels: what is required? *Z. Phys. D 27*, 55, 1993; Pino, G; Rinaldi, CA; Ferrero, JC. Isotopic enrichment of ^{13}C by IR laser photolysis followed by fast bimolecular reaction. *J. Photochem. Photobiol. A 93*, 97–101, 1996.

only on the mass of the sulfur atom. Isotope separation of SF_6 by laser excitation has been demonstrated;[70] unfortunately, SF_6 is such a stable molecule that incorporating the isotopically enriched sulfur into more useful chemical forms proves to be a significant expense.[71]

More speculatively, perhaps the quantum-mechanical phenomenon of vibronic mixing could be exploited for isotopic separation. A vibrational mode whose energy overlaps with that of an electronic transition yields a so-called "mixed state," the degree of mixing depending on the energy difference. Such a state will absorb radiation of the correct energy much more strongly, and so heighten any separation effect. The practicality of this phenomenon will depend even more on molecular details than conventional spectroscopic separation, however. Not only must vibrational modes be minimized, an unusually low-energy electronic transition is required. Most electronic transitions lie in the ultraviolet or visible region, whereas most vibrational transitions lie in the infrared.

Overall, it seems that nanotechnology has little to offer directly to isotope separation except in the general sense of cheap nanofabrication. For example, fabricating conventional isotope separators, such as diffusion membranes, at the nanoscale may yield significant improvements in both cost and speed.

---

[70]E.g., Lin, ST; Lee, SM; Ronn, AM. Laser isotope separation in SF_6. *Chem. Phys. Lett.* 53, 260–5, 1978; Lyman, JL; Rockwood, SD; Freund, SM. Multiple-photon isotope separation in SF_6: Effect of laser pulse shape and energy, pressure, and irradiation geometry. *J. Chem. Phys.* 67, 4545–56, 1977; Del Bello, U; Churakov, V; Fuss, W; Kompa, KL; Maurer, B; Schwab, C; Werner, L. Improved separation of the rare sulfur isotopes by infrared multiphoton dissociation of SF_6. *Appl. Phys. B* 42, 147–53, 1987; D'Ambrosio, C; Fuss, W; Kompa, KL; Schmid, WE. Isotope separation of $^{32}SF_6$ and $^{12}CF_3I$ by a Q-switched CO_2 laser. *Appl. Phys. B* 47, 17–20, 1988.

[71]Caudill, HH; Bond, WD; Collins, ED; Mcbride, LE; Milton, HT; Tracy, JG; Veach, AM; York, JW. Conversion of enriched isotopes of sulfur from SF_6 to a suitable compound for feed to the electromagnetic calutron separators. *Sep. Sci. Technol.* 25, 1931, 1990.

Molecular Separation and Rare Elements

Even though, as discussed below (p. 323), demand for such familiar elements as iron and aluminum is likely to dwindle, at least in their traditional uses, the rest of the Periodic Table is unlikely to become irrelevant. Indeed, nanotechnology-related applications will probably increase the demand for certain rare elements. Although the molecular-scale element separation technologies described above makes it less likely that shortages of such elements will exist, they merit mention. Ironically, a further long-term consequence is likely to be the demise of precious metals as "stores of value." That value has lain in their rarity, which vanishes if they can be extracted at sufficiently low concentrations.

Thermoelectric Materials

Theoretical analyses of the factors that lead to good thermoelectric materials imply that they will be based on heavy (atomic number $Z > \sim40$) elements. Elements such as lead ($Z = 82$), tellurium ($Z = 52$), and bismuth ($Z = 83$), for example, figure prominently in currently known thermoelectric materials.

Such elements are intrinsically rare in the crust. The peak of nucleon binding energy lies in the vicinity of iron-56 (^{56}Fe, $Z = 26$), so elements heavier than this cost energy to make, and this is reflected in these vastly lower abundances of elements with $Z > \sim28$. Hence selective binding agents for these elements may merit attention. Because of its toxicity, selective complexing agents for lead have already been investigated.[72]

Redox–Active Framework Formers

As mentioned (p. 214), a number of elements besides silicon are also capable of forming oxide frameworks, in particular certain

---

[72]E.g., Abu-Dari et al., Ref. 14.

transition metals such as titanium, vanadium, manganese, niobium, molybdenum, tantalum, and tungsten, and these have a number of potential nanotechnological applications. These include such energy-related applications as electrochromic and self-darkening windows (p. 217), intercalation-based batteries (p. 221) and supercapacitors (p. 226), and substrates for reversible, selective intercalation of solutes (p. 295). Tungsten especially is notable for the so-called "tungsten bronzes" (p. 214), which illustrate one useful possibility of such frameworks: their bulk electronic and optical properties can be varied drastically by minor reduction or oxidation of some of the framework atoms, due to the multiple stable oxidation states of a transition element. The same basic framework can thus range from an insulator through a semiconductor to a metal. The potential utility of tungsten in polyoxometalate molecular building blocks was also noted (p. 168).

Many of these framework-forming elements are rare. Tungsten ($Z = 74$) is one of the best but only comprises 0.000017% of the crust on a per-atom basis.[73] It has already been a subject of intense search as a strategic material because of its refractory and alloying properties. Tungsten has been concentrated into ores by geochemical processes, most particularly in the "contact zones" around certain igneous bodies. However, a nontraditional source that has been the subject of research but so far has proved uneconomic is in certain brines, e.g., at Searles Lake, California.[74] Tungstate anions make up a very small (equivalent to ~49 ppm tungsten or ~0.265 millimolar) percentage of such brines, and they are obvious candidates for selective solution-based molecular separation.

---

[73]Ref. 1 in Table 1.1.

[74]Altringer, PB; Brooks, PT; McKinney, WA. Selective extraction of tungsten from Searles Lake brines. *Sep. Sci. Technol. 16*, 1053–69, 1981.

Nuclear Fuels

As indicated in Chapter 5 (pp. 268 and 276), most nuclear fuels are isotopes of rare elements, deuterium being the only exception. Hence, as described there, they should be amenable to molecular separation technologies as for other rare elements.

Molecular Separation and Environmental Remediation

A pollutant is just a misplaced resource.
> —Anonymous

There is little doubt that low-temperature "biomimetic" approaches, capable of switchable binding of solutes (p. 293), will make aqueous-based extraction processes both considerably more efficient and—like biosystems—capable of dealing effectively with considerably lower concentrations. An obvious immediate application of such technologies is in pollution control; indeed, it is a key economic driver, because a great deal of environmental regulation stringently limits the concentration of various substances that can be present in an effluent. This provides a large incentive to remove low concentrations of contaminants cheaply. (Alternatively, of course, it may be possible to redesign the process so that the pollutants aren't generated to begin with. The quest for "greener" syntheses was mentioned briefly in Chapter 3. But such redesign is not always possible.)

Furthermore, the recovered pollutants may themselves turn out to be useful products. After all, copper (say) extracted from a wastewater stream is copper that need not be mined. This is another example of the "waste into resource" theme that has been ongoing in this book, and it is another trend that nanotechnology merely accelerates. Decades ago, for example, economic recovery of manganese from mining waste streams by ion-exchange resins had been demonstrated.[75] For another example, sulfur dioxide is an abundant waste product (or by-product) from most smelting operations, as noted in Chapter 3 (p. 110), and formerly it was just vented directly into the atmosphere. This is no longer permitted due to the serious environmental consequences. The sulfur dioxide is now recovered and converted into sulfuric acid (H_2SO_4) as a by-product. Sulfuric acid is a standard bulk industrial chemical for which a large, well-defined market exists.

It is an obvious step from pollution control to environmental remediation, the cleanup of "legacy" pollutants. Acid-mine drainage (Box 3.7), for example, has led to a number of extremely polluted bodies of water, such as the notorious Berkeley Pit in Butte, Montana, and the Leviathan Mine Superfund site in California near the Nevada border. Not only are such sites toxic in themselves, but they pose an ongoing threat to the environment because of the chance of spilling over and releasing toxic material into a broader area. And yet, these sites are also rich in metals that are already dissolved— indeed, copper is recovered at a low level from the Berkeley pit already. Such sites, therefore, are obvious targets not just for remediation but for resource recovery by molecular-scale separation technologies.

Thus, as molecular separation technologies mature, they will blur the distinction between a "pollutant" and a "resource." Moreover, as mentioned in Chapter 3, a great many aqueous solutions, of both natural and artificial origin, contain potentially useful materials in solution: not just acid-mine drainage waters, but wastewater streams, concentrated natural brines such as those in oil fields or saline lakes, and even seawater.

---

[75]Lotz, P; Green, BR; Fleming, CA. Ion exchange in the treatment of effluent from an electrolytic manganese plant. In *Recent Developments in Ion Exchange*, Williams, PA; Hudson, MJ; eds. Elsevier. pp. 205–12, 1987.

The author suspects that the ~5000 year era of digging up and "cooking" anomalous geologic deposits for their metal content is coming to its end irrespective of whether such deposits will still exist to be exploited. Conventional pyrometallurgy is not only dirty, it is wasteful on a vast scale, and economic considerations are likely to figure in its obsolescence. When (say) zinc can be recovered cheaply from wastewater streams at parts-per-million concentrations, there will be little incentive to process virgin ore.

Indeed, more change in the technology of extractive metallurgy is likely to occur in the next 5 decades than in the last 5 millenia. Ironically, such sweeping changes in *extraction* were explicitly viewed as the least likely alternative in one study of future resources,[76] but those authors ignored biomimetic models completely.

Ultimately, "total resource utilization" is a feasible goal, the complete recycling of matter by using the energy of sunlight to carry out the non-thermal molecular separation of waste materials. Anomalous concentrations—"ores"—will no longer be necessary.

Photocatalysis and Pollutant Destruction

Another environmental application merits mention here as well. In Chapter 5 the potential of semiconductor-based photosynthesis was described, in which the photogenerated hole–electron pair, instead of driving an electric circuit, are used to drive "uphill" chemical reactions, just as natural photosynthesis does. Alternatively, however, these highly active chemical species can react readily with many dissolved compounds. In particular, organic pollutants such as halogenated hydrocarbons, aromatic hydrocarbons, cyanide ion, and so on, can be broken down into simpler and nontoxic compounds such as water and carbon dioxide. Such "photocatalysis" has spawned an extensive literature on using powdered oxide semiconductors, primarily

---

[76]See Ref. 3 in Ch. 3.

titanium dioxide, for pollutant destruction.[77] This application also provides another example of using sunlight directly as a cheap and already distributed source of energy (cf. p. 80).

Change of Materials Mix

> *Some ... argue that supplies of all resources are linked together. The weakest links ... in the chain have never been tested, but they are clearly metals.*
> —Gordon et al., 1987, p. 1[78]

Molecular-level, biomimetic extraction technologies are likely to defer indefinitely any shortages of metals, as described in the previous sections. In ironic contrast to the sentiment expressed above, then, metals may be becoming superabundant just as demand for them plummets. The reason is that as nanotechnology matures, the desired materials mix is likely to change considerably.

---

[77]E.g., Matthews, RW. Kinetics and photocatalytic oxidation of organic solutes over titanium dioxide. *J. Catal.* 111, 264–72, 1988; Wold, A. Photocatalytic properties of titanium dioxide (TiO_2). *Chem. Mater.* 5, 280–3, 1993; Cunningham, J; Al-Sayyed, G; Srijaranal, S. Adsorption of model pollutants onto TiO_2 particles in relation to photoremediation of contaminated water. In *Aquatic and Surface Photochemistry*, Helz, GR; Zepp, RG; Crosby, DG; eds. Lewis, Boca Raton, 317–48, 1994; Linsebigler, AL; Lu, G; Yates, JT, Jr. Photocatalysis on TiO_2 surfaces: Principles, mechanisms, and selected results. *Chem. Rev.* 95, 735–58, 1995; Riegel, G; Bolton, JR. Photocatalytic efficiency variability in TiO_2 particles. *J. Phys. Chem.* 99, 4215–24, 1995; Fujishima, A; Rao, TN; Tryk, DA. Titanium dioxide photocatalysis. *J. Photochem. Photobiol. C* 1, 1–21, 2000. Anpo, M; Kamat, PV; eds. *Environmentally Benign Photocatalysts: Applications of Titanium Oxide-based Materials.* NS&T, 2010; Hu, A; Apblett, A. *Nanotechnology for Water Treatment and Purification.* LNNS&T 2014, esp. Oakes, K; Shan, Z; Kaliaperumal, R; Zhang, SX; Mkandawire, M. Nanotechnology in contemporary mine water issues therein; Chen, W-T; Chan, A; Jovic, V; Sun-Waterhouse, D; Murai, K; Idriss, H; Waterhouse, GIN. Effect of the TiO_2 crystallite size, TiO_2 polymorph and test conditions on the photo-oxidation rate of aqueous methylene blue. *Top. Catal.* 58, 85–102, 2015.

[78]See Ref. 3 in Ch. 3.

The obsolescence of structural metals

Metals are perceived as so important that "ages of civilization" traditionally are described in terms of them: think of the Bronze Age or the Iron Age. Certainly metals pervade modern life, to a degree unimaginable even a couple of centuries ago. The way metal objects are casually discarded as "waste"—think of food or beverage cans—would astonish people even as late as the early 19th century.

Yet this widespread use of metals is another characteristic of the Paleotechnical Era. They are examples of bulk materials (Box 1.1) and are actually surprisingly weak; their dominance in present structural materials is due to their resistance to "Griffith failure" (Box 6.5). Many are also exceptionally vulnerable to corrosion, which is exacerbated by their being electrical conductors. Because electrons can flow in a metal, redox reactions (Box 2.15) occur more easily, and this can hasten corrosion. An everyday example is the crusty stuff on the positive terminal of an automotive battery.

Brittle materials are potentially far stronger than metals, but are subject to catastrophic failure due to crack propagation unless essentially defect-free at the molecular level (Box 6.5). As nanofabrication techniques mature, such flaw-free materials should become accessible and allow the great strength of brittle materials to be exploited routinely. Superstrong materials with covalent chemical bonds, which unlike metallic bonds resist shear, should have strengths approaching the limits set by the chemical bonds themselves.

Of course, the routine nanofabrication of defect-free covalently bound materials, particularly at the scale needed for structural materials, still lies some years away. Composite materials, however, can yield great strength improvements with considerably less perfection in fabrication, and so extreme-strength composites are likely to be available in the nearer term. Composites typically consist of a mat or felt of high-tensile-strength fibers embedded in a matrix. The fibers even may be

Box 6.5 Strength of Materials

The overwhelming importance of materials strength—and in particular *specific* strength, the strength per unit mass—in most applications hardly needs elaboration. Moreover, it proves to be another bulk property that is *not* an average of nanoscopic properties. It's easy to calculate that the chemical bonds in ordinary substances imply strengths far higher than observed.

The strength of conventional materials is dominated by second-order properties: crystal defects and dislocations, grain boundaries, and so on. These flaws illustrate the old adage that a chain is only as strong as its weakest link, as they are the places where cracks initially form. Furthermore, before the First World War, workers such as Inglis and Kolosoff had shown that a crack, once formed, concentrates stress at its tip and thus tends to propagate. Then, in a classic set of papers in the 1920s, Griffith derived a set of criteria for material failure by the propagation of cracks.[1] In particular, he showed that cracks will propagate more readily in brittle materials because of their inability to relieve the stress at the crack tip by plastic flow. Hence, in a brittle material macroscopic failure occurs due to the catastrophic propagation of a vast number of cracks. A water glass shattering on being dropped to the floor is a familiar example.

Metals, however, are capable of plastic flow when sufficiently stressed, and thus crack tips tend to "heal" rather than propagate. Hence, metals retain usable strength even when riddled with microflaws, and also tend to fail "gracefully." As current fabrication techniques are unable to make materials not containing vast numbers of flaws, this robustness has been the overwhelming engineering consideration.

Nonetheless, metals are intrinsically weak. Even in a strong metal, plastic deformation still occurs. It merely requires a large energy input. Indeed, the metallic bond is nondirectional

[1]Ref. 81 in main text, pp. 57 ff.

and so is fundamentally weak. Even if the tensile strength is high; i.e., there is great resistance to pulling adjacent atoms apart, there is little energy barrier to relative shear between them, and hence little barrier to the propagation of dislocations. Approaches to fabricating strong metals thus must involve a tradeoff: they must make the propagation of dislocations energetically expensive, but not so much so as to completely prevent plastic flow and so embrittle the material. Thus the role of alloying elements can largely be viewed as agents to inhibit the movement of dislocations. (Even so, continued flexing can cause a "pile-up" of dislocations and so cause embrittlement. This is the basis of metal fatigue—repeatedly bending a paper clip until it breaks is a familiar example.)

Brittle materials are potentially much stronger, but must be essentially flawless at the molecular level. As discussed in the main text, this has major long-term implications for the demand for raw materials.

oriented, to yield preferred tensile strength in one direction. Often, however, they are randomly oriented to yield an isotropic material. A contemporary example is fiberglass, which is surprisingly strong for being made of glass fibers embedded in polymer resin.

As is shown by biological materials, the improvements in strength can be extraordinary. The example of abalone shell was mentioned in Chapter 3. Furthermore, the fact that the strong elements are fibers helps limit crack propagation. Even if one fiber breaks, the break is not automatically propagated. Finally, fabricating arbitrarily long but essentially defect-free *fibers* is likely to be a considerably easier engineering problem than fabricating two- or three-dimensional objects. In conventional composites such as fiberglass the fiber length is limited, and of course such fibers are hardly defect-free; nonetheless, the improvements in strength are considerable.

In any case, the phrase "structural metal" may well strike future engineers as oxymoronic.

Carbon: The Ultimate Material?

Given a molecular fabrication capability, carbon becomes the most desirable structural material in most applications. This is due to three factors:

- The relatively low mass of the carbon atom. Four out of the atom's six electrons contribute to chemical bonding, which means that the nucleus is contributing relatively little dead weight. In most applications, the figure of merit is not strength per se but *specific* strength, the strength per unit mass, and so small atomic mass is critical.
- The great strength and directionality of carbon–carbon bonds (Box 6.2).
- The carbon atom's ability to bond in two or three dimensions, which means that it can build up two-dimensional sheets and even three-dimensional frameworks (Box 6.2).

Tetrahedrally polymerized (sp^3; Box 6.2) "diamondoid" frameworks, based on the crystalline structure of diamond, are commonly proposed, and even now diamond-based structures are a subject of intense research.[79] For macroscopic structures, however, nanotubes ("buckytubes"; Box 6.2), may be even more valuable. As noted above, one-dimensional objects such as fibers are likely to be considerably easier to fabricate than three-dimensional molecular frameworks, especially in the near term. In macroscopic lengths, such fibers have obvious applications in tethers and cables. Fiber-composite materials, moreover, would combine great strength with a very large resistance to crack propagation. Furthermore, even diamond has well-defined

---

[79]Yang, N; ed. *Novel Aspects of Diamond. From Growth to Applications.* TAP *121*, 2015.

cleavage along the (111) crystal plane, which indicates that large diamond-based objects could fail preferentially in certain orientations.

Silicon, Silicates, and Nanotechnology

Carbon is certainly not the only structural option for nano-technology. And it has disadvantages; for one thing, carbon-based structures oxidize at high temperatures—that is, they burn!

What are possible alternatives to carbon? Silicon (Si) immediately comes to mind, being the next element below carbon in the Periodic Table. But despite all that science fiction about silicon-based life, the chemistry of silicon and carbon is quite different (Box 6.6). A more promising idea looks not at sili*con* but sili*cates*,[80] the compounds of silicon and oxygen with other elements (Box 3.5). The silicon–oxygen bond is strong and partly covalent and hence resistant to *shear*, or bending, as well as to extension. Although not so strong as carbon–carbon bonds, the disiloxy (O–Si–O) link is remarkably strong by everyday standards, the theoretical strength of silica (~16 GPa[81]) comparing well with other theoretical ultimate strengths. (Of course, realizing such strengths will require considerably better nanofabrication capability than exists at present.) Furthermore, silicate-based structures will not oxidize, even at temperatures considerably above ambient. There is a certain irony, of course, that the materials making up ordinary dirt ultimately are likely to become more important than alloy steels.

Indeed, the potential utility of silicates is foreshadowed by such examples as the open aluminosilicate frameworks, the zeolites, that are the prototypes of molecular sieves (Box 6.3) and, as already discussed, have a host of present-day applications.

---

[80]Gillett, SL. Toward a silicate-based molecular nanotechnology, I. Background and review; Toward a silicate-based molecular nanotechnology, II. Modeling, synthesis review, and assembly approaches. At Foresight Institute Web site, www.foresight.org. (Based on 1997 talk) 1998.

[81]Kelly, A; MacMillan, NH. *Strong Solids*, 3rd. ed. Clarendon, 1986, pp. 6, 8, 371; also review in Ref. 80.

Box 6.6 Silicon versus Carbon

Everyone learns in beginning chemistry that silicon is the chemical cousin of carbon, occurring immediately below it in the Periodic Table. That's, of course, also the inspiration for the old science-fiction trope about silicon-based life.

However, the chemistry of silicon and carbon is really quite different. Elemental silicon, to be sure, has the diamond structure, but silicon-silicon bonds are not so strong as are carbon-carbon bonds (Table 6.1), and with the extra mass of the silicon atom the specific strength is much less. Silicon also has little tendency to form the double bonds so characteristic of carbon; whose formation, indeed, largely *defines* the enormous variety of organic chemistry. Because of this, for example, silicon can form no analog of the graphene sheets that make up buckytube walls. The striking contrast in chemistry arises largely from the difference in size of the two atoms. The orbitals where the unshared electrons responsible for bonding are localized can overlap a great deal between adjacent carbon atoms to form a second or even a third chemical bond. The silicon atom is simply too big for that to happen.

One consequence is the stark contrast of silicon dioxide (silica, SiO_2) with its chemical cousin, carbon dioxide (CO_2). Although the atomic ratio—the stoichiometry—is the same, the structures are vastly different. CO_2 consists of individual molecules, each consisting of two oxygens double-bonded to a carbon in the middle. It is not a 3D structure, at least under ordinary conditions.[1] Because of the larger size of the silicon atom, however, double bonds are not stable, so a 3D framework results instead. Indeed, the propensity of silicon to form single bonds with oxygen, as well as the abundance of both elements in Earth's crust, leads to the enormous variety of silicates (Box 3.5), including *zeolites*, the prototypical molecular sieves (Box 6.3), and already an example of "proto-nanotechnology."

---

[1] A 3D polymeric form of CO_2 has been synthesized at extreme pressure: Iota, V; Yoo, CS; Cynn, H. Quartzlike carbon dioxide: An optically nonlinear extended solid at high pressures and temperatures. *Science 283*, 1510–3, 1999.

Table 6.1 Bond energies, silicon and carbon

Important Si and C bond energies, used as a proxy for bond strength. First column is the first element of the bound pair; top header lines are the second element. Energy values (second header line) are given both in kilojoules per mole (kJ/mol) and in attojoules (10^{-18} joules). Silicon values from Barton & Boudjouk[1]; carbon values from Drexler.[2]

	Si		C	
	kJ/mol	aJ	kJ/mol	aJ
Si	322	0.535	372	0.618
O	535	0.889	346	0.575
O (double bond)			808	1.343
C	372	0.618	335	0.556

[1]Barton, TJ; Boudjouk, P. Organosilicon chemistry: A brief overview. *Adv. Chem. Ser.* *224*, 3–46, 1990.
[2]Ref. 3 in Chapter 4, p. 52.

Sheet silicates (Box 3.3), some of which (the micas) can yield atomically flat surfaces over scales of centimeters, are used as the substrate for some self-assembled monolayers (p. 170). So-called pillared clays, formed by swapping out the interstitial cations between the sheets of the clay structure for even larger cations, furnish yet another example. The larger cations act like pillars holding the sheets farther apart. This creates molecular-scale galleries into which other molecules can fit. These now have an extensive literature largely because many make good catalysts, just as with molecular sieves. Such cation exchange is also another example of a topotactic reaction (p. 175).

Dirt and Sewage: Resources of the Future

Terrestrial Carbon Resources, I. Carbon Dioxide Fixation

An obvious distributed source of carbon, already used by photosynthesizing plants, is the carbon dioxide in the atmosphere. As concerns mount about rising concentrations of atmospheric CO_2, extracting CO_2 from the atmosphere as a raw material becomes even more attractive.[82] Such "fixation" of carbon dioxide has already been mentioned in the context of electrosynthesis (p. 228), photosynthesis (p. 260), and hydrogen storage (p. 231; Box 5.7). Of course, carbon dioxide is a fully oxidized compound, so that an energy input is require to reduce it, and that energy must be supplied from an outside source. Other approaches are possible,[83] but approaches requiring heat are obviously less attractive. In any case, catalysts will again be critical.

Parenthetically, another unexpected implication of the value of carbon is that the carbon dioxide atmosphere of Venus becomes of staggering long-term value as the largest off-Earth store of carbon in the inner Solar System.

Terrestrial Carbon Resources, II. The Obsolescence of Petroleum and Coal

As noted in Chapter 3, petroleum is used for more than just fuel: it's also the raw material for petrochemicals, one of whose products is plastics, and it's been widely thought that petroleum is much too valuable to waste as fuel. Conversely, waste biomass includes a whole host of noxious reduced-carbon products whose disposal is currently both expensive and inadequate, such as sewage, agricultural waste, timber slash, fermentation dregs, feedlot manure, and so forth. The irony of the efforts to oxidize

---

[82]E.g., Song, C. Global challenges and strategies for control, conversion and utilization of CO_2 for sustainable development involving energy, catalysis, adsorption and chemical processing. *Catal. Today 115*, 2–32, 2006; Hu, YH, ed., *Advances in CO_2 Conversion and Utilization. ACSSS 1056*, 2010; De Falco, M; Iaquaniello, G; Centi, G; eds. *CO_2: A Valuable Source of Carbon.* Springer, 2013; Jin, F; He, L-N; Hu, Y-HG; eds. *Advances in CO_2 Capture, Sequestration, and Conversion. ACSSS 1194*, 2015.

[83]Jin, F; Huo, Z; Zeng, X; Enomoto, H. Hydrothermal conversion of CO_2 into value-added products: A potential technology for improving global carbon cycle. In Hu, Ref. 82, pp. 31–53; Michiels, K; Peeraer, B; Van Dun, W; Spooren, J; Meynen, V. Hydrothermal conversion of carbon dioxide into formate with the aid of zerovalent iron: the potential of a two-step approach. *Faraday Discuss. 183*, 177–95, 2015.

this material while simultaneously extracting reduced carbon from deep in the Earth was pointed out in Chapter 2 (p. 76).

As outlined in Chapter 5 (p. 242), biomass is also of potential interest for next-generation liquid fuels, and some possible processing approaches were reviewed there. Of course, just as with petroleum itself, the resulting products need not be used as "fuel" but can instead be used as feedstocks for further syntheses. Depending on the end use, the immediate products of biomass conversion will be optimized for their intended end use.

It may be that some demand for coal will remain, not for use as fuel, but as a raw material for carbon-based structures. As with oil, however, the expense of extraction is likely to be an economic barrier, particularly since biomass is already available and must be dealt with in any event.

In any case, some historical perspective may be in order: petroleum's value as a chemical feedstock results merely from historical accident. It derives ultimately from 19th century efforts to use coal tar, a noxious waste from making coke from coal. The efforts were spectacularly successful—they laid the foundation for the modern chemical industry—and they were later readily adapted to oil. Although biowaste is less readily adaptable to the processes developed for petroleum,[84] the inspiration remains.

Terrestrial Carbon Resources, III. The Significance of Carbonates

There seems to be a widespread perception that near-surface carbon reservoirs on Earth are very limited. This is untrue. What *is* true is that only a tiny percentage of Earth's carbon resides in the atmosphere. Reduced carbon in the crust constitutes a much larger percentage (Table 6.2), this carbon representing biological material that was buried in sedimentary rocks over geologic time. Fossil fuels (coal, oil, and natural gas) constitute in turn a

---

[84]Rinaldi, R; Schuth, F. Design of solid catalysts for the conversion of biomass. *Energy Environ. Sci.* 2, 610–26, 2009.

Table 6.2 Terrestrial carbon reservoirs

Total carbon in units of 10^{17} kilograms ("geograms"[1]). All carbon is expressed as CO_2 equivalent regardless of actual chemical state. Abundances from Walker,[2] except that subdivision of crustal carbon is from Hunt[3]; his crustal total agrees with Walker's.

Reservoir	Amount
Atmosphere	0.025
Ocean	1.4
Crust:	
reduced carbon	700
carbonates	2600
Total	3300
Upper mantle	6600
Grand total	9900

[1]Garrels, RM; Mackenzie, FT. *Evolution of Sedimentary Rocks*, Norton, 1971.
[2]Walker, JCG. *Evolution of the Atmosphere*, Macmillan, 1977, pp. 200, 233.
[3]Hunt, JM. Distribution of carbon in crust of Earth, *AAPG Bull.* 56, 2273-7, 1972.

tiny percentage of this material (p. 41). It is all ultimately derived from carbon dioxide via photosynthesis.

By far the largest percentage of Earth's carbon, however, resides in carbonate rocks (Table 6.2), nearly all consisting of limestone (composed of calcite, rhombohedral calcium carbonate, $CaCO_3$) and dolostone (composed of dolomite, rhombohedral calcium-magnesium carbonate, $CaMg(CO_3)_2$.) Indeed, carbonate rock can be regarded as "petrified" carbon dioxide, being formed by the reaction of carbon dioxide with water and dissolved metal ions via the so-called Urey reaction (Box 6.7). It may be, then, that carbonate rocks, which comprise the "backstop" terrestrial carbon reservoir, can ultimately be expected to become considerably more valuable than traditional metal ores (Fig. 6.8). This is not a scenario typically envisioned in

Box 6.7 The Urey Reaction

The Urey reaction, named after the pioneering geochemist and Nobel prize winner Harold Urey, and sometimes termed the "carbonate-silicate cycle," acts as the global "thermostat" by regulating the carbon dioxide content of the atmosphere over geologic time. Metal ions, mostly calcium, released by the weathering of rock precipitate carbonate rocks such as limestone by reactions like the following. First, carbon dioxide dissolves in water to form the weak acid "carbonic acid:"

$$CO_2 + H_2O \leftrightarrow \text{"}H_2CO_3\text{"}.$$

"Carbonic acid" is in quotes because much of the carbon dioxide remains merely dissolved rather than reacting with the water molecules. In the absence of other reactions, natural waters are distinctly acidic due to this reaction. (Rainwater has a pH of ~5.6; in contrast, seawater is slightly basic, pH ~8, because of additional reactions.)

The carbonic acid then releases hydrogen ions and free carbonate ion:

$$H_2CO_3 \leftrightarrow H^+ + HCO_3^-;$$
$$HCO_3^- \leftrightarrow H^+ + CO_3^{2-}.$$

Carbonate ion then reacts with dissolved calcium ion to precipitate calcium carbonate, the mineral making up limestone:

$$Ca^{++} + CO_3^{2-} \rightarrow CaCO_3.$$

This last reaction is often called the "limewater" reaction, because a solution saturated with calcium ions will turn milky, due to precipitated $CaCO_3$ particles, if carbon dioxide is blown into it. This is a classic test for CO_2.

Thus, increases in atmospheric carbon dioxide are unstable and will ultimately be drawn down. The problem in the case of anthropogenic CO_2 emissions is the short timescale on which they're occurring, as the Urey reaction typically takes thousands to tens of thousands of years.

Figure 6.8 Limestone outcrop.

Limestone cliffs of Late Paleozoic age in Lee Canyon, Spring Mountains, Nevada. Such rocks are composed of >90% calcium carbonate (calcite, rhombohedral $CaCO_3$) and represent the bulk of the carbon inventory in Earth's crust. Photograph by the author.

traditional speculations on the future of nonrenewable resources and economic geology.

Silicate resources

The overwhelming abundance of silicate minerals, on Earth and other rocky bodies (Box 3.5), may seem to make discussion of silicate "resources" superfluous. When ordinary dirt becomes more valuable than metals, so that things like (say) dry lake sediments become valuable feedstocks, the immense exploration infrastructure of present-day mining is hardly required. As outlined in Chapter 3 (Box 3.2), however, the *waste* from conventional mining operations is largely silicate debris that also is commonly a source of environmental problems. Due to the great strength of the silicon-oxygen bond, silicates are seldom attractive as ore minerals. Instead, they constitute the bulk of the gangue (p. 91). Hence, it seems particularly attractive to use such

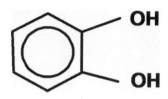

Figure 6.9 Catechol.

The spacing between the OH groups is well-matched to the size needed to "grab" a silicon atom. See Fig. 3.7 for a description of the conventions used in such molecular diagrams.

waste as the raw material for a silicate nanotechnology, which is also in keeping with the "waste as resource" theme of this book.

Of course, as silicates are so chemically stable, the question now arises as to how they might be dealt with for nanotechnological applications, especially in bulk. As with dealing with carbon-based waste (p. 242), processing that involved large amounts of heating, as for melting or pyrolysis, would hardly fit with the emphasis on non-thermal processing that's also been a theme of this book. Remarkably, however, certain organic reagents are capable of dissolving silicates. An example is *catechol* (1-2 dihydroxybenzene, Fig. 6.9); a basic catechol solution will even dissolve quartz.[85] The catechol complexes formed are also particularly reactive, which is an advantage in using them for further syntheses.[86] Ethylene glycol, $C_2H_4(OH)_2$, familiar as the active ingredient in automotive antifreeze, is also capable of reacting with silicates. An alkaline "cocktail" containing ethylene glycol can even dissolve beach sand.[87] Such "digestion" of silicate raw materials has been investigated for

[85]Bach, R., Sticher, H. Abbau von Quarz mit Brenzkatechin [in German]. *Experientia 22*, 515–6, 1966.

[86]Boudin, A; Cerveau, G; Chuit, C; Corriu, RJP; Reye, C. Reactivity of hypervalent silicon derivatives. One step synthesis of mono- and dihydrosilanes. *J. Organomet. Chem. 362*, 265–72, 1989.

[87]Laine, RM, Blohowiak, KY; Robinson, TR; Hoppe, ML; Nardi, P; Kampf, J; Uhm, J. Synthesis of pentacoordinate silicon complexes from SiO$_2$. *Nature 353*, 642–4, 1991.

zeolite synthesis,[88] and as an alternative route to raw materials for sol–gel applications (p. 154).[89] These complexes formed by this digestion are even more reactive than the complexes derived from catechol, a further advantage when used as feedstocks for syntheses. Hence a "hydrometallurgy of silicates" seems within reach.

Indeed, the direct synthesis of organic silicates—that is, silicon bonded directly to organic side groups—from minerals has been demonstrated, of alkoxides in particular.[90] An "alkoxide" can be visualized as compound of an alcohol with a metal, with the hydrogen in the −OH group of the alcohol replaced by a metal atom. Tetraethoxysilane (called tetraethyl silicate in the older literature), for example, with formula $Si(OC_2H_5)$, consists of a silicon atom bonded to four ethoxy groups, which can be formally derived from ethanol (C_2H_5OH) by omitting the hydrogen on the −OH group. Furthermore, the resulting alkoxides largely reflect the structure of the original silicate anion; hence, the syntheses also provide unexpected control over the products obtained. In particular, a complex chain siloxane was synthesized directly from $K_2CuSi_4O_{10}$, a "tube" chain silicate, with preservation of

[88]Herreros, B; Carr, SW; Klinowski, J. 5-Coordinate Si compounds as intermediates in the synthesis of silicates in nonaqueous media. *Science*. 263, 1585–7, 1994; Herreros, B; Klinowski, J. Synthesis of zeolites from sodium silicoglycolate, a 5-coordinate silicon compound. *Chem. Phys. Lett.* 220, 478, 1994; Hydrothermal synthesis of zeolites from 5-coordinate silicon compounds. *J. Phys. Chem.* 99, 1025–8, 1995.

[89]Kansal, P; Laine, RM. Pentacoordinate silicon complexes as precursors to silicate glasses and ceramics. *J. Am. Ceram. Soc.* 77, 875–82, 1994; Group II tris(glycolato)silicates as precursors to silicate glasses and ceramics. *J. Am. Ceram. Soc. 78*, 529–38, 1995.

[90]Kenney, ME; Goodwin, GB. Silicate esters and organosilicon compounds. Patent (US) 4717773, 1988; Goodwin, GB; Kenney, ME. A new approach to the synthesis of alkyl silicates and organosiloxanes. ACSSS 360, 238–48, 1988; A silicate substitution route to organosilicon compounds. *Adv. Chem. Ser. 224*, 251–63, 1989; A new route to alkoxysilanes and alkoxysiloxanes of use for the preparation of ceramics by the sol-gel technique. *Inorg. Chem. 29*, 1216–20, 1990; Kemmitt, T; Henderson, W. A new route to silicon alkoxides from silica. *Aust. J. Chem. 51*, 1031–5, 1998.

Figure 6.10 Tailings.

Comminuted silicate debris, the material remaining after beneficiation and removal of the ore minerals. This is part of an enormous tailings apron around the site of the abandoned copper mine near Yerington, Nevada, USA. The apron covers several square kilometers, and a sense of its thickness may be gathered by comparing with the barbed-wire fence in front of the scarp (arrow). The fence posts are about 4 feet (1.2 m) high.

the intricate tubular silicate "backbone."[91] Alkoxides are reactive compounds and so again are useful feedstocks for further syntheses. They can be used directly in sol–gel synthesis, for example, or further reduced to siloxanes. Sodium methoxide is used in the conventional synthesis of biodiesel (Box 5.12).

Of mining wastes, the most promising raw material for such digestion is probably the tailings, the material that's left behind after the crushing and grinding during beneficiation (Box 3.2; Fig. 6.10). Dissolution of such fine-grained material will be easier, avoiding further crushing and grinding steps that would probably be required with other waste materials such as slag, and would also lend itself to further separation. Because they are so fine-grained, too, tailings often are a source of particularly noxious environmental impact, such as through blowing dust and from their high chemical reactivity, so using them is particularly desirable.

As an aside, silicate-based nanotechnology is likely to prove critical on rocky bodies on which carbon is essentially absent,

---

[91]Harrington, BA; Kenney, ME. The preparation and properties of a fibrous tube alkoxysiloxane derived from a tube silicate. *Coll. Surf.* *63*, 121–9, 1992.

such as the Moon. Lunar regolith, the surface layer of rock comminuted by eons of meteorite impact, would be an ideal feedstock for a silicate nanotechnology.

A Note on Space Resources and Nanotechnology

Off-earth resources of both energy and materials have come under increasingly serious scrutiny over the last few decades.[92] They have many potential advantages, including minimal environmental impact and vast potential abundance. Their overwhelming disadvantage, of course, has been the high cost of access to space, and this factor has dominated the economics. (Returning material from space is *not* expensive, however. Hence the economics are dominated by the "payback" possible for a given payload mass.)

Nanotechnology promises to decrease greatly this access cost. First, the favorable economics of high materials strengths on transportation costs are nowhere more pronounced than in the case of space. Even with conventional chemical fuels, the payload increase is highly significant due to the diminished vehicle mass.[93] Indeed, single-stage to orbit (SSTO) may become practical. Second, the payload mass itself will decrease drastically. Both these factors will have a dramatic effect on the economics of space operations.

In fact, rockets may be supplanted as the routine means for getting off Earth. The enormous potential strength of "buckytube" cables has motivated serious looks at a "space

---

[92]Gillett, SL. Extraterrestrial resources. Paper presented at the 90th Annual Meeting of the Northwest Mining Association, December 8, 1984—available from NWMA, 1984; Lewis, JS; Lewis, RA. *Space Resources: Breaking the Bonds of Earth*. Columbia University Press, 1987; Lewis, JS; Matthews, MS; Guerrieri, ML; eds. *Resources of Near-Earth Space*. University of Arizona Press, Space Science Series, 1993.

[93]McKendree, T. Planning scenarios for space development. In *Space Manufacturing 10: Pathways to the High Frontier*, Faughnan, B; ed. 254–64, 1995.

elevator" or "skystalk."[94] These have been a staple of science fiction for a generation,[95] but known materials had been far too weak to merit serious study until now. An extremely strong cable would connect a point on Earth's surface with a satellite in geosynchronous orbit, effectively making a suspension bridge to space. In theory, a buckytube has a support length—the length of cable that could hang freely, under one Earth gravity, without snapping—of over 5000 kilometers. It follows that the "taper factor"—the ratio of the cross section of the cable at geosynchronous orbit, where tensional loads are greatest, to that at the surface—of such a cable is less than 3, which is certainly reasonable. Because the cable would extend to a counterweight beyond geosynchronous orbit, at which point the cable is traveling faster than orbital velocity at that height, material could also be flung off into space with no expenditure of reaction mass at all. Such a system would have a more profound effect on space access than did the transcontinental railroad for access to the US west in the late 19th century.

Near-Earth Materials

Nonetheless, near-earth sources of *materials* are more problematic. So-called near-earth asteroids, a population of small objects with orbits crossing the inner planets', have excited interest as resources.[96] These objects are widely thought to constitute the parent material for most of the present meteorite flux on the terrestrial planets. For terrestrial resource purposes, objects with compositions typical of nickel–iron meteorites have

---

[94]Cowen, R. Ribbon to the Stars. *Science News 162*, 218–9 (plus sidebar), 5 Oct 2002.

[95]Clarke, AC; *The Fountains of Paradise*. 305 pp., Ballantine Books, 1980 Softbound; Sheffield, C. *The Web Between the Worlds*. Ace Books, 1979.

[96]Kuck, DL. Near-Earth extraterrestrial resources. In *Space Manufacturing III*, Grey, J; Krop, C; eds., Fourth Princeton/AIAA Conference on Space Manufacturing Facilities, AIAA #79–1377, Princeton, May 14–17, 1979.

been particularly favored.[97] Such objects also contain platinum-group and precious metals at concentrations comparable to or greater than the best terrestrial ores. Indeed, mining such objects has become a staple of much serious futurism, and has recently been the subject of high-profile business proposals.[98]

An ironic consequence of maturing nanotechnology, however, is that asteroidal nickel-iron becomes of merely historical interest, at least for terrestrial use. Not only are molecular separation technologies likely to make asteroidal retrieval of Ni-Fe considerably less attractive, but availability of superstrong materials, probably carbon-based, will make Ni-Fe alloy unimportant.

The relative abundance of platinum-group metals, however, may yet prove an economic motivator if they maintain their present importance in catalysts, even with the availability of molecular separation for extracting platinum-group metals from terrestrial materials. It seems problematic, however, whether they will remain so valuable once molecular-scale understanding and fabrication of catalysts is achieved

In the last chapter, some more general philosophical considerations will be addressed.

---

[97]Drexler, KE. The Asteroidal Manifesto. *L-5 News*. L-5 Society, Tucson, AZ, Feb 1983; Lewis, JS; Nozette, S. Extraction and purification of iron-group and precious metals from asteroidal feedstocks (AAS 83–236). Space Manufacturing 1983, edited by J.D. Burke and A.S. Whitt. (proceedings of a conference held May 9–12, 1983 at Princeton University), *Adv. Astronaut. Sci.* 53, 351–354. Univelt, Inc., San Diego, 1983.

[98]E.g., Deep Space Industries, deepspaceindustries.com.

Chapter 7

A View from the Paleotechnical Era

History is what happens while you're making other plans.
—Anonymous

This book has been a message of technological optimism. The author makes no apology for that. In fact, he finds the notion that such an apology might be necessary impossible to take seriously. So it's perhaps appropriate to return for a re-examination of the "doom" scenarios touched on in Chapter 1.

Scenarios of "overshoot"—resource exhaustion followed by the collapse of civilization—are nothing new. One such, *The Limits to Growth*,[1] was published over 40 years ago, and became one of the most influential books of the 20th century. Others before and since, though, have echoed the same theme. The influence of *Limits to Growth* was probably enhanced by its predictions' being based on a computer model. In those long-ago days, "the computer says so" had a cachet that is difficult to recapture now. But in fact the computer models were seriously

---

[1]Meadows, DH; Meadows, DL; Randers, J; Behrens, WW. *The Limits to Growth: A Report for the Club of Rome's Predicament of Mankind.* Universe Books, New York, (Signet paperback), 1972.

Nanotechnology and the Resource Fallacy
Stephen L. Gillett
Copyright © 2018 Pan Stanford Publishing Pte. Ltd.
ISBN 978-981-4303-87-3 (Hardcover), 978-0-203-73307-3 (eBook)
www.panstanford.com

deficient, consisting of simplistic, ad hoc linkages between "inputs" and "outputs" that were not based on—nor even constrained by—fundamental physical laws. For example, the *Limits to Growth* model found that even "unlimited" energy supplies led to ultimate collapse, through rising pollution. This is just silly. Thermodynamic misconceptions notwithstanding, element separation is not intrinsically energy-expensive, as thoroughly discussed in Chapter 3. So it would be easy enough to extract pollutants with "unlimited" energy. Hence pollution control is not a fundamental problem; for a repeatedly mentioned counterexample, the biosphere manages to completely recycle matter indefinitely, using only the energy of sunlight. Hence a model that assumes that a certain amount of economic output (of energy or whatever) "must" generate a certain amount of pollution is just wrong.

A common thread running through similar analyses is that technical advances can't help in moving to an industrial future beyond oil. The high energy density, or engineering convenience, or some such property is supposed to make oil "irreplaceable."

Again, though, this is wrong. When not utterly disingenuous, these notions are based on confounding engineering limitations with physical laws (Boxes 2.16 and 7.2). They are often even buttressed with misapplied physical laws, the long-suffering second law of thermodynamics (Box 2.4) being the usual prop.

In fact, however, a correct thermodynamic appreciation of most current technology shows that it squanders energy, because the energy is in the form of *heat*. This is such an important point that I make no apology for repeating it throughout this book: the high energy density of current fuels is merely a brute-force compensation for the inefficiency with which they are used. When fuels are "burned"—converted into heat—a great deal of the energy must be thrown away. It must be dissipated into the environment, as described in Chapter 2.

We didn't have (and don't have) the "energy" crisis. We have had the "heat" crisis.

Thus, the fundamental laws of nature are much more permissive than most of the "limits" literature recognizes. We can talk about engineering *timescales*, over which technology might be available. We can also talk about deleterious effects of current or future technologies, such as pollution or its potential for misuse.

But the bald statement that "nothing can replace oil" is simply, utterly false.

Wishful Thinking: "It's All for the Best Anyway"

The neo-Luddites imagine a future of organic vegetables and hand-woven rugs; the reality of their policies would be a mad-dog scramble between nations, classes and races over scraps of water and fuel and food ... Followed by restabilization right back in the post-Neolithic norm of human history, a society of starving peasants ruled by bandits, perhaps with a few high-tech trinkets for the elite.

—S.M. Stirling, 1991[2]

It is in some ways more naïve to dream of 'making people so good that technologies do not need to be perfected.'

—AtKisson, 1998[3]

One school of thought views the "inevitable" resource crunch as a godsend, an opportunity to return to a "simpler" life. It's thought that a more "balanced" world will ensue, with people content to accept substantially lowered expectations in exchange for a life "closer to the land," typically in conjunction with some sort of local agriculture. A heightened sense of community and self-sufficiency—perhaps with more leisure time, though how *that* would follow from a greater reliance on traditional agriculture is hardly obvious—are thought to be the rewards for

[2]See Ref. 1 in Ch. 1.
[3]AtKisson, A. *Believing Cassandra: An Optimist Looks at a Pessimist's World.* Chelsea Green Publ. Co., 1998, p. 125.

a drastically lowered standard of living. Vacca has a tart response to such fantasies: "The alleged amenities permitted by the slower rhythms of rural life often coincide with an inescapably sluggish and deprived existence."[4] It also is commonly asserted that such a lifestyle might be a lot better for the environment, although the degree of environmental degradation associated with low-technology agriculture worldwide might give one pause—a point discussed further below. Such writers also typically do not note that a world where travel is rare and reserved only for an elite is an invitation to rampant tribalism and war—again reflecting a return to the post-Neolithic norm of history.

To summarize it gently, such scenarios suffer from a gross naïveté on political realities; indeed, even biological realities—not to mention a hazy and romanticized view of the squalor that attended the "simpler" societies of the past.[5] Do these authors *really* think that people are going to forgo power, status, and leisure, not to mention striving for a better life for themselves (and especially) for their children? Ironically, such folk also often sneer at the "unrealism" of "technoutopians," too.

In fact, if poverty *does* get forced on people by external events, whether political or natural, they will do their best to downsize someone else's lifestyle instead. The stability—the "social contract"—in a modern democracy depends critically on a healthy economy, and indeed a minimum level of income, as has been repeatedly underscored historically.[6] The crash of the 1930s brought Hitler to power in Germany.[7] Even in the US, more

---

[4]Vacca, R. *The Coming Dark Age*, rev. electronic ed., 2000, p. 21. He then continues with, "Those precious human relationships that the life of a rural peasantry is supposed to foster have to rest in fact on a basis of absolute cultural poverty ..." Vacca, be it noted, is also *not* optimistic about the future of industrial civilization.

[5]E.g., Manchester, W. *A World Lit Only by Fire: The Medieval Mind and the Renaissance*. Little, Brown & Co., 1993.

[6]E.g., Chua, A. *World on Fire*, Doubleday, 2003; cf. King, SD. *When the Money Runs Out: The End of Western Affluence*, Yale University Press, 2013.

[7]E.g., Fulbrook, M. *A Concise History of Germany*, updated ed. Cambridge University Press, 1992.

than one contemporary observer thought the social strains of the Great Depression would lead to revolution in the US.

The real consequence of such resource-limited scenarios would be increasingly vicious and desperate fighting over the dwindling resource base, as more realistic writers acknowledge.[8] In such a Hobbesian world, only the most ruthless would survive. Even if complete apocalypse is avoided, the result will be a highly stratified society; think feudal Europe with a few guns and electronics for the elite.

That's the real lesson of history.

Perhaps an even more chilling possibility is "eco-fascism." An overarching global authority might be set up to enforce "equitable" distribution and forbid "dirty" or "dangerous" technologies. Even with the highest ideals and the best of intentions, such an elite would soon degenerate into a corrupt aristocracy. Look at the appalling history of the Soviet Union for an example of the inevitable and pervasive corruption of "utopian" elites! Despite the communal rhetoric,[9] the Communist Party in the Soviet Union became the most finely graded aristocracy the world has ever seen. George Orwell,[10] of course, had anticipated such an outcome decades before in *Animal Farm*.

Farfetched? Unfortunately, no. Even now some people who advocate "massive downsizing of our lifestyles" point to Cuba, evidently in all seriousness, as an example for the world to emulate. Since the withdrawal of the Soviet subsidies, ordinary Cubans have been forced back into grinding poverty because Castro's regime refuses to yield any power. Yes, the oil consumption of ordinary Cubans is a fraction of the developed countries'. This,

---

[8]E.g., Duncan, RC. World energy production, population growth, and the road to the Olduvai Gorge. *Popul. Environ.* 22, 503–22, 2001. Also online at http://dieoff.com/page234.htm.

[9]E.g., Smith, H. *The Russians*. Quadrangle/NY Times Book Co., ca. 1974, pp. 25–52; Shipler, DK. *Russia: Broken Idols, Solemn Dreams*. Times Books, 1983, pp. 163–248, esp. pp. 192 ff.; Heller, M; Nekrich, AM. *Utopia in Power: The History of the Soviet Union from 1917 to the Present*. Summit Books, 1986. pp. 606–8, 696–99.

[10]Orwell, G. *Animal Farm*.

though, is hardly a reflection of "community spirit," but simply of the naked power of Castro's *nomenklatura*, which remains a privileged elite. A better example of Stirling's "return to the post-Neolithic norm of history," cited in the quote above, would be hard to find.

Complexity and Its Discontents

Others suggest that industrial civilization is doomed due to a fragility stemming from its complexity, the idea being that the host of specializations necessary for its operation could render it vulnerable to even minor disruptions.[11] In a now classic study, for example, Tainter[12] studied the collapse of complex societies. In pre-industrial cultures, the ultimate energy supply comes from agriculture; that is, from sunlight converted into useful forms of chemical energy by natural, self-replicating nanosystems. Agriculture requires (well-watered; cf. Box 5.14) soil; it requires people to work that soil; and it requires organization to build infrastructure, direct planting and harvesting, and so on. This is nothing more than the classic economic triad of land, labor, and capital. (Toward the dawn of industrialization in Europe and elsewhere, the energy supply was supplemented by such things as windmills and water wheels. And, of course, fires for cooking, smelting, and firing pottery are ancient, but, like agricultural products, are derived from biomass.) In Tainter's view, societies become prone to collapse when investment in additional social complexity, to manage the agricultural system, yields diminishing returns. In other words, the societies become overwhelmed by their overhead. According to Tainter industrial society has managed to stave off this fate, so far, by exploiting alternative energy sources, fossil fuels in particular.

---

[11]Cf. Ref. 4.
[12]Tainter, A. *The Collapse of Complex Societies*. Cambridge University Press, 250 p., 1988.

The thesis is not, to be sure, utterly unreasonable. Nonetheless, it's not at all clear that industrial civilization is quite so fragile. Ecology furnishes illuminating counterexamples: an ecological truism is that complex ecosystems are more stable. It's simple ecosystems that are most subject to wild fluctuations of overshoot and collapse: wasps feeding on beetles in grain elevators, lynxes feeding on hares. Indeed the entire biosphere furnishes a counterexample: it is very complex indeed, far more so than human systems, yet it is extraordinarily robust. Hence complexity can be a source of strength rather than weakness.

Why can this be? Because of the way the complexity is organized. In Tainter's scenarios, complexity implies centralized information overhead. In a traditional civilization, for example, you need accountants to count the harvest, tax collectors to seize a portion for the government, architects and builders to construct public works such as roads, irrigation works, and so forth, and soldiers to make sure the peasants follow orders. And, last but not least, you need some sort of organizing ideology, whether religion (many examples), nationalism (cf. Rome), or whatever. Priests, astrologers, or other such specialists may even have to decide when to plant and when to harvest. The late Roman Empire fell in the west, in Tainter's view, because the agricultural base could no longer support the infrastructure of empire. In some cases, landowners and peasants even welcomed the barbarians as a relief from crushing taxes!

Furthermore, all this activity is largely if not completely centrally directed. Sure, the (say) proconsul of a province had a great deal of local authority, but he nonetheless had to report to the Emperor, and execute the orders returned.

Ecosystems and modern civilization aren't organized like that.[13] They are not a hierarchy but a web. It's the difference between the centralized computer systems of not so long ago versus the Internet. Information doesn't merely flow upward

---

[13]Page, SE. *Understanding Complexity* [video lecture series], esp. Lecture 9, The Great Courses, Chantilly, Virginia, USA 2009.

from local authorities to the center, and then back in the form of commands or directives. Vast numbers of feedback loops transmit information sideways, downward, upward, every which way. Consider a market economy: prices, of course, are information signals about relative scarcity and abundance, and they are not set by any central authority. They are epiphenomena that result from lots of individual transactions. The mass media (and not-so-mass media; cf. trade journals, and these days, social media) not only disseminate information downward but also receive it upward, and that in turn influences the actions of the leaders as well as the mass of citizenry. Even elections can be regarded as a mechanism by which the leaders receive feedback(!) from that mass of citizens. Tainter acknowledges that participatory societies may have evolved as one response to forestall collapse. By providing the "masses" with a bigger stake in the system, they are more likely to actively support it rather than cut their losses.

Perhaps the Soviet Union provided an example of Tainter collapse of an industrial society. A highly hierarchal, inflexible, centralized organization that suppressed information flows: as many have noted, it paid the price for that inflexibility.[14]

Another weakness of the Tainter thesis is that it's highly linear. To use an overworked word, it ignores synergies among different activities in a highly complex industrial society. For example, he presents data suggesting that there are at present diminishing returns on research and development (R&D), but he ignores the fact that research ramifies in unexpected ways. A modern automobile contains dozens of computer chips, but automotive manufacturers didn't invest in the R&D to invent those chips! Microchips were developed for very different purposes, using very different sources of funding, but once developed, they proved to have a staggering number of applications.

---

[14]Shane, S. *Dismantling Utopia: How Information Ended the Soviet Union.* Ivan R. Dee, Chicago: 325 pp., 1994.

There's nothing new here, either. Steam engines were invented to pump water out of mines, not to revolutionize transportation—but that was a far more important consequence.

Of course, all this implies that science-based technology is critical to the health and survival of industrial civilization. It ensures that alternatives are available and that re-invention of infrastructure can continually occur. By contrast, a preindustrial "hydraulic" (i.e., irrigation-based) civilization, once it's irrigated all it can irrigate, has boxed itself into a corner. It can't suddenly start investing in solar power instead, particularly with no clear idea of what "energy" is or how it's critical in maintaining a society! Once the traditional agricultural route to energy is exhausted, that's it.

To be sure, paleotechnical civilization—the one we are living in now—probably isn't very robust because its resource base, fossil fuels used pathetically inefficiently as heat, is seriously unsustainable, as belabored at some detail in this book. But, as also reviewed in this book, alternative energy sources are certainly abundant.

The Population Bust

Of course, if human populations rise indefinitely, no amount of new technology will stave off ultimate disaster. The "population bomb" has been a buzzword for over half a century now, and certainly the striking rise of human populations through the 20th century made the scenario plausible. As even a cursory online search will show, the coming, allegedly inevitable, "great die-off" remains a potent idea even now.

The fact that a massive collapse of population had been predicted to occur by now, however, might indicate that this model is nearly not so clear-cut nor so inevitable as it seems. In fact, recent studies show declining fertility worldwide, such that even most countries in the developing world will be below replacement level by midcentury.[15] The "population bomb" might still explode, but it is not inevitable. Indeed, a declining fertility rate brings new problems, at least from the viewpoint of traditional economics, because a smaller working population is supporting many more retirees. Of course, *that* scenario in turn makes no allowance for increased productivity from new technologies, such as those described in this book.

The Preservation of Nature

Preservation of the natural world, or at least a significant part of it, is another argument advanced against the further development of industrial civilization, and certainly the legacy of polluted soil, water, and air, the devastation of abandoned mines, the degradation of agricultural lands, and so on provides a powerful incentive to halt further destruction. But, as has been repeatedly noted in this book, the highly polluting nature of present technology merely shows how primitive it is. No laws of nature demand that pollution, certainly at this scale, must occur. And, as noted above, that "simpler" life is likely to be catastrophic for the natural world. Only wealthy societies can have such things as wilderness areas, national parks, and Endangered Species acts, and have the resources to enforce them. Parks and wilderness areas would be overrun by squatters if there were a breakdown of the social order, and endangered species would rapidly become extinct, through subsistence hunting, competition with domestic stock, or both. By contrast, the looming obsolescence of much of conventional resource deposits due to the nanotechnologies elaborated in this book means that concerns about "locking away future resources" are highly exaggerated, so that more natural areas could be preserved.

Keeping some perspective is also important. "Mother" Nature is not, in fact, very maternal (Box 7.1) and has inflicted

---

[15]Morgan, SP; Miles G. Taylor, MG. Low fertility at the turn of the twenty-first century. *Annu. Rev. Sociol. 32*, 375–99, 2006.

Box 7.1 "Mother" Nature

Tertiary eruptions of the Great Basin would compare with those of modern times as the explosion of a hydrogen bomb with the bursting of a firecracker.

—J. Hoover Mackin[1]

There seems to be a widespread misperception, especially in popular culture, that all natural changes, if not necessarily completely benign, are measured and gradual, and do not lead to long-term, cataclysmic changes.

Nothing could be further from the truth. In fact, the notion of "Mother" Nature as always kind and beneficent is a peculiarly modern flavor of nonsense, an outgrowth of the Romantic myth of the late 18th and early 19th centuries. Not only is nature always changing, always in flux, but some of those changes are catastrophic indeed, and reset the whole subsequent course of Earth history.

For one thing, volcanic eruptions far, far larger than anything modern humans have dealt with have tormented the biosphere in the recent geologic past. Flood basalts are an example. Basalt is an extremely common type of lava— Hawaiian eruptions are basalt, for example—but a "flood" basalt means just that: an eruption so large that a vast tract of land is flooded with fluid, molten rock. Southeastern Washington State and adjacent Oregon are covered with dozens of basalt lava flows hundreds of square miles in extent. These eruptions, moreover, are only about 15 million years old—i.e., less than one-fourth as old as the youngest dinosaurs.

Giant ignimbrite (Latin for "fire-cloud") eruptions were even worse. These are igneous rocks, also known as "welded tuffs," that were laid down from molten rock particles

---

[1]Mackin, JH. Structural significance of Tertiary volcanic rocks in southwestern Utah. *Am. J. Sci. 258*, 81–131, 1960. Usually phrasing like that doesn't survive the peer review in a technical journal article!

dispersed in hot gas, somewhat like the froth on a soda. When the particles settled out, they were still hot enough to weld together to form rocks as solid as lava flows.

In the Great Basin of the western United States are dozens to hundreds of vast ignimbite sheets, each the result of a single gigantic eruption. Some of these ignimbrites cover hundreds of square kilometers—and that, of course, is just where they're preserved, where the rock was still hot enough to weld when it settled out. Hot ash, but not hot enough to weld, must originally have extended for hundreds of kilometers farther from the original eruption.

Such eruptions are so large—so beyond current and historical experience—as to be almost inconceivable (cf. quote above). Under the name "supervolcanoes," however, they have become the subject of serious study as "megadisaster" scenarios. Three eruptive centers within the contiguous United States—Long Valley, California, Yellowstone, Wyoming, and Valles, New Mexico—have generated such eruptions in the recent geologic past.

Of course, the catastrophic results of a large asteroid impact on the Earth are now not only the stuff of serious scientific research,[2] with an impact now widely accepted as the cause of the end-Cretaceous mass extinction,[3] but have also entered popular culture. Indeed, sky surveys to locate potentially hazardous objects have been going on for the last couple of decades, and serious attention has been paid to possible amelioration strategies, such as deflecting an object with a nuclear explosive.

---

[2]E.g., McLaren, DJ; Goodfellow, WD. Geological and biological consequences of giant impacts. *Annu. Rev. Earth Planet. Sci. 18*, 123–71, 1990; Gibson, RL; Reimold, WU; eds. *Large Meteorite Impacts and Planetary Evolution IV. Geol. Soc. Amer. Spec. Pap. 465*, 2010.

[3]E.g., Smit, J. Extinctions at the Cretaceous-Tertiary boundary: The link to the Chicxulub impact. In *Hazards Due to Comets and Asteroids*, Gehrels, T; Matthews, MS; Schumann, A; eds. Space Science Series, University of Arizona Press, 1994, p. 1191.

Biologically induced disasters have occurred, too, the emergence of oxygen-releasing photosynthesis some time before 2.5 billion years ago remaining the most catastrophic pollution event in Earth's history. ("Oxygen-releasing" isn't redundant, by the way: early versions of photosynthesis, which are still used by some bacteria even on the modern Earth, did not release free oxygen.) Among other things, the build-up of highly reactive oxygen caused sweeping changes in ocean chemistry. Before oxygen, dissolved iron seems to have been a major component of ocean chemistry, but oxidized iron is nearly insoluble. The consequences of this change in marine chemistry linger to this day. In the modern ocean iron is often a limiting nutrient, a nutrient whose scarcity limits organisms' growth.

This air-pollution event also caused global refrigeration through draw-down of atmospheric CO_2, causing the first known major glaciation about 2.2 billion years ago,[4] a planetary Fimbulwinter that may have extended pole-to-pole and lasted several hundred million years.

By the late Precambrian, about 600–800 million years ago, glaciations had returned in a big way. An intimate association of glacial and tropical rock types is found all over the world, in which a stack of glacial strata is overlain by a thick dolomite layer, a carbonate rock related to limestone (Box 6.7) that indicates very warm (and possibly CO_2-rich) conditions. The association of these wildly different rock types is now commonly thought to reflect "icehouse/greenhouse" oscillations, in which global glaciations alternated with globally warm conditions.[5] During the warm times dolomite was laid down under an atmosphere rich in CO_2 while photosynthesizers grew furiously. In so doing they drew down the CO_2, which eventually triggered glaciations

---

[4]Kopp, RE; Kirschvink, JL; Hilburn, IA; Nash, DZ. The Paleoproterozoic snowball Earth: a climate disaster triggered by the evolution of oxygenic photosynthesis. *Proc. Natl. Acad. Sci. USA 102*, 11131–6, 2005.
[5]Hoffman, PF; Kaufman, AJ; Halverson, GP; Schrag, DP. A Neoproterozoic snowball Earth. *Science, 281*, 1342–44, 1998.

that then grew through a positive feedback of their own. Because ice is white, it reflects solar energy very effectively; so things get cooler, so that more snow can fall and more ice can grow ... etc. As the globe then becomes enshrouded in ice most of the photosynthesizers die. Once the Earth is completely covered with ice, moreover, the ice won't melt on its own, because the Earth's reflectivity—"albedo"—is now so high. However, the freeze-over isn't permanent. With the photosynthesizers killed back, carbon dioxide from ongoing volcanism accumulates, leading eventually to a new greenhouse. A global greenhouse, warm pole to pole—it's paradise for those few surviving photosynthesizers, who now start growing furiously again....

Evidently the cycle occurred at least twice, and probably several times. The oscillations probably stopped with emergence of grazers: if the photosynthesizers are eaten fast enough, they can't draw down the CO_2 fast enough. Although there have been severe glacial ages since, the Earth has never been completely ice-covered again.

Who would have thought the havoc wreaked by green growing scum could be so profound? Yet we now think of the Earth as "green." Our fundamental characterization of our planet as a "living" world is intimately tied up with the consequences and elaborations of this biologically self-inflicted catastrophe.

catastrophes on the biosphere that beggar anything done by humans. This is not a matter of "two wrongs make a right"—the fact that nature can be destructive does not excuse humans' choosing to be so—but it underscores that relinquishing high technology merely leaves humans, and the biosphere, vulnerable. Sooner or later one of those megacatastrophes will happen. With the proper technologies in place, humans would be in a position to ameliorate or even avoid them completely, the deflection of a "dinosaur killer" asteroid being one example.

It is no doubt ironic to think of humankind as becoming the guardian of the biosphere.

That's Buck Rogers Stuff!

There has been a great deal said about a 3,000 miles high-angle rocket. In my opinion such a thing is impossible for many years. The people who have been writing these things that annoy me, have been talking about a 3,000 mile high-angle rocket shot from one continent to another, carrying an atomic bomb and so directed as to be a precise weapon which would land exactly on a certain target, such as a city.

I say, technically, I don't think anyone in the world knows how to do such a thing, and I feel confident that it will not be done for a very long period of time to come. . . . I think we can leave that out of our thinking. I wish the American public would leave that out of their thinking.

—Dr. Vannevar Bush (science adviser to US President Franklin Delano Roosevelt), Senate testimony, December 3, 1945

(The Soviets flew Sputnik I atop just such a "high-angle" rocket less than a dozen years later, in October 1957. And the existence of "high-angle rockets" that could land an atomic bomb on a city was a major issue in the US Presidential election of 1960.)

Anything that is theoretically possible will be achieved in practice, if it is desired greatly enough.
—Arthur C. Clarke[16]

After the surrealistic technical advances in the last century or so, it's difficult to believe in the early 21st century that people still seriously dismiss proposals not because they flout the laws of nature but simply because of a gut feeling of personal incredulity—what Arthur C. Clarke has called the "failure of nerve." I'm sitting here writing this at my personal computer, which has vastly more power than even the Pentagon had in the

---

[16]Clarke, AC. The failure of nerve. In *Profiles of the Future*, Bantam, 1962.

1960s—and remember again how preposterous the notion of your *personal* computer would have seemed then![17]

Moreover, few modern technologies are as "Buck Rogers" as those associated with oil production and refining. Sound waves and intense computer processing are used to image the subsurface miles deep, to identify likely exploration targets that may be a few dozen meters across. Drill bits are guided to these targets through kilometers of solid rock (often after passing through a kilometer or so of water), "driven" by an operator at a monitor, like a spaceship in a video game. Enormous drilling platforms are placed in water hundreds of feet deep in some of the stormiest seas on the planet—and maintenance just on these platforms requires technology that would put a space station to shame. After all, seawater is corrosive, and under pressure—the tiniest leak can lead to catastrophe. Even space is more benign. Taking oil extraction for granted, while poking fun at "far out" technology, simply shows that these critics have no appreciation of contemporary petroleum technology.[18]

Certainly one factor in such denial is simple disorientation: things have changed so much, so fast, surely they can't *keep* doing so! But they can—and in fact, *because* of the oncoming resource crunch they are likely to change even faster in the next few decades. Furthermore, evaluations of "frontier" technologies are especially plagued by arguments purportedly based on physical laws that merely confuse limitations of particular engineering approaches with laws of nature (Box 7.2).

---

[17]In 1971 the science-fiction(!) magazine *Analog* carried an article about a computer game at the Massachusetts Institute of Technology, which students played in the wee hours when the computer was free ("Spacewar," AW Kuhfeld, July). In his marveling about this system, then-editor John W. Campbell regretted that computers were far too expensive ever to be practical just for games. Barely six years before the Apple II, home computer games were too far out even for science fiction! Of course, your typical early 21st century teenage computer gamer would now find "Spacewar" extremely lame.

[18]Cf. Ch. 5 in Ref. 15 (Chapter 2).

Box 7.2 "Limits" and Limits

Real limits *do* exist. Natural laws place fundamental limits on human activity, but the key words are "natural laws:" not misconceptions about physical laws, nor personal incredulity, nor naïve assumptions based on particular engineering approaches. In fact, the confounding of engineering limits based on current paradigms and capabilities with laws of nature is what has given "limits" a bad name. As with the boy who cried wolf, when so many phony "limits" have been seen to be nonsense, the real ones will have difficulty being taken seriously.

As already noted (Box 2.16), "net energy" analyses are often marred by such errors. Indeed, a number of energy-related misconceptions degenerate to the level of straw men. It's been claimed, for example, that ontrolled fusion is impossible, one naïve argument stating that nothing can contain the extreme temperatures required. But by this argument the Sun can't exist! Of course, the Sun is not confining the reaction with a physical "container" at all. The force of gravity keeps the Sun together, even against the extreme temperatures in its core. Technological approaches typically envision using forces far stronger than gravity[1]—magnetic and electrostatic—to contain the fusion reaction.

As noted in the main text, too, another questionable assumption is that fusion requires heat in the first place. But—again—heat's a clumsy way of doing things. Why not slam the nuclei together directly, using (say) electrostatic acceleration?[2]

Energy technologies are not the only subjects of such stacked analyses, however. Historical examples include Simon

---

[1]Gravity is actually an exceedingly weak force. The simplest toy magnet holding up a paper clip is outpulling the entire Earth. Gravity seems overwhelming in everyday life only because it doesn't come in positive and negative varieties; it always adds up. The far stronger electromagnetic forces, by contrast, largely cancel out.
[2]Ref. 154 (Chapter 5) in main text.

Newcomb's—a noted astronomer of the day—"proof" that heavier-than-air flight was impossible, despite the example of birds.[3]

A more recent example is illustrative: in the early 1950s it was seriously suggested that there was an upper limit to how complex computers could be, because the human assemblers of the electronics, envisioned as soldering individual connections among thousands to millions of separate components, would become overwhelmed by the number of connections that needed to be made. At some point mistakes become statistically inevitable. This "inteconnection problem," however, was not a fundamental issue at all. It was an engineering challenge that led directly to the integrated circuit.[4]

---

[3]Ref. 15 in main text. Clarke also gleefully lists some other bad analyses.
[4]Reid, TR. *The Chip*, 2nd ed., Random House, 2001, p. 15.

Another factor no doubt is ideological blinders—after all, some people profess to like the idea that civilization will "inevitably" collapse (cf. section above). In fact, a great deal of the purported incredulity is disingenuous. Such critics, instead of insisting that "nothing can replace oil," realize that in fact something can. They really (and again, naïvely) want that low-tech future. Ehrlich has said that giving humanity cheap energy would be like "giving an idiot child a machine gun."[19] Another book[20] consists largely of hand-wringing about the alleged "loss of our humanity" that would follow from technologies such as genetic engineering and nanotechnology. (Left unspoken is the issue of deciding what's "really" human. The fact that many perfectly human activities are very unattractive indeed—cf. Stirling's quote above—is also not addressed.) So the ridicule

---

[19]Ehrlich, P. An ecologist's perspective on nuclear power. *Federation of American Scientists Public Issue Report*. May/June 1978.
[20]McKibben, B. *Enough*. Times Books, 2003.

("that's Buck Rogers stuff!") is often merely a substitute for saying "we *shouldn't.*"

In turn, too, a dark, unspoken thread weaves through these scenarios of "relinquishing" technologies—who enforces that relinquishment? The enforcement would have to be strict and total, because of the military implications of misuse. Such considerations lead straight back to the eco-fascism scenarios described above. Moreover, the ruling elite must have access to those technologies, to forestall their misuse. Of course.

An all-powerful tyranny suppressing "dangerous" technologies so that we can be "fully human"—can people *really* take such an argument seriously?

Other writers acknowledge that a resource crunch will not lead to some sort of Ecotopia. They in fact view the Hobbesian scenarios as inevitable. But they are just as wrong—not because the Hobbesian scenarios couldn't happen, but because they don't have to happen.

In either case, too, the biosphere will be the loser. If industrial civilization collapses, it will take a goodly chunk of the natural world with it.

There's a further, albeit someone delicate, point. Although "gloom and doom" prognostications can hardly be blamed for all social stresses in the contemporary world, they have been a message of significant parts of the intelligentsia for some forty years—at least since the publications of the Club of Rome. Whether based on naïve scientific analyses (including bogus thermodynamics), simplistic "world dynamics" models, or on a particular ideological position, and however earnest and well meant, such predictions are unlikely to foster social stability. If young folk are told they are doomed, and must forego any hope of bettering themselves or their children, the reaction will be eat, drink, and be merry, for tomorrow will be worse, or else a retreat into drugs and apathy. The message is particularly disheartening because it comes from largely established people who will *not* bear the burden.

Or, alternatively, people can better their conditions with violence. This is, alas, the traditional response (cf. Stirling's quote above again): after all, *someone* can always manage to live better, even in a zero-sum world.

The message falls even harder on the developing world, who see the lifestyle in developed countries, but who are told that they can never aspire to it. Indeed, they are told in all seriousness by some commentators that they should be "triaged"; they can't be saved anyway, and the world shouldn't fritter away resources on them.[21]

It's perhaps a wonder that things have not blown up more than they have! One reason, no doubt, is that a lot of people simply don't believe it. Asian societies, in particular, haven't bought into this worldview, but there are still entrepreneurs and future-lookers even in the western world. Look at the personal computer revolution, unanticipated when *Limits to Growth* came out.

The tragedy of a collapse would be heightened because it would be based on bad analysis and a perverse wishful thinking: will savage, warring tribes squabbling over a declining Earth become self-fulfilling? Such an outcome would furthermore represent a broken promise. Even at the dawn of the industrial revolution in the early 19th century, the vastly increased material productivity it promised already led some observers to suggest that the biblical pronouncement that "there will be poor always" might not always be true.[22] Although that promise remains unfulfilled, there are vastly more people living better now than ever before in history. And the promise will *remain* unfulfilled if industrial civilization collapses.

The alternative, of course, is to work toward a future of abundance, of possibilities—of opportunities on and off Earth.

---

[21]Cf. Abernethy, V. *The Atlantic Monthly*, Dec 1994.

[22]Johnson, P. *The Birth of the Modern: World Society 1815–30*, HarperCollins, 1991, esp p. 360 ff.

The Prosperous Future

For I dipt into the future, far as human eye could see,
Saw the vision of the world, and all the wonder that would be.
—"Locksley Hall," Alfred, Lord Tennyson, 1842

We are currently living in the Paleotechnical Era, otherwise known as the Early Industrial. Its defining characteristics include the widespread use of heat and bulk materials, particularly metals. Motive power is furnished almost exclusively by that apotheosis of 19th-century technology: the heat engine. Due to the clumsiness of resource extraction and assembly technology, moreover, vast transportation network spans the entire globe for moving raw materials in and finished products out. Non-agricultural raw materials are extracted from geologic enhancements ("ores"), while agricultural products are grown on lands suited to the vagaries of particular crops. And raw materials are taken to massive matter organizers—"factories"—from which the products are exported to markets that may be halfway around the planet. Moreover, these arrangements are not only taken for granted, they are justified by appeals to "economies of scale" and "relative advantage."

Because of this, conventional technology is depleting its resource base at an accelerating rate and creating an increasingly unlivable mess in doing so, even as rising expectations around the world heighten demand still further. We simply cannot maintain the standard of living in the industrialized world much longer with conventional technology, much less raise the rest of the world's.

Nonetheless, it is critical to recognize that the dwindling resource base is not set by the laws of nature, despite common assertions to the contrary. Typically such assertions are based on misconceptions about the laws of thermodynamics, and—unsurprisingly—are unbuttressed with any calculations of limiting thermodynamic costs. In fact, present technology is grossly inefficient and gratuitously dirty, and this results from an overwhelming reliance on heat—the "Promethean paradigm." Nanotechnology will allow us to supplant the Promethean paradigm and move beyond the Paleotechnical.

This new paradigm, with an increasing emphasis on information and control rather than raw power,[23] will have roots in the past as well, most obviously in the rise of electronics ever since the 1920s. The Edison effect, the basis of vacuum tubes and hence of "first-generation" electronics, was discovered by the inventor Thomas Edison in 1880, and his patent[24] using the effect for a voltage regulator is considered the first US patent for an electronic device. Edison, however, was not interested in delicate electronic responses but in large-scale electrical devices, and so it was left to others to develop the vacuum tube. Meanwhile, Guglielmo Marconi and others were developing what was to become radio from experiments on so-called Hertzian waves, and the marriage of these efforts with the vacuum tube set the stage for the wireless communications that so dominated the 20th century, and continue into the 21st. Electronics proved to have many applications beside communication, of course, as the development of computer and recording technology demonstrated.

Nanotechnology, as already noted, also continues a long-standing trend toward miniaturization, with the further implication that the micro- and nanoscale *organization* of a material is key to its value (Box 7.3).

Industrial civilization is not ending. It is transforming. We are not returning to that Neolithic norm of starving peasants ruled by bandits. It is going to be more high-tech than ever, with more people wealthier than ever before. It will also, however, be more decentralized, less energy intensive, and considerably

---

[23]Cf. "Von Neumann's work ... moved in the general direction of the post-war scientific disciplines, which had a decreased emphasis on motion, force, energy, and power and an increased emphasis on communication, organization, programming, and control." Aspray, W. *John von Neumann and the origins of modern computing*. MIT Press, pp. 211, 376, 1990.
[24]US patent 307,031, 1883.

Box 7.3 In Defense of The Economists:
The Economy as Information

"Anyone who believes exponential growth can go on forever in a finite world is either a madman or an economist."
—Kenneth Boulding[1]

"The world of mind and pattern ... holds room for endless evolution and change."
—Eric Drexler[2]

Like many people with a natural science background, the author has done some economist-bashing in this book. Physical reality places constraints on human activity irrespective of market forces. What's not there to be found can't be found, for example, however strong the market incentives.

That said, though, it is commonly felt that economics must involve the transfer of *matter*, as implied by Boulding's quote above, and that the greater the nominal value in the economy, the more material must be involved.

But in fact there is little correlation between "value" and actual physical "stuff." Consider, for example, a "blockbuster" movie series, say the *Lord of the Rings* or *Harry Potter*. The value of these films, in terms of the money they've made, beggars the value of the actual material tied up in film, DVDs, and other media. Even the resources used in *making* the movies—paying the actors, building the sets, the transportation, and so on—is inconsequential by comparison. So hundreds of millions of dollars were added to the GNP for minimal materials. Even as far back as the 1980s the decoupling of energy growth from GNP was noted.[3]

---

[1]Attributed to Boulding in the hearings for the Energy Reorganization Act of 1973, US Congress House of Representatives.
[2]Ref. 3 (Chapter 1) in main text, p. 166.
[3]Ehrlich, AH; Ehrlich, PR. *Earth*. Franklin Watts, 1987, p. 222.

Such "dematerialization" of the economy leads to the generalization of economics as *information*-intensive. Indeed, some of the highest value "goods" are the most intangible; consider software! The characterization of ever more of the workforce as "knowledge workers" or "symbolic analysts,[4]" concerned with the manipulation of information, is now a cliché. Of course, the recognition that the skills and knowledge of the workforce (now often called "human capital") is critical to production is not new. Even "in-person service workers[5]" are largely information handlers; after all, a custodian doesn't manufacture physical goods but organizes matter!

More than this, though, the increasing *organization* of matter, at smaller and smaller scales, is disproportionately valuable. The value of the pure silicon in a computer chip is trivial; it is the organization of that silicon, and the vanishingly small proportion of impurities in it, that gives it value. Indeed, the "organization" can transcend the physical hardware itself: consider, for example, compression algorithms that can boost data transmission rates with the same transmission hardware. In any case, the different *arrangements* of atoms are far, far vaster than the (already vast) number of atoms themselves. By organizing matter literally at the atomic level, nanotechnology is the apotheosis of this trend.

In any case, *information* growth—as opposed to physical growth—can continue indefinitely, with no growth in materials, but specifically growth in knowledge, culture, art, and so on. Person-centuries of work and creativity will add billions to the economy, yet the physical "consumption" involved is minimal. This is especially the case with the infrastructure in place to transmit information electronically.

Finally, although "primary" production—agriculture, energy (fossil fuels and renewables such as hydropower and wind), even mining—clearly remain of fundamental importance, equally clearly they are a small part of the modern economy

---

[4]Reich, R. *The Work of Nations*, Vintage Books, 1992, pp. 177–80.
[5]Ref. 4 (Chapter 4), pp. 176–7.

overall. Indeed, as technologies like those described in this book render conventional fossil fuels and ores obsolete, such production will continue to dwindle. This is particularly true because so many novel materials consist of common elements, just in highly organized forms. It is hardly necessary to explore for silicon deposits, for example, as it's the second most common element in the crust (Table 1.1)!

less reliant on a planet-spanning transportation infrastructure. Not only will the five-millennium era of locating anomalous deposits to "dig up and cook" for metals extraction come to its close, but energy resources will become more localized as well. Furthermore, over the long term as the "distributed fabrication" promised by nanotechnology becomes realized, global energy consumption for transportation will also drop massively. If almost any artifact can be fabricated locally, no longer will it be necessary to ship raw materials and finished goods halfway around the world, with the enormous energy consumption this entails.

Toward a New Stone Age?

Finally, we think of technological history as a series of Ages of Metals, but with a mature nanotechnology metals become obsolete. The earliest materials used by humanity were wood (a carbon-based composite) and stone (silicates). Ironically, therefore, the technology of materials seems to be coming full circle. Metal, as well as fire, may prove to be a stigma of the Paleotechnical Era.

Notes and References Cited

Journal and Book Series Abbreviations

ACSSS	American Chemical Society, Symposium Series
GE&T	Green Energy & Technology, Springer
KOECT	Kirk-Othmer Encyclopedia of Chemical Technology
LNE	Lecture Notes in Energy, Springer
LNNS&T	Lecture Notes in Nanoscale Science & Technology, Springer
LNP	Lecture Notes in Physics, Springer
NATO-ASI	NATO Advanced Study Institute
NS&T	Nanostructure Science and Technology, Springer
RSC	Royal Society of Chemistry
SSMS	Springer Series in Materials Science, Springer
TAP	Topics in Applied Physics, Springer
TC	Topics in Catalysis, Springer
TCC	Topics in Current Chemistry, Springer

Index

PGSTL 03/02/2018